Topology Now!

Library of Congress Catalog Control Number 2005937270

ISBN 0-88385-744-8

Printed in the United States of America

Current Printing (last digit):
10 9 8 7 6 5 4 3 2 1

Topology Now!

by

Robert Messer
Albion College

and

Philip Straffin
Beloit College

Published and Distributed by
THE MATHEMATICAL ASSOCIATION OF AMERICA

CLASSROOM RESOURCE MATERIALS

Classroom Resource Materials is intended to provide supplementary classroom material for students—laboratory exercises, projects, historical information, textbooks with unusual approaches for presenting mathematical ideas, career information, etc.

101 Careers in Mathematics, 2nd edition edited by Andrew Sterrett

Archimedes: What Did He Do Besides Cry Eureka?, Sherman Stein

Calculus Mysteries and Thrillers, R. Grant Woods

Combinatorics: A Problem Oriented Approach, Daniel A. Marcus

Conjecture and Proof, Miklós Laczkovich

A Course in Mathematical Modeling, Douglas Mooney and Randall Swift

Cryptological Mathematics, Robert Edward Lewand

Elementary Mathematical Models, Dan Kalman

Environmental Mathematics in the Classroom, edited by B. A. Fusaro and P. C. Kenschaft

Essentials of Mathematics, Margie Hale

Exploratory Examples for Real Analysis, Joanne E. Snow and Kirk E. Weller

Fourier Series, Rajendra Bhatia

Geometry From Africa: Mathematical and Educational Explorations, Paulus Gerdes

Historical Modules for the Teaching and Learning of Mathematics (CD), edited by Victor Katz and Karen Dee Michalowicz

Identification Numbers and Check Digit Schemes, Joseph Kirtland

Interdisciplinary Lively Application Projects, edited by Chris Arney

Inverse Problems: Activities for Undergraduates, Charles W. Groetsch

Laboratory Experiences in Group Theory, Ellen Maycock Parker

Learn from the Masters, Frank Swetz, John Fauvel, Otto Bekken, Bengt Johansson, and Victor Katz

Mathematical Connections: A Companion for Teachers and Others, Al Cuoco

Mathematical Evolutions, edited by Abe Shenitzer and John Stillwell

Mathematical Modeling in the Environment, Charles Hadlock

Mathematics for Business Decisions Part 1: Probability and Simulation (electronic textbook), Richard B. Thompson and Christopher G. Lamoureux

Mathematics for Business Decisions Part 2: Calculus and Optimization (electronic textbook), Richard B. Thompson and Christopher G. Lamoureux

Ordinary Differential Equations: A Brief Eclectic Tour, David A. Sánchez

Oval Track and Other Permutation Puzzles, John O. Kiltinen

A Primer of Abstract Mathematics, Robert B. Ash

Proofs Without Words, Roger B. Nelsen

Proofs Without Words II, Roger B. Nelsen

A Radical Approach to Real Analysis, David M. Bressoud

Real Infinite Series, Daniel D. Bonar and Michael Khoury, Jr.

She Does Math!, edited by Marla Parker

Solve This: Math Activities for Students and Clubs, James S. Tanton

Student Manual for Mathematics for Business Decisions Part 1: Probability and Simulation, David Williamson, Marilou Mendel, Julie Tarr, and Deborah Yoklic

Student Manual for Mathematics for Business Decisions Part 2: Calculus and Optimization, David Williamson, Marilou Mendel, Julie Tarr, and Deborah Yoklic

Teaching Statistics Using Baseball, Jim Albert

Topology Now!, Robert Messer and Philip Straffin

Understanding our Quantitative World, Janet Andersen and Todd Swanson

Writing Projects for Mathematics Courses: Crushed Clowns, Cars, and Coffee to Go, Annalisa Crannell, Gavin LaRose, Thomas Ratliff, Elyn Rykken

MAA Service Center
P. O. Box 91112
Washington, DC 20090-1112
1-800-331-1MAA FAX: 1-301-206-9789
www.maa.org

Contents

Preface

Topology is a branch of mathematics packed with intriguing concepts, fascinating geometric objects, and ingenious methods for studying them. The authors have written this textbook to make this material accessible to undergraduate students who may be at the beginning of their study of upper-level mathematics and who may not have covered the extensive prerequisites required for a traditional course in topology. Our preference is to cultivate the intuitive ideas of continuity, convergence, and connectedness so that students can quickly delve into knot theory, surfaces, fixed-points, and even obtain a taste of algebraic topology. We believe that students should see the exciting geometric ideas of topology now (!) rather than later.

We acknowledge the danger in building on less than a solid foundation, and one of our goals is to provide adequate reinforcement. Principles we introduce without a rigorous development are supported with numerous examples and explicit statements. We have also provided a selection of careful proofs. For example, we include a complete proof that the Alexander polynomial is a well-defined invariant for knots—a rare feat for a text at this level. After working with the material in this text, students will be well equipped to study the intricacies and abstractions of more advanced courses in point-set and algebraic topology. We hope that many of them will be motivated to take these courses.

The geometric approach to topology in this text also exposes students to the interrelation among the various branches of mathematics. Continuity and geometry are of course at the heart of the matter. This is a natural preparation for courses in real analysis, geometry, and further work in topology. Students will also see strong ties with linear algebra in the dimension of various objects, with abstract algebra in the fundamental group of a space, and with discrete mathematics in a variety of combinatorial and counting arguments.

The prerequisite for this approach to topology is some exposure to the geometry of objects in higher-dimensional Euclidean spaces, together with appreciation of precise mathematical definitions and proofs. We recommend courses in multivariable calculus and linear algebra, and one further proof-oriented course.

Organization. The first chapter introduces various ways objects can be considered to be the same. We begin with set-theoretical concepts and introduce the concept of continuity for preserving the geometrical properties of a space. For an object that is a subset of a larger space, we also consider a continuous family of deformations of the embedded object.

In the second chapter we apply these concepts to knots and links. Although we rely on intuitive ideas about polyhedral structure of sets and subsets, we point out the difficulties and how to address them before sweeping them under the rug. We develop some useful ways of distinguishing among knots and sample some polynomials of recent discovery that are quite powerful invariants of knots.

Chapters 3 and 4 present some basic examples of surfaces and their three-dimensional analogs. These classical results are among the most beautiful ideas of geometric topology. The proofs are honest, lacking only some of the technical details.

Chapter 5 takes a side tour into the theory of fixed-points. This provides an interesting application of topology that in turn has useful applications in disciplines beyond mathematics.

Chapter 6 introduces an algebraic system of dealing with loops in a space. Although we only scratch the surface of algebraic topology, students have an opportunity to work with the basic concepts and develop a sense of the power of algebraic techniques.

Chapter 7 presents material that is often the starting point in traditional texts. Our idea is that the abstractions of point-set topology will make more sense after the students have seen the geometric examples that motivate these abstractions. For example, the general concept of a quotient topology is nicely motivated by the geometric technique of gluing together edges of a polygonal disk to form a closed surface.

A course covering Chapters 1 through 4 would provide a geometric introduction to topology. The later chapters are somewhat independent. Chapter 5 connects topology with analysis, Chapter 6 introduces the algebraic aspects of topology, and Chapter 7 introduces abstract topological spaces.

We have included some harder material that will challenge students but which is not necessary to the flow of ideas. For example, the proofs that the Alexander polynomial is a well-defined knot invariant can easily be omitted from the end of Section 2.5 without loss of continuity. Likewise, the discussion of Heegaard splittings in Section 4.4, the Contraction Mapping Theorem in Section 5.2, and the material on words and relations in Section 6.5 can be regarded as supplemental topics.

For over twenty years, the second author has taught a junior-level course at Beloit College based on the material in Chapters 1 through 4 (omitting the Alexander polynomial proofs at the end of Section 2.5 and the material in Section 4.4) and most of Chapter 6, and tying everything together with Chapter 7.

Features. Each chapter begins with an informal introduction. This gives students a historical or mathematical context for topics contained in the chapter. The individual sections and topics within the sections are also linked with transitional comments that place the topics into a larger context.

An abundance of examples illustrate the new concepts. These are stated as problems that the students might encounter as homework or on exams. The solutions provide models for dealing with the concepts as well as illustrations of the level of rigor expected in deriving results.

Each section contains a rich variety of exercises. Many exercises give students practice with the definitions and theorems of the section. Other exercises relate the current material to previous topics or provide motivation for future developments. Exercises frequently ask students to fill in gaps in arguments given in the section. The most challenging exercises extend topics in new directions, offering possibilities for independent study or undergraduate research.

A Web site is maintained at http://www.albion.edu/math/ram/TopologyNow! to provide additional support material for this text. Students and instructors are invited to visit this site and to submit comments, suggestions, questions, exercises, sample syllabi, and supplementary course material that might be of value to others.

Acknowledgments. The authors hope both students and teachers will enjoy this text. We have worked hard to make this material clear and comprehensible while maintaining a standard of honest mathematics. We encourage readers to contact us with suggestions and comments. The following students and professors have made comments and suggestions for improvements in preliminary material for this book. We gratefully acknowledge their contributions.

Albion College: Elizabeth Chen, Brad Emmons, Robert Gray, Miles Horak, Frederick Horein, Jamie Kucab, Martha O'Kennon, David Reimann, Timothy Schafer, Aaron St. John, David Tollefson, Matthew Woods

Beloit College: students in Phil Straffin's topology courses for over twenty years

Cranbrook Kingswood School: Yuyin Chen

University of Akron: Lisa Lackney, Benjamin Marko, Ian Deters, Lori McDonnell, Tom Price, Joel Rabe, Charles Williams

University of Detroit Mercy: Gillian Carney

University of Wyoming: Sylvia Hobart

Valparaiso University: Elizabeth Brondos, Ross Corliss, Steven Klee, Kimberly Pearson, Paul Schmid, Dan Tesch, Philip Whaley

Robert Messer
Albion College
ram@albion.edu

Philip Straffin
Beloit College
straffin@beloit.edu

1

Deformations

Topology is the study of geometric objects as they are transformed by continuous deformations. To a topologist the general shape of the objects is of more importance than distance, size, or angle. This chapter introduces some typical topological spaces and ways of measuring the similarity between various pairs of spaces.

1.1 Equivalence

Which of the following geometric objects are the same as triangle A?

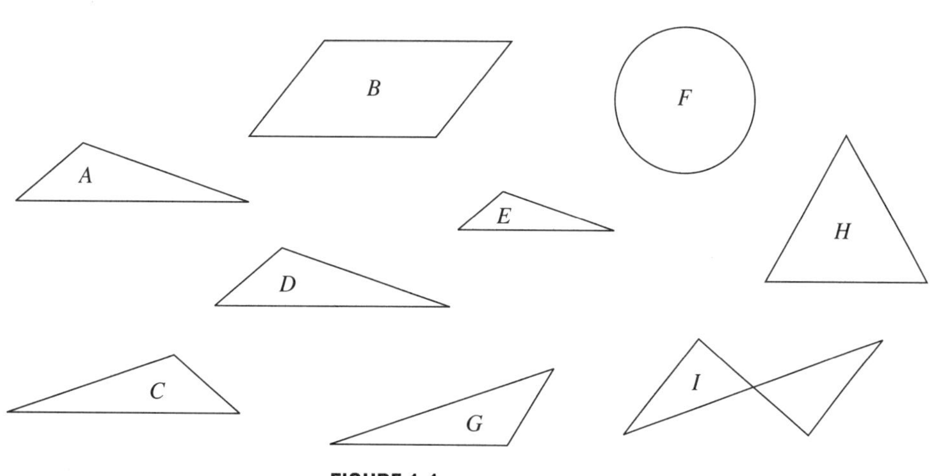

FIGURE 1.1
A variety of geometric figures

One person might say that since triangles A, C, D, and G are congruent to A, they are the only ones among the figures here that are the same as A. Another person might point out that G requires A to be rotated and C requires A to be reflected, so these are not really the same as A. Even D is a translation of A, so it is not exactly the same as A. Heraclitus, the philosopher of ancient Greece who questioned whether you can step twice into the same river, might object to saying that A is the same as itself.

On the other hand, E is a triangle similar to A (side lengths are proportional). Even H is like A in that they are both triangles. But B is also a polygon, and F is at least a simple closed curve. With a little twisting we might even bring I into the family.

The point is that there are many notions of equivalence among geometric figures. In surveying a plot of land, length and angle as well as location are important; for constructing a rigid framework, any triangle will do; to enclose a region of the plane, we can use any simple closed curve. We can choose the meaning of sameness to fit our purposes. This is also true when we consider the more flexible equivalences of topology. Nevertheless, all of the notions of sameness will satisfy three basic properties listed in the following definition. A straightforward verification is all you usually will need to confirm that a given relation satisfies these three properties.

Definition 1.2 *A binary relation $\sim$ on a set X is an* **equivalence relation** *if and only if for all $x, y, z \in X$ it satisfies*

1. $x \sim x$ *(reflexivity)*
2. *if $x \sim y$, then $y \sim x$* *(symmetry)*
3. *if $x \sim y$ and $y \sim z$, then $x \sim z$* *(transitivity)*

An equivalence relation on a set partitions the set into distinct classes. Each of these equivalence classes consists of objects all equivalent to the other objects in the class. Thus, any object in the class can be taken as a representative of the entire class. Exercise 2 at the end of this section asks you to establish the connection between an equivalence relation on a set and a partition of the set.

Example 1.3 Let $\mathbb{Z} = \{\ldots, -2, -1, 0, 1, 2, \ldots\}$ denote the set of integers. Consider the relation on the set

$$X = \{(a, b) \mid a \in \mathbb{Z} \quad \text{and} \quad b \in \mathbb{Z},\ b \neq 0\}$$

defined by $(a, b) \sim (c, d)$ if and only if $ad = bc$. Show that this is an equivalence relation, and interpret the equivalence classes as rational numbers.

Solution. The defining condition of the relation makes short work of verifying the three conditions for any ordered pairs of integers (a, b), (c, d), and (e, f) with $b \neq 0$, $d \neq 0$, and $f \neq 0$.

1. $(a, b) \sim (a, b)$ because $ab = ba$.
2. Suppose $(a, b) \sim (c, d)$. Then $ad = bc$. This can be rewritten as $cb = da$, which means that $(c, d) \sim (a, b)$.
3. Suppose $(a, b) \sim (c, d)$ and $(c, d) \sim (e, f)$. Then $ad = bc$ and $cf = de$. Since $b \neq 0$, $d \neq 0$, and $f \neq 0$, we can rewrite these equations as $\frac{a}{b} = \frac{c}{d}$ and $\frac{c}{d} = \frac{e}{f}$. Thus, $\frac{a}{b} = \frac{e}{f}$. Or, $af = be$. That is, $(a, b) \sim (e, f)$.

Since $(a, b) \sim (c, d)$ if and only if $\frac{a}{b} = \frac{c}{d}$, we can match up the equivalence class of all ordered pairs equivalent to (a, b) with the rational number $\frac{a}{b}$. This establishes a one-to-one correspondence between equivalence classes and rational numbers. ❄

Example 1.4 Consider the relation on the set $Y = \{(a, b) \mid a \in \mathbb{Z} \text{ and } b \in \mathbb{Z}\}$ defined by $(a, b) \sim (c, d)$ if and only if $ad = bc$. Show that this extension of the relation defined in Example 1.3 to the set of all ordered pairs of integers (with no restriction on the second coordinate) fails to be an equivalence relation.

Solution. The verifications of the reflexive and symmetric properties proceed just as in Example 1.3. We run into trouble with the transitive property. In particular, $(1, 1) \sim (0, 0)$ and $(0, 0) \sim (2, 1)$, yet $(1, 1)$ is not related to $(2, 1)$ as is required by transitivity. Thus, this relation is not an equivalence relation on Y. ❄

Notice how the failure of the transitive property in Example 1.4 prevents us from partitioning Y into disjoint equivalence classes. Every element in Y is related to $(0, 0)$, but not every element is related to an element such as $(1, 1)$. This illustrates one of the problems with trying to extend the rational numbers to allow for division by 0.

The granddaddy of all equivalence relations is of course the relation of equality. Although most people feel they understand this relation fairly clearly, it does present some philosophical difficulties. To begin with, isn't it rather silly to talk about two objects being equal? To say two objects are equal is to say they are the same object. That is, there is only one object, which of course is equal to itself.

Gottlob Frege brought this problem to the attention of philosophers in the late nineteenth century. Here is a specific example involving simple arithmetic. We are all familiar with the equality $3^2 + 4^2 = 5^2$. It brings to mind the Pythagorean Theorem of geometry and suggests searching for other identities involving the sums of squares of integers. Even Fermat's Last Theorem that $x^n + y^n = z^n$ has no positive integer solutions for $n > 2$ is related to this fact. But in stating that $3^2 + 4^2$ is identical to 5^2, aren't we just saying that $25 = 25$? If so, this trivial example of the reflexive property of equality certainly lacks the information content of the original equation.

Do we dare claim that $3^2 + 4^2$ is not equal to 5^2? Perhaps we should distinguish between different representations of numbers and claim that $3^2 + 4^2$ and 5^2 are only equivalent in the sense of having the same numerical value. This merely pushes the problem to another level. We still want to say that the numerical value of the expression $3^2 + 4^2$ is equal to the numerical value of 5^2. Why does this statement seem to have a greater information content than stating that the numerical value of 5^2 is equal to the numerical value of 5^2?

Nathan Salmon discusses various attempts to resolve this issue in his book *Frege's Puzzle* [9]. He suggests that statements about equality actually involve three terms. Two of the terms are the entities whose equality we are asserting. The third component is a reason or justification for the assertion. This justification is often suppressed. But to say $5^2 = 5^2$, we implicitly acknowledge the role of the reflexive property of equality. This allows us to contrast the equality $3^2 + 4^2 = 5^2$ whose verification involves arithmetic computations.

Exercises 1.1

1. Consider the set $D = \{0, 1, 2, 3, 4, 5, 6, 7, 8, 9\}$ of decimal digits.

(a) We want to count the number of possible ordered lists (permutations) of three distinct elements chosen from D. Find a systematic way to generate all such permutations. Without writing them all out, determine the number permutations of three distinct elements of D.

(b) Consider the relation $\sim$ on the set of these permutations defined by $(a, b, c) \sim (d, e, f)$ if and only if $\{a, b, c\} = \{d, e, f\}$; that is, the same three digits occur (in any order) in each list. Show that this is an equivalence relation.

(c) How many permutations are equivalent to $(0, 1, 2)$?

(d) How many permutations are equivalent to any ordered list (a, b, c) of three distinct digits?

(e) Among all the permutations of three distinct digits, how many distinct combinations (unordered lists) are there?

(f) Generalize these results to develop a formula for the number of combinations of k distinct digits for $k = 0, 1, \ldots, 10$.

(g) Generalize further to develop a formula for the number of combinations of k distinct elements chosen from a set of n elements.

2. Suppose $\sim$ is an equivalence relation on a set X. For any $x \in X$ let $C_x = \{y \in X \mid y \sim x\}$ be the **equivalence class** represented by x.

(a) Show that if $C_a \cap C_b \neq \emptyset$, then $C_a = C_b$.

(b) Show that the union $\bigcup_{a \in X} C_a$ of all the equivalence classes is equal to X.

(c) Suppose for each element α in some index set A that X_α is a nonempty subset of a set X. Suppose these subsets have the properties that if $X_\alpha \cap X_\beta \neq \emptyset$, then $X_\alpha = X_\beta$, and that $\bigcup_{\alpha \in A} X_\alpha = X$. Define an equivalence relation on X that has the sets X_α as equivalence classes.

(d) Show that your relation is an equivalence relation and that it has the indicated equivalence classes.

3. The equivalence relation defined in Example 1.3 can be used to define rational numbers as equivalence classes of ordered pairs of integers (with nonzero second coordinates). Of course, it would not be appropriate to use the concept of fractions prior to defining the rational numbers. Verify that this is an equivalence relation with a proof that does not use fractions. Suggestion: Consider two cases depending on whether $c = 0$ or

$c \neq 0$. In the case that $c \neq 0$, use the two equations $ad = bc$ and $cf = de$ to derive the equation $(af - be)dc = 0$. Apply the principle that a product of integers cannot equal 0 unless one of the factors is 0.

4. Exercise 3 shows how to define rational numbers as equivalence classes of ordered pairs of integers. In a similar fashion, we can define the integers (including zero and negative numbers) as equivalence classes of ordered pairs of positive integers. Let $\mathbb{N} = \{1, 2, \ldots\}$ denote the set of natural numbers. Consider the relation on the set $\{(a, b) \mid a \in \mathbb{N} \text{ and } b \in \mathbb{N}\}$ defined by $(a, b) \sim (c, d)$ if and only if $a + d = b + c$.

(a) Show that this is an equivalence relation. Give a proof that does not use the concept of negative numbers.

(b) Give a one-to-one correspondence between the equivalence classes and the set of integers.

(c) Define an operation of addition on the equivalence classes that is compatible with ordinary addition of the corresponding integers.

(d) Verify that your operation of addition is well-defined. That is, prove that the sum of two equivalence classes does not depend on the particular ordered pairs you used to define the sum.

(e) Which equivalence class is the additive identity?

(f) Show that each equivalence class has an additive inverse.

5. Consider the three binary relations on $\mathbb{Z}$ defined by $a \leq b$, $a \sim b$ if and only if $ab \neq 0$, and $a \approx b$ if and only if $|a - b| < 1$.

(a) Determine which of these relations is reflexive and symmetric, but not transitive.

(b) Determine which of these relations is reflexive and transitive, but not symmetric.

(c) Determine which of these relations is symmetric and transitive, but not reflexive.

6. Find interesting examples other than those given in Exercise 5 of binary relations that satisfy two of the properties required to be an equivalence relation but not the third.

7. A student argues that any relation that is symmetric and transitive will automatically be reflexive. After all, the symmetric law says that if we consider x and y with $x \sim y$, then $y \sim x$. Applying the transitive law to these two conditions yields that $x \sim x$.

(a) Explain how Exercise 5 indicates that something is wrong with this argument.

(b) Pinpoint the flaw in the student's argument.

(c) Give a weaker alternative to the reflexive condition that, together with symmetry and transitivity, will yield the reflexive property.

8. Lois Lane would be surprised to learn that Clark Kent can leap tall buildings in a single bound. She would not have the slightest hesitation accepting the fact that Superman can leap tall buildings in a single bound (except perhaps to wonder why anyone was telling her such a well-known bit of information). But since Clark Kent is Superman, how are these two revelations different? If two things are equal, we should be able to substitute one for the other in any statement about them.

9. Consider the relation $\sim$ defined on the set of polynomials by $p \sim q$ if and only if $p(t) = \pm t^k q(t)$ for some integer k.

(a) Verify that this is an equivalence relation.

(b) Which equivalence class has only one element?

(c) Show that every equivalence class, other than the one you found in part (b), can be uniquely represented by a polynomial with a positive constant term.

10. Two sequences $a_1, a_2, \ldots$ and $b_1, b_2, \ldots$ are **eventually equal** if and only if there is a natural number N such that if $n \geq N$, then $a_n = b_n$. Show that this defines an equivalence relation on the set of sequences of real numbers.

11. Let $\mathbb{R}$ denote the set of all real numbers (positive, negative, zero, rational, and irrational). A subset X of $\mathbb{R}$ has **measure zero** if and only if for every $\varepsilon > 0$ there is a sequence of intervals (a_n, b_n) such that

$$X \subseteq \bigcup_{n=1}^{\infty}(a_n, b_n) \qquad \text{and} \qquad \sum_{n=1}^{\infty}(b_n - a_n) < \varepsilon.$$

Informally, a set has measure zero if it can be covered by intervals of arbitrarily small total length. Two functions $f : \mathbb{R} \to \mathbb{R}$ and $g : \mathbb{R} \to \mathbb{R}$ are **equal almost everywhere** if and only if the set $\{x \in \mathbb{R} \mid f(x) \neq g(x)\}$ has measure zero.

(a) Show that the union of two sets of measure zero has measure zero.

(b) Show that the relation of being equal almost everywhere is an equivalence relation on the set of functions $f : \mathbb{R} \to \mathbb{R}$.

12. Let $\mathbb{R}$ denote the set of real numbers, and let $\mathbb{Q}$ denote the set of rational numbers. Consider the relation $\sim$ defined on $\mathbb{R}$ by $x \sim y$ if and only if $x - y \in \mathbb{Q}$.

(a) Show that $\sim$ is an equivalence relation on $\mathbb{R}$.

(b) Show that each equivalence class intersects $[0, 1)$.

(c) For each equivalence class, choose one of its elements that lies in $[0, 1)$. Let E be the set of all chosen elements. For any $r \in \mathbb{R}$, let $E_r = \{e + r \mid e \in E\}$. Show that if $x \in [0, 1)$, then $x \in E_r$ for some $r \in \mathbb{Q} \cap (-1, 1)$.

(d) Show that if $E_r \cap E_s \neq \emptyset$ for some $r, s \in \mathbb{Q}$, then $r = s$.

(e) Let

$$S = \bigcup_{r \in \mathbb{Q} \cap (-1,1)} E_r.$$

Show that $[0, 1) \subseteq S \subseteq (-1, 2)$.

(f) Show that if E has a linear measure of 0, then the linear measure of S is 0.

(g) Show that if E has positive linear measure, then S has infinite linear measure.

(h) What do you conclude about the measure of E?

1.2 Bijections

Equivalence relations between geometric objects are most commonly defined in terms of functions between objects. For example, two triangles are congruent if and only if there is a function from one onto the other that preserves the distances between pairs of points.

Likewise, two triangles are similar if and only if there is a function that preserves the ratios of distances between points. In these examples and many others, the functions preserve the set-theoretical structure of the geometric objects. That is, such a function establishes a correspondence between the points of the equivalent objects. You are probably familiar with the following definition.

Definition 1.5 *A function $f : X \to Y$ is an* ***injection*** *(or a* ***one-to-one*** *function) if and only if for any $x_1, x_2 \in X$ we have*

$$f(x_1) = f(x_2) \quad \Longrightarrow \quad x_1 = x_2.$$

The function is a ***surjection*** *(or an* ***onto*** *function) if and only if for every $y \in Y$ there is $x \in X$ with $f(x) = y$. The function is a* ***bijection*** *if and only if it is an injection and a surjection.*

Example 1.6 Find a bijection from the interval $(0, 1)$ to the real line $\mathbb{R}$.

Solution. The tangent function is a familiar bijection from $(-\frac{\pi}{2}, \frac{\pi}{2})$ onto $\mathbb{R}$. With a simple change of variable, we can cook up a function $f : (0, 1) \to \mathbb{R}$ defined by

$$f(x) = \tan\left(\pi x - \frac{\pi}{2}\right).$$

Based on the result of Exercise 23 at the end of this section, we can verify that f is a bijection simply by writing down the formula for the inverse:

$$f^{-1}(x) = \frac{1}{\pi}\arctan(x) + \frac{1}{2}.$$

❄

Example 1.7 Find a bijection from the interval $(0, 1)$ to the interval $[0, 1)$.

Solution. The inclusion $i : (0, 1) \to [0, 1)$ defined by $i(x) = x$ gives an injection, but the image does not include the endpoint 0. We can arbitrarily choose a point, say $\frac{1}{2}$, in $(0, 1)$ and remap it to 0. But then $\frac{1}{2}$ is no longer in the image. So choose another point, say $\frac{1}{3}$, and remap it to $\frac{1}{2}$. This fills one gap in the image but again creates another. It seems that this game will never end. But that is fine; after all, we are working with an infinite number of points. This idea leads to the function $g : (0, 1) \to [0, 1)$ defined by

$$g(x) = \begin{cases} 0 & \text{if } x = \dfrac{1}{2}, \\ \dfrac{1}{n-1} & \text{if } x = \dfrac{1}{n} \text{ for } n = 3, 4, 5, \ldots, \\ x & \text{otherwise.} \end{cases}$$

Once you see what is going on here, you can easily check that the inverse is defined by

$$g^{-1}(x) = \begin{cases} \dfrac{1}{2} & \text{if } x = 0, \\ \dfrac{1}{n+1} & \text{if } x = \dfrac{1}{n} \text{ for } n = 2, 3, 4, \ldots, \\ x & \text{otherwise.} \end{cases}$$

❄

Since topology is a branch of geometry, we will often forsake analytical formulas in favor of defining functions geometrically. Although it is theoretically possible to interpret a geometric description in terms of an analytical formula, the following example illustrates the advantage of sticking with geometry.

Example 1.8 Find a bijection from the unit circle $S^1 = \{(x, y) \in \mathbb{R}^2 \mid x^2 + y^2 = 1\}$ to the square $X = \{(x, y) \in \mathbb{R}^2 \mid \max\{|x|, |y|\} = 1\}$.

Solution. We define a function $r : S^1 \to X$ by radial projection as follows. For each point $p \in S^1$, draw a ray from the origin through p. Define $r(p)$ to be the point at which this ray intersects the square. It is geometrically clear that r is an injection and a surjection.

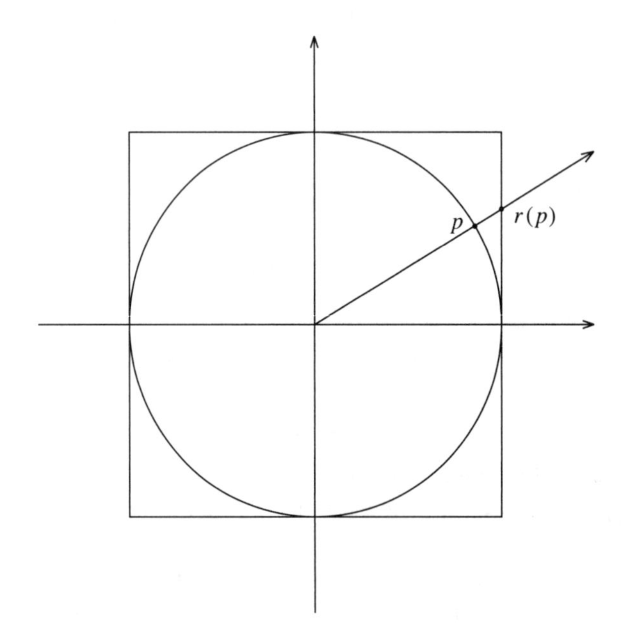

FIGURE 1.9
Radial projection from a circle onto a square.

The analytical description of r is slightly more complicated. The radial projection onto the square scales each point $p = (x, y)$ so the coordinate with the larger absolute value becomes ± 1. Thus,

$$r(x, y) = \begin{cases} \left(\dfrac{x}{|x|}, \dfrac{y}{|x|}\right) & \text{if } |x| \geq |y|, \\ \left(\dfrac{x}{|y|}, \dfrac{y}{|y|}\right) & \text{if } |y| \geq |x|. \end{cases}$$

Don't even think of deriving a formula for r^{-1} from the formula for r. However, from the geometric definition of r, notice that r^{-1} just shrinks each vector from the origin to a point of X to have unit length. Thus, $r^{-1}(p) = \frac{p}{||p||}$, or in terms of the coordinates of $p = (x, y)$,

$$r^{-1}(x, y) = \left(\frac{x}{\sqrt{x^2 + y^2}}, \frac{y}{\sqrt{x^2 + y^2}}\right).$$

❄

The following example treats a geometric object as a single point in the space of all such objects. We will frequently encounter these **configuration spaces**. The ability to find a bijection between the configuration space and a simpler set often sheds light on the organization of objects in the space.

Example 1.10 Find a subset of $\mathbb{R}^3$ and a bijection from the subset to the congruence classes of triangles.

Solution. Based on results of high-school geometry, we select from among the three sides and three angles a set of parameters that will specify the congruence class of a triangle. Perhaps the simplest is to use the lengths a, b, c of the three sides. Not all triples of three numbers determine lengths of sides of a triangle; we need the largest number to be less than the sum of the other two numbers. There is one additional problem. For example, $(2, 3, 4)$ and $(2, 4, 3)$ determine congruent triangles. Once we rule out this kind of duplication, we have a region $\{(a, b, c) \mid 0 < a \leq b \leq c < a + b\}$ so that each point corresponds to a unique congruence class of triangles and so that each triangle corresponds to a unique point in the region. ❄

You are undoubtedly familiar with the Cartesian coordinate plane as the set of ordered pairs of real numbers. The following definition extends this idea to the product of any two sets.

Definition 1.11 *The **Cartesian product** of two sets X and Y is the set, denoted $X \times Y$, consisting of all ordered pairs (x, y) with $x \in X$ and $y \in Y$. That is,*

$$X \times Y = \{(x, y) \mid x \in X \quad \text{and} \quad y \in Y\}.$$

Example 1.12 Describe a geometric bijection between a torus (the surface of a doughnut) and the Cartesian product $S^1 \times S^1 = \{((a, b), (c, d)) \in \mathbb{R}^2 \times \mathbb{R}^2 \mid a^2 + b^2 = 1 \text{ and } c^2 + d^2 = 1\}$ of two circles.

Solution. Imagine the torus sitting in $\mathbb{R}^3$ with the origin at the center of the doughnut hole and the xy-plane passing through the widest part of the surface. Consider the points of the torus whose polar angle (around the z-axis) is a certain value θ. This will be the circle formed by the intersection of the torus with a half-plane emanating from the z-axis at the angle θ. In this circle, measure a second angle φ with $\varphi = 0$ at the widest part of the torus and positive values of φ in the counterclockwise direction when you view the half-plane with the z-axis on your left.

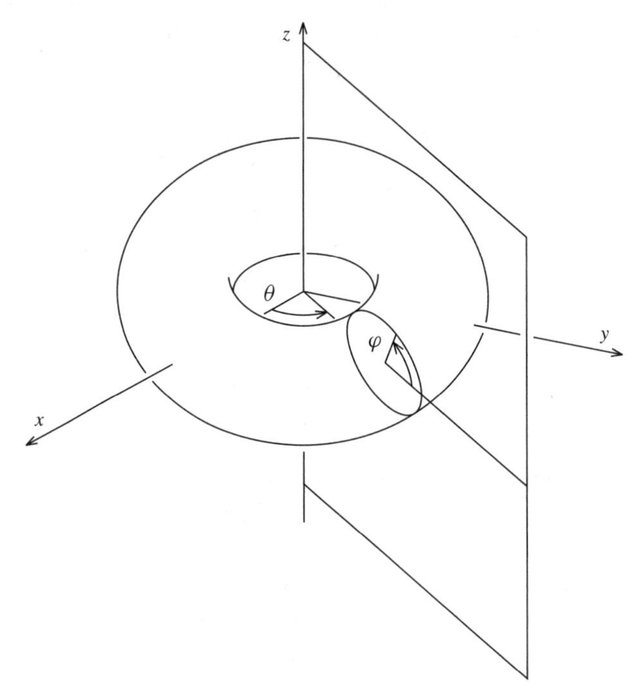

FIGURE 1.13
Toroidal coordinates θ and φ

Each of the angles θ and φ determines a point on a circle. Thus, each point of the torus determines an ordered pair (θ, φ) of angles, and this ordered pair determines a point of the Cartesian product $S^1 \times S^1$ of two circles.

Although a point on the torus only determines the angles θ and φ to within an integer multiple of 2π, such an addition to θ and φ will not change the point determined in $S^1 \times S^1$. Thus, the correspondence between the torus and $S^1 \times S^1$ is well-defined. Two points on the torus will either be in different half-planes from the z-axis and thus have different values of $\theta \in [0, 2\pi)$, or they will lie at different positions on the circle in a given half-plane and thus have a different value of $\varphi \in [0, 2\pi)$. Thus, the correspondence is injective. Any pair of angles (θ, φ) determines a point on the torus. Thus, the correspondence is surjective. ❄

Here is a summary of some basic concepts related to functions. Exercises 22 and 23 ask you to prove the two theorems.

Definition 1.14 *Suppose $f : X \to Y$ and $g : Y \to Z$ are functions. The* ***composition*** *of f followed by g is the function $g \circ f : X \to Z$ defined by $(g \circ f)(x) = g(f(x))$ for all $x \in X$.*

Definition 1.15 *For any set X, the* ***identity function*** *on X is the function $\mathrm{id}_X : X \to x$ defined by $\mathrm{id}_X(x) = x$ for all $x \in X$.*

Definition 1.16 *Consider a function $f : X \to Y$. A function $g : Y \to X$ is an* ***inverse*** *of the function f if and only if $g \circ f = \mathrm{id}_X$ and $f \circ g = \mathrm{id}_Y$.*

Theorem 1.17 *For any function $f : X \to Y$, we have $f \circ \mathrm{id}_X = f$ and $\mathrm{id}_Y \circ f = f$.*

Theorem 1.18 *Consider a function $f : X \to Y$. Then f is a bijection if and only if f has an inverse function.*

Exercises 1.2

1. Some people regard trigonometric functions as moderately exotic. Find a rational function (one whose formula only involves the four operations of arithmetic applied to real numbers and the independent variable) that is a bijection from $(0, 1)$ to $\mathbb{R}$.
2. Find a bijection from $(0, 1)$ to $[0, 1]$.
3. **(a)** Prove that the composition of injections is an injection.
   (b) Prove that the composition of surjections is a surjection.
   (c) Prove that the composition of bijections is a bijection.
4. Consider two functions $f : X \to Y$ and $g : Y \to Z$.
   (a) Suppose $g \circ f$ is an injection. Prove that f must be an injection. Give an example to show that g need not be an injection.
   (b) Suppose $g \circ f$ is a surjection. Prove that g must be a surjection. Give an example to show that f need not be a surjection
   (c) Suppose $g \circ f$ is a bijection. What can you say about f and g?
5. Give a geometric description of a bijection from the Cartesian product $S^1 \times [0, 1]$ to the annulus $\{(x, y) \in \mathbb{R}^2 \mid 1 \le x^2 + y^2 \le 4\}$.
6. Give a geometric description of a bijection from the sphere $S^2 = \{(x, y, z) \in \mathbb{R}^3 \mid x^2 + y^2 + z^2 = 1\}$ to the cube $X = \{(x, y, z) \in \mathbb{R}^3 \mid \max\{|x|, |y|, |z|\} = 1\}$.

7. Give a geometric description of a bijection between the sphere with the north pole removed $\{(x, y, z) \in \mathbb{R}^3 \mid x^2 + y^2 + z^2 = 1,\ z < 1\}$ and the plane $\mathbb{R}^2$.

8. Give a geometric description of a bijection between the circular disk

$$D^2 = \{(x, y) \in \mathbb{R}^2 \mid x^2 + y^2 \leq 1\}$$

and the square disk

$$[-1, 1] \times [-1, 1] = \{(x, y) \in \mathbb{R}^2 \mid \max\{|x|, |y|\} \leq 1\}.$$

9. Give a geometric description of a bijection between the lines in $\mathbb{R}^2$ through the origin and a subset of the unit circle S^1.

10. Give a geometric description of a bijection between the lines in $\mathbb{R}^3$ through the origin and a subset of the unit sphere $S^2 = \{(x, y, z) \in \mathbb{R}^3 \mid x^2 + y^2 + z^2 = 1\}$.

11. Find a subset of $\mathbb{R}^2$ and a bijection from the subset to the congruence classes of rectangles. Which points correspond to squares?

12. Find a subset of $\mathbb{R}^2$ and a bijection from the subset to the similarity classes of triangles. Which points correspond to isosceles triangles? Which point corresponds to the class of equilateral triangles?

13. Investigate the correspondence between the set of ordered pairs of real numbers (a, b) and the roots of the polynomials $x^2 + ax + b$.

14. Television sets and computer monitors display colors by combining light with various intensities of the three colors red, green, and blue. Thus, the color of any pixel can be described as an ordered triple $(r, g, b) \in [0, 1] \times [0, 1] \times [0, 1]$ where the coordinates specify the fraction of full intensity for each particular color. With this coordinate system, $(0, 0, 0)$ gives black, $(1, 0, 0)$ gives red, $(1, 0, 1)$ gives magenta, and $(1, 1, 1)$ gives white, for example. Alternatively, colors can be specified by their hue (across the spectrum of the rainbow), saturation (from washed out to the full intensity of the hue), and luminosity (from black to white).

(a) Investigate the correspondence between the red-green-blue coordinate system and the hue-saturation-luminosity coordinate system for colors.

(b) Look into other possible coordinate systems for defining colors (such as the cyan-magenta-yellow-black system), and describe how the different systems correspond.

15. Suppose $g : A \to C$ and $h : B \to D$ are bijections. Use g and h to define a bijection between the Cartesian products $A \times B = \{(a, b) \mid a \in A \text{ and } b \in B\}$ and $C \times D = \{(c, d) \mid c \in C \text{ and } d \in D\}$.

16. Consider the equivalence relation defined on the square region $B^2 = \{(x, y) \in \mathbb{R}^2 \mid |x| \leq 1 \text{ and } |y| \leq 1\}$ by $(x_1, y_1) \sim (x_2, y_2)$ if and only if either $(x_1, y_1) = (x_2, y_2)$, or $x_1 = x_2$ and $|y_1| = |y_2| = 1$.

(a) Draw a picture showing some typical equivalence classes.

(b) Illustrate a simple bijection between the set of equivalence classes and the points of the cylinder $\{(x, y, z) \in \mathbb{R}^3 \mid x^2 + y^2 = 1 \text{ and } 0 \leq z \leq 1\}$.

17. Consider the equivalence relation defined on the square region $B^2 = \{(x, y) \in \mathbb{R}^2 \mid |x| \le 1 \text{ and } |y| \le 1\}$ by $(x_1, y_1) \sim (x_2, y_2)$ if and only if $(x_1, y_1) = (x_2, y_2)$, or $x_1 = x_2$ and $|y_1| = |y_2| = 1$, or $|x_1| = |x_2| = 1$ and $y_1 = y_2$, or $|x_1| = |x_2| = 1$ and $|y_1| = |y_2| = 1$.

(a) Draw a picture showing some typical equivalence classes.

(b) Illustrate a simple bijection between the set of equivalence classes and the points of a torus.

18. Define an equivalence relation on the closed interval $[0, 1]$ so there is a simple bijection between the set of equivalence classes and the circle $S^1 = \{(x, y) \in \mathbb{R}^2 \mid x^2+y^2 = 1\}$.

19. Consider the equivalence relation defined on the unit disk $D^2 = \{(x, y) \in \mathbb{R}^2 \mid x^2 + y^2 \le 1\}$ by $(x_1, y_1) \sim (x_2, y_2)$ if and only if either $(x_1, y_1) = (x_2, y_2)$ or else $x_1^2 + y_1^2 = x_2^2 + y_2^2 = 1$, $x_1 = x_2$, and $y_1 = -y_2$.

(a) Draw a picture showing some typical equivalence classes.

(b) Illustrate a simple bijection between the set of equivalence classes and the points of the sphere $S^2 = \{(x, y, z) \mid x^2 + y^2 + z^2 = 1\}$.

20. **(a)** Define an equivalence relation on the union of two disjoint intervals that pairs up the left endpoints and pairs up the right endpoints. Show that there is a simple bijection between the set of equivalence classes and a circle.

(b) Define an equivalence relation on the union of two disjoint disks that pairs up corresponding points of the boundaries of the disks. Show that there is a simple bijection between the set of equivalence classes and a sphere.

(c) What happens if you extend this construction to the disjoint union of two three-dimensional balls?

(d) Can you generalize your results to higher dimensions?

21. A Möbius band is formed from a rectangular strip of paper by twisting one of the ends 180° and gluing it to the other end. Notice that the Möbius band has a single edge.

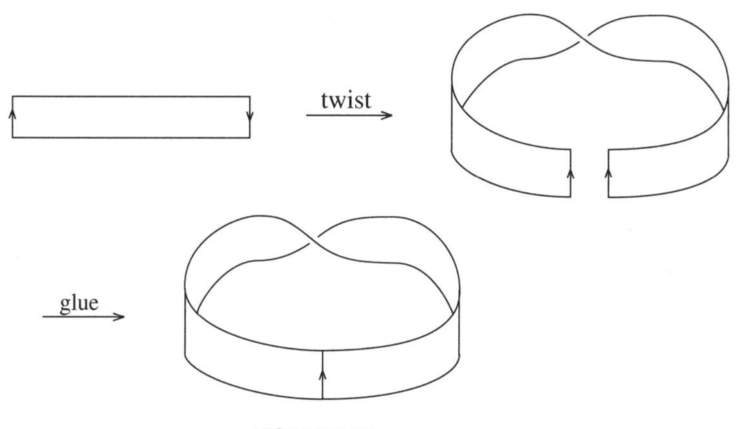

FIGURE 1.19
Forming a Möbius band.

This boundary is a simple closed curve that seems to go twice around the Möbius band. Define an equivalence relation on the rectangular strip so that there is a simple bijection between the set of equivalence classes and the points of the Möbius band.

22. Prove Theorem 1.17.

23. **(a)** Show that a function $f : X \to Y$ is a surjection if and only if there is a function $g : Y \to X$ such that $f \circ g = \mathrm{id}_Y$.

(b) Show that a function $f : X \to Y$ with nonempty domain X is an injection if and only if there is a function $g : Y \to X$ such that $g \circ f = \mathrm{id}_X$. How does this result break down if $X = \emptyset$?

(c) Show that $f : X \to Y$ is a bijection if and only if f has an inverse function.

(d) Consider a function $f : X \to Y$. Suppose $g : Y \to X$ is an inverse of the function f, and suppose $h : Y \to X$ is also an inverse of the function f. Prove that $g = h$.

1.3 Continuous Functions

As you look over the example of bijections in the previous section, you may notice that some functions did better than others at preserving the integrity of their domains. For example, although the bijection $\tan : (-\frac{\pi}{2}, \frac{\pi}{2}) \to \mathbb{R}$ does an infinite amount of stretching, points close enough together in the domain are mapped close to each other in the range. On the other hand, the bijection $g : (0, 1) \to [0, 1)$ defined in Example 1.7 by

$$g(x) = \begin{cases} 0 & \text{if } x = \dfrac{1}{2}, \\ \dfrac{1}{n-1} & \text{if } x = \dfrac{1}{n} \text{ for } n = 3, 4, 5, \ldots, \\ x & \text{otherwise} \end{cases}$$

tears $x = \frac{1}{2}$ away from nearby points in the sense that values of x near $\frac{1}{2}$ are mapped far away from $g(\frac{1}{2}) = 0$.

Continuity is the key property here. A continuous function preserves closeness of points. A discontinuous function maps arbitrarily close points to points that are not close. The precise definition of continuity involves the relation of distance between pairs of points. We can let $d(x, y)$ denote the distance between points x and y. For points in $\mathbb{R}$, for example, the usual measure of distance is $d(x, y) = |x - y|$. This extends to the distance between points $x = (x_1, \ldots, x_n)$ and $y = (y_1, \ldots, y_n)$ in $\mathbb{R}^n$:

$$d(x, y) = ||x - y|| = \sqrt{(x_1 - y_1)^2 + \cdots + (x_n - y_n)^2}.$$

The definition of continuity is rather technical, although you may recognize it from your study of calculus.

Definition 1.20 *A function* $f : X \to Y$ *is* ***continuous*** *at* $x_0 \in X$ *if and only if for every* $\varepsilon > 0$ *there is* $\delta > 0$ *such that for all* $x \in X$ *we have the implication* $d(x, x_0) < \delta \implies d(f(x), f(x_0)) < \varepsilon$. *A function is* ***continuous*** *if and only if it is continuous at each point of its domain.*

The implication $d(x, x_0) < \delta \implies d(f(x), f(x_0)) < \varepsilon$ is saying that if points x and x_0 are close together in X, then their images $f(x)$ and $f(x_0)$ are close together in Y. The choice of $\varepsilon > 0$ specifies the size of a target about $f(x_0)$ in which the image $f(x)$ must lie. Although a continuous function may do an enormous amount of stretching (consider $\tan : (-\frac{\pi}{2}, \frac{\pi}{2}) \to \mathbb{R}$ near the endpoints of its domain), we require for a specified target around $f(x_0)$ that all points sufficiently close to x_0 have images within this target region. The number ε represents the size of the target, and the number δ specifies how close points must be to x_0.

Example 1.21 Let $s : S^1 \to S^1$ be defined by $s(z) = z^2$ where we represent a point (x, y) of the unit circle S^1 as a complex number $z = x + iy$ and use multiplication of complex numbers to compute $z^2 = (x + iy)(x + iy) = x^2 + i^2y^2 + 2ixy = (x^2 - y^2) + i(2xy)$. Give a geometric argument that s is continuous.

Solution. We can write any point of S^1 as $(x, y) = (\cos\theta, \sin\theta)$ where θ is the angle from the positive x-axis to the ray from $(0, 0)$ through (x, y). In terms of complex numbers, this point can be written $z = x + iy = \cos\theta + i\sin\theta$. Notice that

$$\begin{aligned} s(z) &= (x^2 - y^2) + i(2xy) \\ &= (\cos^2\theta - \sin^2\theta) + i(2\cos\theta\sin\theta) \\ &= \cos 2\theta + i\sin 2\theta. \end{aligned}$$

Thus, the ray from 0 through $s(z)$ makes an angle of 2θ with the positive x-axis. Consequently, the angle between any two points z_0 and z is doubled when s is computed. In particular, the distance between $s(z)$ and $s(z_0)$ is never more than twice the distance between z and z_0. So if we want $d(s(z), s(z_0)) < \varepsilon$, we simply choose z within $\delta = \frac{\varepsilon}{2}$ of z_0. Then $d(z, z_0) < \delta \implies d(s(z), s(z_0)) \le 2d(z, z_0) < 2\delta = 2(\frac{\varepsilon}{2}) = \varepsilon$. ❄

Example 1.22 Let $D^2 = \{(x, y) \in \mathbb{R}^2 \mid x^2 + y^2 \le 1\}$ be the unit circular disk in $\mathbb{R}^2$ and let $B^2 = \{(x, y) \in \mathbb{R}^2 \mid \max\{|x|, |y|\} \le 1\}$ be the unit square disk. Let $r : D^2 \to B^2$ be the function that expands D^2 radially onto B^2. That is, any line segment from the origin to the boundary of D^2 is stretched linearly to a segment in the same direction from the origin to the boundary of B^2. Show that r is continuous.

Solution. The maximum stretching occurs near points that map to the corners of B^2. Even here the maximum amount of stretching is bounded by a constant factor K. That is, for any two points $x, x_0 \in D^2$ we have $d(r(x), r(x_0)) \le Kd(x, x_0)$. A person who was good with polar coordinates and had a fair amount of patience could even compute a value

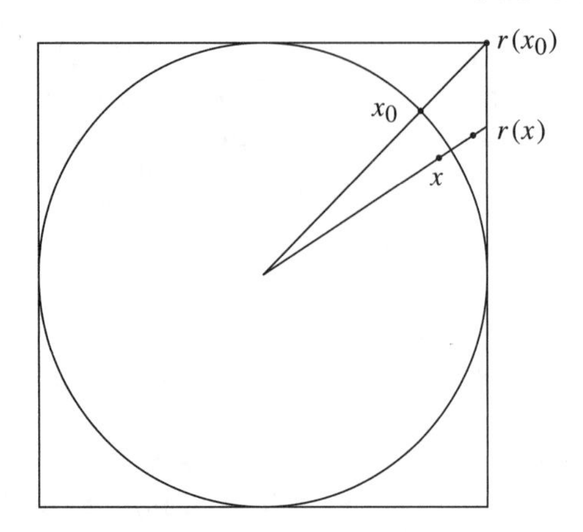

FIGURE 1.23
Radial stretching is continuous.

for K. However, the existence of such a bound is more important to the continuity of the function than its value. For any $\varepsilon > 0$ we can let $\delta = \frac{\varepsilon}{K}$. Now if $d(x, x_0) < \delta$, then $d(r(x), r(x_0)) \leq Kd(x, x_0) < K\delta = \varepsilon$. ❄

Example 1.24 Let M be a region of $\mathbb{R}^2$ with a finite area. Define $A : \mathbb{R} \to \mathbb{R}$ by letting $A(x)$ denote the area of the portion of M to the left of the vertical line at position x along a horizontal axis. Show that A is continuous.

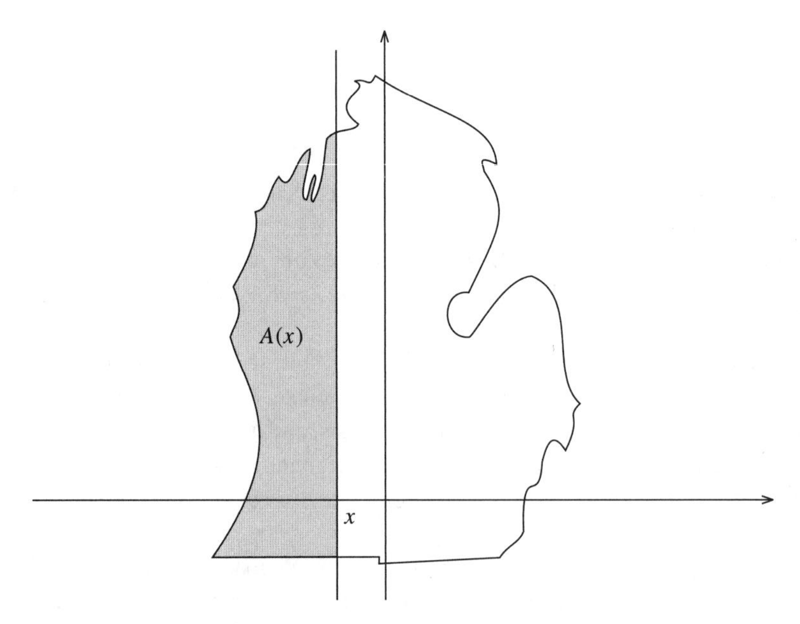

FIGURE 1.25
$A(x)$ is the area of the shaded region.

Solution. If M is bounded (see Figure 1.25), the argument is fairly simple. Let h be the difference between an upper bound and a lower bound on the y-coordinates of points in M. Then the area of the portion of M within a vertical strip of width δ is no larger than the area of the rectangle of width δ and height h. Thus, given $\varepsilon > 0$, we let $\delta = \varepsilon/(h+1)$. Then

$$|x - x_0| < \delta \Longrightarrow |A(x) - A(x_0)| \leq \delta \cdot h < \varepsilon.$$

If M is unbounded (see Figure 1.26), we need to do a little more work. The idea is to find a bounded subregion of M that contains enough of the area of M that we can make the above argument work with only minor adjustments. As always, we let $\varepsilon > 0$ be given. Choose a number $b > 0$ so that the portion of M between the horizontal lines $y = -b$ and $y = b$ is within $\frac{\varepsilon}{2}$ of the entire area of M. See Exercise 4 at the end of this section for an idea of how to find b. Now let $\delta = \frac{\varepsilon}{4b}$. The area of the portion of M bounded between $y = -b$ and $y = b$ and lying within a vertical strip of width δ is less than $\delta \cdot (2b) = \frac{\varepsilon}{2}$. The area of the rest of M lying within the vertical strip is less than $\frac{\varepsilon}{2}$. Thus, $|x - x_0| < \delta \Longrightarrow |A(x) - A(x_0)| < \frac{\varepsilon}{2} + \frac{\varepsilon}{2} = \varepsilon$. ❄

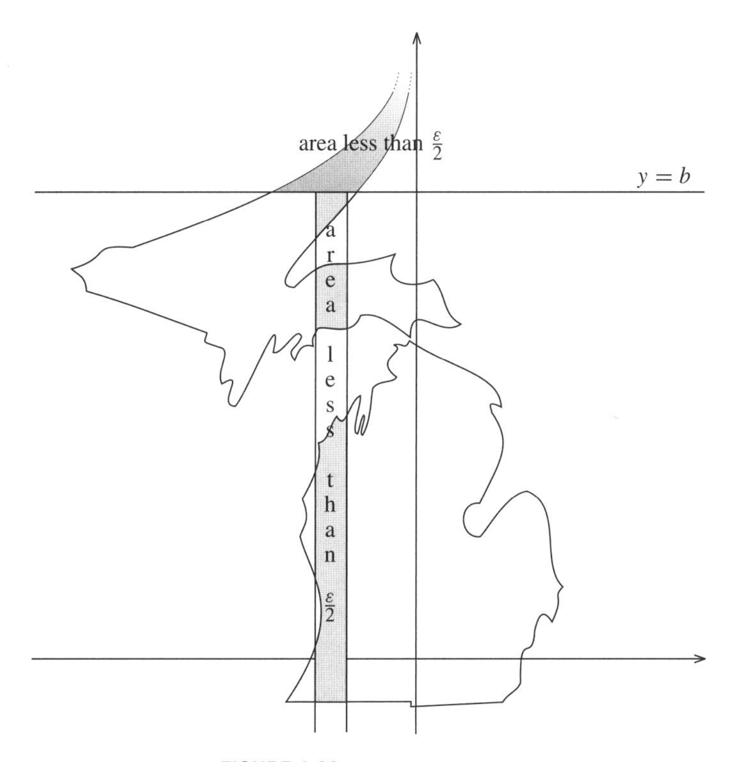

FIGURE 1.26
The area function is continuous.

The following theorem is useful in a variety of situations where we want to paste together some continuous functions to form a new continuous function. The proof involves a little good, old-fashioned ε and δ chasing.

Theorem 1.27 *Suppose A and B are regions of $\mathbb{R}^2$ that are bounded by polygons. Suppose $f : A \to Y$ and $g : B \to Y$ are continuous functions such that $f(x) = g(x)$ for all $x \in A \cap B$. Then the function $h : A \cup B \to Y$ defined by*

$$h(x) = \begin{cases} f(x) & \text{if } x \in A, \\ g(x) & \text{if } x \in B \end{cases}$$

is continuous.

Proof. We need to show that h is continuous at each point x_0 of its domain. First consider the case that $x_0 \in A$, but $x_0 \notin B$. Because x_0 is outside the polygonal region B, it will be some positive distance r from the nearest point of B. Since $f : A \to Y$ is continuous, for a given $\varepsilon > 0$, there is $\delta > 0$ such that for all $x \in A$ we have $d(x, x_0) < \delta \implies d(f(x), f(x_0)) < \varepsilon$. Modify the choice of δ if necessary to make sure it is less than r. Then any point $x \in A \cup B$ with $d(x, x_0) < \delta$ is guaranteed not to be in B. Thus, for all $x \in A \cup B$, we have $d(x, x_0) < \delta \implies d(h(x), h(x_0)) < \varepsilon$. A similar argument shows that h is continuous at all points with $x_0 \in B$ but $x_0 \notin A$.

Now consider the case that $x_0 \in A \cap B$. Let $\varepsilon > 0$ be given. Since $f : A \to Y$ is continuous, there is $\delta_1 > 0$ such that for any $x \in A$ we have $d(x, x_0) < \delta_1 \implies d(f(x), f(x_0)) < \varepsilon$. Since $g : B \to Y$ is continuous, there is $\delta_2 > 0$ such that for any $x \in B$ we have $d(x, x_0) < \delta_2 \implies d(g(x), g(x_0)) < \varepsilon$. Let δ be the smaller of the two numbers δ_1 and δ_2. Then for all $x \in A \cup B$, we have

$$\begin{aligned} d(x, x_0) < \delta &\implies d(x, x_0) < \delta_1 \quad \text{and} \quad d(x, x_0) < \delta_2 \\ &\implies d(f(x), f(x_0)) < \varepsilon \text{ if } x \in A, \quad \text{and} \quad d(g(x), g(x_0)) < \varepsilon \text{ if } x \in B \\ &\implies d(h(x), h(x_0)) < \varepsilon. \end{aligned}$$

❄

Exercises 1.3

1. Suppose that $f : X \to Y$ is a constant function. That is, for all $x_1, x_2 \in X$ we have $f(x_1) = f(x_2)$. Prove that f is continuous.

2. Suppose $f : X \to Y$ is continuous and $g : Y \to Z$ is continuous. Prove that the composition $g \circ f : X \to Z$ is continuous.

3. Suppose $f : X \to Y$ is a continuous function. Let A be a subset of X. Show that the restriction $f|_A : A \to Y$ is continuous.

4. Suppose M is a region of $\mathbb{R}^2$ with finite area. For any nonnegative integer n let A_n be the area of the portions of M between the horizontal lines $y = n$ and $y = n + 1$ and between the horizontal lines $y = -n$ and $y = -(n + 1)$.
   (a) Draw a picture to show that the area of M is equal to $\sum_{n=0}^{\infty} A_n$.
   (b) For any $\varepsilon > 0$, show that there is an integer N such that $\sum_{n=N+1}^{\infty} A_n < \varepsilon$.

5. Show that $K = 2$ is a bound for the stretching factor in Example 1.22.

6. Let X be the rectangular region $[-4, 4] \times [-2, 2]$ with the square region $(-3, -1) \times (-1, 1)$ removed. Let Y be the rectangular region $[-4, 4] \times [-2, 2]$ with the square region $(1, 3) \times (-1, 1)$ removed. Let $s : X \to Y$ be the bijection obtained by sliding the square region $(-3, -1) \times (-1, 1)$ to $(1, 3) \times (-1, 1)$ and letting segments from the outer boundary to the inner boundary stretch or shrink linearly.

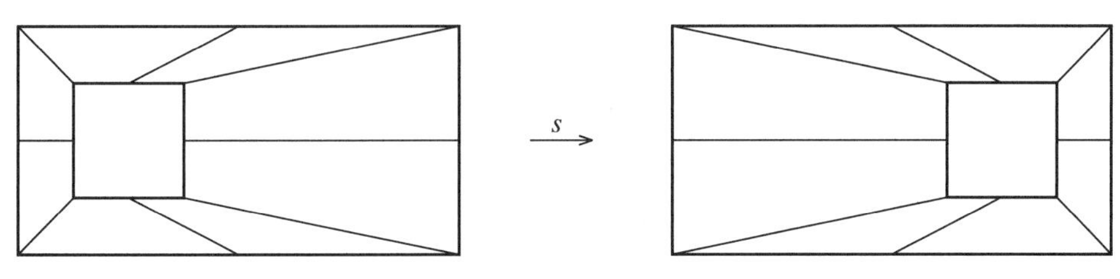

FIGURE 1.28
A sliding square

(a) Where does the greatest stretching take place?

(b) What is an upper bound for the stretching factor?

(c) Given $\varepsilon > 0$, choose δ in terms of ε and a bound for the stretching factor.

(d) Show that $d(x, x_0) < \delta \Longrightarrow d(s(x), s(x_0)) < \varepsilon$.

7. Consider the reciprocal function $r : (0, 1) \to \mathbb{R}$ defined by $r(x) = \frac{1}{x}$.

(a) Where does the greatest stretching take place?

(b) Show that there is no upper bound for the stretching factor. That is, for any $K > 0$ show there are points $x, y \in (0, 1)$ with $|r(x) - r(y)| > K|x - y|$.

(c) Show that r is continuous.

8. (a) Compare the length of the arc of the unit circle with the length of the straight line between $(\cos a, \sin a)$ and $(\cos b, \sin b)$ to show that $|\cos a - \cos b| \leq |a - b|$ and $|\sin a - \sin b| \leq |a - b|$ for all $a, b \in \mathbb{R}$.

(b) Conclude that $\sin : \mathbb{R} \to \mathbb{R}$ and $\cos : \mathbb{R} \to \mathbb{R}$ are continuous functions.

9. Consider a function $f : X \to Y$ that satisfies $d(f(x_1), f(x_2)) \leq d(x_1, x_2)$ for all points $x_1, x_2 \in X$. Prove that f is continuous.

10. Find a bijection from the open interval $(0, 1)$ to the unit circle $S^1 = \{(x, y) \in \mathbb{R}^2 \mid x^2 + y^2 = 1\}$. Is your function continuous? Is its inverse continuous?

11. The definition of continuity depends on the measure of the distances between points in the domain and in the range of the function. It is interesting to see how different ways of measuring distance result in different notions of continuity. For example, consider the **discrete metric** defined on any set X by the simple rule

$$d(x, y) = \begin{cases} 0 & \text{if } x = y, \\ 1 & \text{if } x \neq y. \end{cases}$$

With this measure of distance, any two distinct points are exactly one unit apart. This leads to two extreme situations involving continuity.

(a) Suppose X is a space with the discrete metric. Show that any function $f : X \to Y$ is continuous.

(b) Suppose Y is a space with the discrete metric. Show that if a function $f : \mathbb{R} \to Y$ is continuous, then f must be a constant function.

12. Recall from your calculus course that a sequence $a_1, a_2, \ldots$ **converges** to a limit L if and only if for all $\varepsilon > 0$ there is a positive integer N such that if $n \geq N$, then $d(a_n, L) < \varepsilon$. The use of the distance function d to measure how close a_n is to the limit L shows that this definition of convergence can be extended to Euclidean spaces of any dimension. In fact, it can be adapted to any space that has a reasonable way to measure distance.

(a) Suppose $f : A \to B$ is continuous and $a_1, a_2, \ldots$ is a sequence of points in A that converges to a point L in A. Show that in B the sequence $f(a_1), f(a_2), \ldots$ converges to $f(L)$.

(b) Suppose $f : A \to B$ has the property that whenever a sequence $a_1, a_2, \ldots$ converges to a limit L in A, then the image sequence $f(a_1), f(a_2), \ldots$ converges to $f(L)$ in B. Prove that f is continuous.

1.4 Topological Equivalence

In this chapter we have looked at bijections, those functions that preserve set-theoretical structure. We have also looked at continuity, a property of functions that allows stretching, shrinking, and folding, but preserves the closeness relation among points. We now want to combine these two concepts and introduce the principal notion of equivalence in topology.

In the following definition, X and Y will denote sets with a concept of distance. The ability to measure distances is used to define continuity. You should think of X and Y as subsets of Euclidean spaces with distances measured as usual. The Euclidean distance function is a special case of the more general concept of a metric. We will explore this useful generalization in Section 7.1 on metric spaces. In fact, the concept of continuity can be further extended with the general idea of topological spaces. In Section 7.2 we will see how a notion of nearness can be defined without a specific measure of distance.

Definition 1.29 *A* ***homeomorphism*** *(or* ***topological equivalence****) is a bijection $h : X \to Y$ such that both h and h^{-1} are continuous. The spaces X and Y are* ***homeomorphic*** *(or* ***topologically equivalent****) if and only if there is a homeomorphism from X to Y.*

Example 1.30 Show that the radial stretching function $r : D^2 \to B^2$ of Example 1.22 is a homeomorphism.

Solution. We first check the two conditions for r to be a bijection. Because radial line segments of D^2 are stretched to lie along line segments in the same direction from the origin, points with the same image must lie on the same radius of D^2. Because these segments are stretched linearly, such points must in fact be identical. Therefore, r is an injection. Any point of B^2 is on a line segment from the origin to the boundary of B^2. The radius of D^2 in the same direction contains a point at a distance from the origin proportional to the distance the given point lies along the line segment of B^2. This point maps to the given point. Therefore, r is a surjection.

In Example 1.22 we verified that r is continuous. Since r^{-1} moves points closer together, Exercise 9 of Section 1.3 shows that the inverse of r is also continuous. Putting these facts together, we conclude that r is a homeomorphism. ❄

The following example illustrates the importance of requiring continuity for both a function and its inverse in the definition of homeomorphism. We certainly would not want to consider an interval to be topologically equivalent to a circle.

Example 1.31 Consider the continuous bijection $w : [0, 2\pi) \to S^1$ that wraps the interval $[0, 2\pi)$ around the unit circle S^1 according to the formula $w(t) = (\cos t, \sin t)$. Show that $w^{-1} : S^1 \to [0, 2\pi)$ is not continuous.

Solution. Since $w(0) = (\cos 0, \sin 0) = (1, 0)$, we have $w^{-1}((1, 0)) = 0$. But w^{-1} maps points near $(1, 0)$ (those below the x-axis, in particular) to points near the right end of $[0, 2\pi)$. So let $\varepsilon = 1$. For any $\delta > 0$ there will be points below the x-axis within δ of $(1, 0)$ that w^{-1} maps to points of $[\pi, 2\pi)$. That is, there are points $(x, y) \in S^1$ with $d((x, y), (1, 0)) < \delta$ and $d(w^{-1}(x, y), w^{-1}(1, 0)) \geq \pi > 1 = \varepsilon$. Therefore, w^{-1} is not continuous at $(1, 0)$. ❄

We are now in a position to define geometric objects from a topologist's point of view. Among a collection of homeomorphic objects, we choose one that is particularly simple or easy to define and designate it as the standard object. We then call all topologically equivalent spaces by the same name. It is usually easy to tell from the context of a discussion whether we are referring to the standard object or to an arbitrary object in the equivalence class. The following definition gives some typical examples of this kind of language.

Definition 1.32 *The **standard disk** is the set* $\{(x, y) \in \mathbb{R}^2 \mid x^2 + y^2 \leq 1\}$. *A **disk** is any topological space homeomorphic to the standard disk.*

*The **standard n-dimensional ball** (or more simply, the **standard n-ball**) is the set* $\{(x_1, x_2, \ldots, x_n\}) \in \mathbb{R}^n \mid x_1^2 + x_2^2 + \cdots + x_n^2 \leq 1\}$. *An **$n$-ball** (or **$n$-cell**) is any topological space homeomorphic to the standard n-ball.*

*The **standard n-dimensional sphere** (or more simply, the **standard n-sphere**) is the set* $\{(x_1, x_2, \ldots, x_{n+1}) \in \mathbb{R}^{n+1} \mid x_1^2 + x_2^2 + \cdots + x_{n+1}^2 = 1\}$. *An **$n$-sphere** is any topological space homeomorphic to the standard n-sphere.*

Example 1.33 Show that the circle $S^1 = \{(x, y) \in \mathbb{R}^2 \mid x^2 + y^2 = 1\}$ is homeomorphic to the knot K illustrated in the figure.

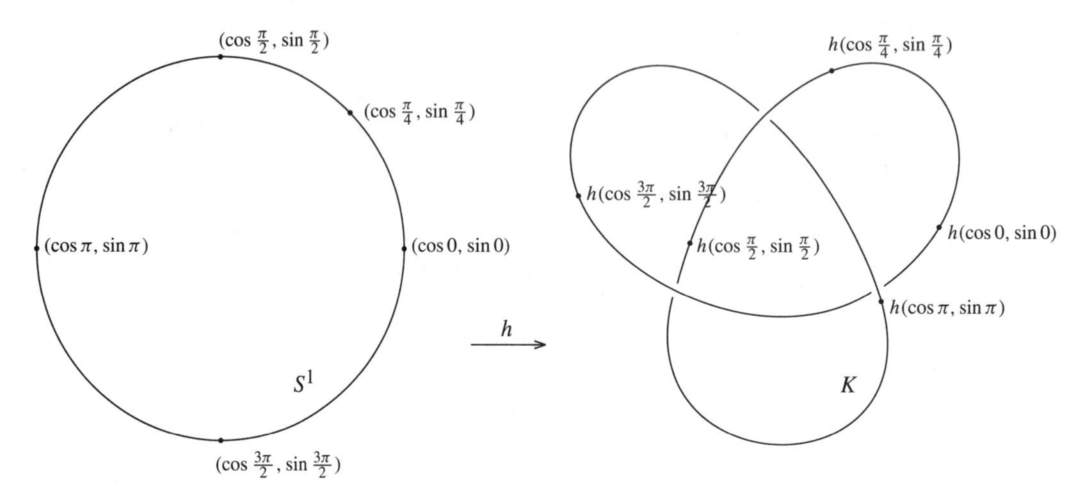

FIGURE 1.34
The circle and knot are homeomorphic

Solution. Break the circle S^1 open at $(0, 1)$. Map one endpoint of the resulting arc to an arbitrary point of the knot K. Lay the arc along the knot with a constant amount of stretching so that when you get to the end of the arc you are back to your starting point on K. This will result in a well-defined bijection $h : S^1 \to K$. Because there is a bound on the amount of stretching h does to the distances between pairs of points on S^1, it follows that h is continuous. Similarly, h^{-1} is continuous. Thus, h is a homeomorphism. ❄

Example 1.35 Show that a 2-sphere with a point removed is homeomorphic to the plane.

Solution. By translation and scaling, we can assume the sphere is the standard sphere $S^2 = \{(x, y, z) \in \mathbb{R}^3 \mid x^2 + y^2 + z^2 = 1\}$. By rotation, we can assume that the point removed is the north pole, $(0, 0, 1)$. A ray from the north pole through another point (x, y, z) on the sphere will intersect the xy-plane P in a point we can denote by $T(x, y, z)$. The mapping $T : S^2 - \{(0, 0, 1)\} \to P$ is called stereographic projection.

It is fairly clear that T is a homeomorphism. But it is also reassuring to be able to derive formulas for T and T^{-1} to confirm that T is a continuous bijection with a continuous inverse. Let r denote the distance $T(x, y, z)$ is from the origin. The similar triangles in Figure 1.36 give that

$$\frac{r}{1} = \frac{\sqrt{x^2 + y^2}}{1 - z}.$$

Note that $T(x, y, z)$ is a scalar multiple of $(x, y, 0)$. These two facts give

$$T(x, y, z) = \left(\frac{x}{1 - z}, \frac{y}{1 - z}, 0\right)$$

as the point in P in the right direction at the right distance from the origin. Exercise 5 asks you to check that the formula for $T^{-1} : P \to S^2 - \{(0, 0, 1)\}$ is

$$T^{-1}(u, v, 0) = \left(\frac{2u}{u^2 + v^2 + 1}, \frac{2v}{u^2 + v^2 + 1}, \frac{u^2 + v^2 - 1}{u^2 + v^2 + 1} \right).$$

Thus T is a bijection, and T and T^{-1} are both continuous since their formulas use only the operations of arithmetic. Therefore, T is a homeomorphism from $S^2 - \{(0, 0, 1)\}$ to the plane P. ❄

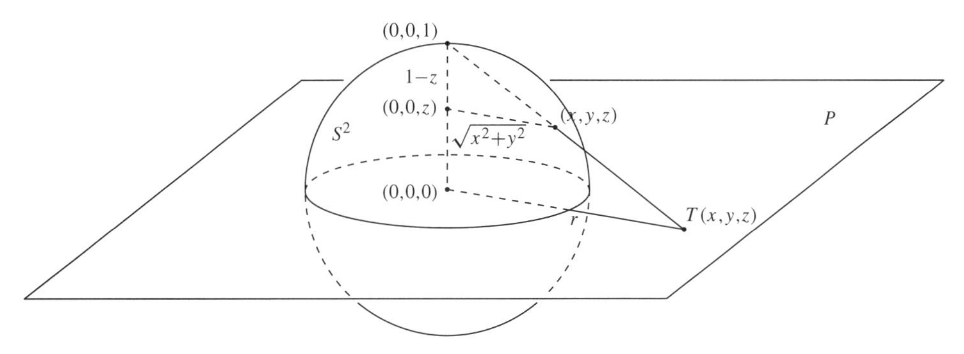

FIGURE 1.36
Stereographic projection from $S^2 - \{(0, 0, 1)\}$ to the xy-plane P.

Exercises 1.4

1. Show that the homeomorphism relation is an equivalence relation.

2. Show that the half-open interval $[0, 1)$ is homeomorphic to the closed ray $[0, \infty)$. Suggestion: Cook up a simple algebraic expression for an increasing function that maps 0 to 0 and has a vertical asymptote at 1, or modify one of your favorite trigonometric functions.

3. **(a)** Show that any open interval (a, b) is homeomorphic to the open interval $(0, 1)$.
   (b) Show that any open ray (a, ∞) or $(-\infty, b)$ is homeomorphic to the open interval $(0, 1)$.
   (c) Show that the real line $\mathbb{R}$ is homeomorphic to the open interval $(0, 1)$.
   (d) Conclude that any two open intervals of $\mathbb{R}$ are homeomorphic.

4. Consider the wrapping function $w : [0, 2\pi) \to S^1$ defined by the formula $w(t) = (\cos t, \sin t)$ as in Example 1.31.
   (a) Show that w is a bijection.
   (b) Use the continuity of the sine and cosine functions to show that w is continuous.

5. Consider the function T from $S^2 - \{(0, 0, 1)\}$ to the xy-plane P defined by

$$T(x, y, z) = \left(\frac{x}{1 - z}, \frac{y}{1 - z}, 0 \right)$$

   as in Example 1.35.

(a) For any $(u, v, 0) \in P$, show that

$$\left(\frac{2u}{u^2+v^2+1}, \frac{2v}{u^2+v^2+1}, \frac{u^2+v^2-1}{u^2+v^2+1}\right) \in S^2 - \{(0,0,1)\}.$$

(b) Show that

$$T^{-1}(u, v, 0) = \left(\frac{2u}{u^2+v^2+1}, \frac{2v}{u^2+v^2+1}, \frac{u^2+v^2-1}{u^2+v^2+1}\right)$$

by verifying that $T \circ T^{-1} = \mathrm{id}_P$ and $T^{-1} \circ T = \mathrm{id}_{S^2-\{(0,0,1)\}}$.

(c) Conclude that T is a bijection.

6. Homeo, homeo, wherefore art thou homeo?

7. **(a)** Give a formula for a function $f : [-2, 2] \to [-2, 2]$ that stretches the interval $[-2, -1]$ linearly onto $[-2, 1]$ and shrinks the interval $[-1, 2]$ linearly onto $[1, 2]$.

(b) Show that f is a homeomorphism of $[-2, 2]$ onto itself.

(c) Show that f extends to a homeomorphism $F : \mathbb{R} \to \mathbb{R}$ by using the identity function outside the interval $(-2, 2)$.

8. We often want to slide points around inside a set. This exercise illustrates the technical details needed to move the point $(-1, 0)$ to the point $(1, 0)$ inside a rectangle surrounding these two points. Once you see what is involved in this example, you should be willing to forgo the details involved in more general point-sliding. Consider the rectangular region $R = [-2, 2] \times [-1, 1]$. Let $R_0 = \{(x, y) \in R \mid |x| + |y| \leq 1\}$ be the diamond-shaped region in the center of R. Let $R_- = \{(x, y) \in R \mid -x + |y| \geq 1\}$ be the portion of R to the left of R_0. And let $R_+ = \{(x, y) \in R \mid x + |y| \geq 1\}$ be the portion of R to the right of R_0.

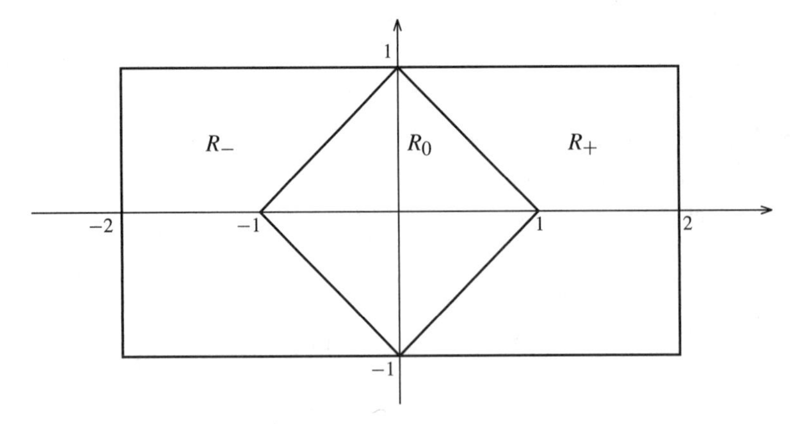

FIGURE 1.37
Stretching R_- onto $R_- \cup R_0$

(a) Give a formula for a function $f : R \to R$ that moves points horizontally mapping R_- onto $R_- \cup R_0$ and $R_0 \cup R_+$ onto R_+.
(b) Show that f is a homeomorphism of R onto itself.
(c) Show that f extends to a homeomorphism $F : \mathbb{R}^2 \to \mathbb{R}^2$ by using the identity function outside of R.

9. Consider a circular disk $\{(x, y) \in \mathbb{R}^2 \mid x^2 + (y-1)^2 \leq 1\}$ of unit radius centered at $(0, 1)$. We want to slide the vertical chords of this disk down so their lower endpoints are on the x-axis.
(a) Show that the image of the disk will be the upper half of an elliptical region

$$\left\{(x, y) \in \mathbb{R}^2 \,\middle|\, x^2 + \frac{y^2}{4} \leq 1, y \geq 0\right\}.$$

(b) In terms of the coordinates of a point (x, y) in the disk, how far must we slide the point?
(c) Write a formula for the sliding function

$$s : \{(x, y) \in \mathbb{R}^2 \mid x^2 + (y-1)^2 \leq 1\} \to \left\{(x, y) \in \mathbb{R}^2 \,\middle|\, x^2 + \frac{y^2}{4} \leq 1, y \geq 0\right\}.$$

(d) Show that s is a homeomorphism.

10. Show that the open disk $\{(x, y) \in \mathbb{R}^2 \mid x^2 + y^2 < 1\}$ is homeomorphic to $\mathbb{R}^2$. Suggestion: Use the homeomorphism from Exercise 2 to perform radial stretching.

11. Show that the disk with its center removed, $\{(x, y) \in \mathbb{R}^2 \mid 0 < x^2 + y^2 \leq 4\}$, is homeomorphic to the disk with a hole, $\{(x, y) \in \mathbb{R}^2 \mid 1 < x^2 + y^2 \leq 4\}$.

12. Suppose $g : A \to C$ and $h : B \to D$ are homeomorphisms. Use g and h to define a homeomorphism between the Cartesian products $A \times B = \{(a, b) \mid a \in A \text{ and } b \in B\}$ and $C \times D = \{(c, d) \mid c \in C \text{ and } d \in D\}$.

13. Use Definition 1.32 as a guide in creating your own definitions.
(a) Give a topologist's definition of a circle.
(b) Give a topologist's definition of a torus.
(c) Give a topologist's definition of a theta curve (a geometric figure that looks like θ, the Greek letter theta).

14. **(a)** Give a geometric description of a homeomorphism from the quarter plane $\{(x, y) \in \mathbb{R}^2 \mid x \geq 0 \text{ and } y \geq 0\}$ to the half plane $\{(x, y) \in \mathbb{R}^2 \mid y \geq 0\}$.
(b) Give a geometric description of a homeomorphism from the quarter space $\{(x, y, z) \in \mathbb{R}^3 \mid y \geq 0 \text{ and } z \geq 0\}$ to the half space $\{(x, y, z) \in \mathbb{R}^3 \mid z \geq 0\}$.

15. Give a geometric description of a homeomorphism from $[0, 1) \times [0, 1)$ to $[0, 1] \times [0, 1)$.

1.5 Topological Invariants

The easiest way to show two figures are homeomorphic is often to construct an explicit homeomorphism between them. But what if two figures are not homeomorphic? Surely we cannot be expected to check every function between the sets and show that it is not a homeomorphism. One of the goals of the field of topology is to discover easier ways of detecting the differences between spaces that are not homeomorphic.

Intuitively, two spaces that are homeomorphic have the same general shape in spite of possible deformations of distance and angle. Thus, if two spaces are not homeomorphic, they will tend to look distinctly different. Our job is to specify the difference. To do this rigorously, we need to define some property of topological spaces and show that the property is preserved under transformations by any homeomorphism. Then if one space has the property and the other one does not have the property, there is no way they can be homeomorphic.

Definition 1.38 *A property that is preserved under homeomorphisms is called a* ***topological invariant****.*

The following example illustrates how a topological invariant enables us to distinguish spaces that are not homeomorphic.

Example 1.39 Show the subsets $\{1, 2, 3\}$ and $\{1, 2, 3, 4\}$ of $\mathbb{R}$ are not homeomorphic.

Solution. Consider the property of being a three-element set. Since homeomorphisms are bijections, they preserve the number of elements in a set. Thus, the property of having three elements is a topological invariant. Since $\{1, 2, 3\}$ has the property and $\{1, 2, 3, 4\}$ does not, we know these two sets are not homeomorphic. ❄

This simple example can be generalized to the result that any two homeomorphic sets will have the same cardinality. Thus, cardinality is the first invariant to check if you suspect two figures are not homeomorphic.

Fine. However, most interesting geometric sets have the cardinality of the set of real numbers. So this counting argument will not work directly on the points of the set. But we can play a similar game with the connected pieces of the figures. One way to decide whether two points are in the same connected piece of a set is to try to draw a path within the set from one point to the other. The following definition makes this idea precise. Exercise 1 at the end of this section asks you to verify that the relation described in this definition is in fact an equivalence relation.

Definition 1.40 *A* ***path*** *in a space X is a continuous function $\alpha : [0, 1] \to X$. Consider the equivalence relation between pairs of points in a set X defined by $x \sim y$ if and only if there is a path $\alpha : [0, 1] \to X$ with $\alpha(0) = x$ and $\alpha(1) = y$. The equivalence classes under this relation are called the* ***path components*** *of X. A set such that every two points are joined by a path is said to be* ***path-connected****.*

Theorem 1.44 will facilitate our ability to identify path components of a space. Its proof relies on a fundamental property of the real numbers. So we first have to introduce some concepts that you may have seen (or skipped over) in your calculus course.

Definition 1.41 *A number c is an* ***upper bound*** *for a set $A \subseteq \mathbb{R}$ if and only if $x \leq c$ for all $x \in A$. A number is a* ***least upper bound*** *for A if and only if it is an upper bound for A and no smaller number is an upper bound for A.*

Completeness Property of the Real Numbers 1.42 *Any nonempty subset of $\mathbb{R}$ with an upper bound has a least upper bound.*

The Completeness Property of the Real Numbers is the basis for many results that we usually take for granted.

Example 1.43 Show that there is a real number c whose square is 2.

Solution. Consider the set $S = \{x \in \mathbb{R} \mid x^2 \leq 2\}$. Since $1^2 = 1 \leq 2$, we know that $1 \in S$, and hence S is nonempty. Notice that if $x > 2$, then $x^2 > 4 > 2$, so $x \notin S$. That is, 2 is an upper bound for S. By the Completeness Property, S has a least upper bound c. Notice that $c \geq 1$. We want to prove that $c^2 = 2$.

Well, if $c^2 < 2$, then $2 - c^2 > 0$. So we can choose a real number $r < 1$ with

$$0 < r < \frac{2 - c^2}{2c + 1}.$$

Then

$$\begin{aligned}
(c + r)^2 &= c^2 + 2cr + r^2 \\
&< c^2 + 2cr + r \qquad \text{(since } 0 < r < 1\text{)} \\
&= c^2 + (2c + 1)r \\
&< c^2 + (2c + 1)\frac{2 - c^2}{2c + 1} \qquad \left(\text{since } r < \frac{2 - c^2}{2c + 1}\right) \\
&= c^2 + 2 - c^2 \\
&= 2.
\end{aligned}$$

Thus, $c + r$ is an element of S that is greater than c. This contradicts the fact that c is an upper bound for S.

Likewise, if $c^2 > 2$, then $c^2 - 2 > 0$. So we can choose a real number r with

$$0 < r < \frac{c^2 - 2}{2c}.$$

Then

$$\begin{aligned}(c-r)^2 &= c^2 - 2cr + r^2\\ &> c^2 - 2cr\\ &> c^2 - 2c\,\frac{c^2-2}{2c}\\ &= c^2 - c^2 + 2\\ &= 2.\end{aligned}$$

If $x > c - r$, then $x^2 > (c-r)^2 > 2$. Thus, $c - r$ is an upper bound for S that is less than c. This contradicts the fact that c is the least upper bound for S.

The only remaining possibility is that $c^2 = 2$. ❄

Theorem 1.44 *Suppose* $\alpha : [0, 1] \to A \cup B$ *is a path with* $\alpha(0) \in A$ *and* $\alpha(1) \in B$. *Then there is a sequence of points of* A *that converges to a point of* B *or else there is a sequence of points of* B *that converges to a point of* A.

Proof. Let c be the least upper bound for the set $\{t \in [0, 1] \mid \alpha(t) \in A\}$. Intuitively, c is the cutoff point beyond which α maps no points into A. Choose a sequence of points that converges to c and that α maps to A. See Exercise 14 for a technique for constructing such a sequence. Choose another sequence of points that converges to c and that α maps to B. Since α is continuous, the result of Exercise 12 of Section 1.3 yields that images of these two sequences converge to $\alpha(c)$. Thus, whether $\alpha(c) \in A$ or $\alpha(c) \in B$, we have a sequence in one set that converges to a point in the other set. ❄

Example 1.45 Show that $[-1, 0) \cup (0, 1]$ has two path components.

Solution. For any points $x, y \in [-1, 0)$ and any $t \in [0, 1]$, the point $x + t(y - x) = y + (1 - t)(x - y)$ will lie in the closed interval with endpoints x and y (or if $x = y$ it will equal the common value). In particular, it will be contained in $[-1, 0)$. Thus, $\alpha : [0, 1] \to [-1, 0) \cup (0, 1]$ defined by $\alpha(t) = x + t(y - x)$ is a path between x and y. Similarly, if $x, y \in (0, 1]$, then the same formula again defines a path in $[-1, 0) \cup (0, 1]$ between x and y.

Finally, we use Theorem 1.44 to show that points in $[-1, 0)$ cannot be connected by a path to points in $(0, 1]$. Observe that no sequence of negative numbers can converge to a positive limit, and no sequence of positive numbers can converge to a negative limit. Thus, the gap between $[-1, 0)$ and $(0, 1]$ prevents a sequence of points in either of these intervals from converging to a point in the other interval.

We conclude that $[-1, 0)$ and $(0, 1]$ are the two path components of $[-1, 0) \cup (0, 1]$. ❄

The definition of homeomorphism was motivated by the idea of preserving the general shape or configuration of a geometric figure. Since path components are significant characteristics of a space, it is certainly reasonable that a homeomorphism will preserve the decomposition of a space into path components. The following theorem and corollary show how this happens.

Theorem 1.46 *Suppose $f : X \to Y$ is a continuous function. Then f maps all the points of a path component of X into a single path component of Y.*

Proof. Let a and b be any two points in a path component of X. Let $\alpha : [0, 1] \to X$ be a path from $a = \alpha(0)$ to $b = \alpha(1)$. Because the composition of continuous functions is continuous, $f \circ \alpha : [0, 1] \to Y$ is a path from $(f \circ \alpha)(0) = f(a)$ to $(f \circ \alpha)(1) = f(b)$. Thus, f maps these two points to the same path component of Y. ❄

Example 1.47 Determine the path components of $\mathbb{R}^2 - S^1$.

Solution. Let $A = \{(x, y) \in \mathbb{R}^2 \mid x^2 + y^2 < 1\}$ be the region inside the unit circle S^1, and let $B = \{(x, y) \in \mathbb{R}^2 \mid x^2 + y^2 > 1\}$ be the region outside the circle. Any two points in A are the endpoints of a straight line segment in A. This line segment gives a path in A between the two points. Any two points in the region B can be joined in B by a the arc of a circle followed by a straight line segment. This gives a path in B between the two points.

The function $f : \mathbb{R}^2 - S^1 \to \{-1, 1\}$ defined by

$$f(x, y) = \frac{x^2 + y^2 - 1}{|x^2 + y^2 - 1|}$$

is continuous since it is defined in terms of the continuous operations of addition, subtraction, multiplication, division by nonzero numbers, and the absolute value function. It maps the region A to -1 and the region B to 1. Since $\{-1\}$ and $\{1\}$ are distinct path components of $\{-1, 1\}$, Theorem 1.46 gives that A and B are not in the same path component of $\mathbb{R}^2 - S^1$.

It follows that A and B are the two path components of $\mathbb{R}^2 - S^1$. ❄

The previous theorem can be restated to say that a continuous function $f : X \to Y$ induces a function $f_* : P(X) \to P(Y)$ where $P(X)$ is the set of path components of X and $P(Y)$ is the set of path components of Y. The function f_* is defined by the rule that it maps the path component C_a of X containing the point a to the path component $C_{f(a)}$ of Y containing the point $f(a)$. The theorem guarantees that f_* is well-defined; that is, if b is another point in C_a, then $f_*(C_a) = C_{f(a)} = C_{f(b)} = f_*(C_b)$. With this notation, we have the following corollary.

Corollary 1.48 *A homeomorphism $h : X \to Y$ induces a bijection $h_* : P(X) \to P(Y)$. In particular, the number of path components of a space is a topological invariant.*

Proof. For any point $r \in Y$, consider C_r, the path component of Y that contains r. Since h is a surjection, there is a point $a \in X$ such that $h(a) = r$. Thus, $h_*(C_a) = C_{h(a)} = C_r$. It follows that h_* is a surjection.

For any $a, b \in X$, suppose the path components C_a and C_b of X containing these two points satisfy $h_*(C_a) = h_*(C_b)$. This condition can be rewritten as $C_{h(a)} = C_{h(b)}$. Thus, $h(a)$ and $h(b)$ are in the same path component of Y. Since h^{-1} is continuous, $h^{-1}(h(a)) = a$ and $h^{-1}(h(b)) = b$ are in the same path component of X. That is, $C_a = C_b$. It follows that h_* is injective. ❄

Suppose we are given two geometric figures that we suspect are not topologically equivalent. If both of the figures are path-connected, counting components will not distinguish the spaces. However, we might be able to remove a special subset of one of the figures and count the number of components of the remainder. If no comparable set can be removed from the other space to leave the same number of components, we will then know that the two spaces are not homeomorphic.

Example 1.49 Show that the intervals $(0, 1)$ and $[0, 1)$ are not homeomorphic.

Solution. The interval $[0, 1)$ contains the endpoint 0 with the property that it does not separate the space. That is, when 0 is removed from the set $[0, 1)$, the resulting subset $(0, 1)$ is still path-connected. A homeomorphism $h : (0, 1) \to [0, 1)$ would map some point a to 0. The argument given in Example 1.45 can be adapted (see Exercise 5) to show that $(0, a) \cup (a, 1)$ has two path components. The restriction of h to the subset $(0, a) \cup (a, 1)$ would be a homeomorphism onto the path-connected subset $(0, 1)$ of the range. Since homeomorphisms preserve the number of path components of a space, this is impossible. It follows from this contradiction that the original intervals $(0, 1)$ and $[0, 1)$ are not homeomorphic. ❄

We saw in Example 1.47 that the standard circle separates $\mathbb{R}^2$ into two path components. The Jordan Curve Theorem is the topological version of this result. It states that any simple closed curve (the continuous injective image of the circle S^1) in $\mathbb{R}^2$ or S^2 separates the space into two path components. This theorem was first stated by Camille Jordan in his *Cours d'analyse* published in 1887. His proof was incomplete, and it was not until 1913 that Oswald Veblen supplied the first rigorous proof. See Michael Henle's book *A Combinatorial Introduction to Topology* [5] for a modern proof of the Jordan Curve Theorem.

Related to the Jordan Curve Theorem is the Schönflies Theorem. This theorem, dating from 1908, states that a simple closed curve in $\mathbb{R}^2$, together with the bounded path component of the complement of the curve, is a topological disk (homeomorphic to $\{(x, y) \in \mathbb{R}^2 \mid x^2 + y^2 \leq 1\}$) with the simple closed curve as its boundary.

These theorems are useful in a variety of applications. In particular the Jordan Curve Theorem allows us to distinguish between a sphere and a torus.

Example 1.50 Show that the torus T^2 and the sphere S^2 are not homeomorphic.

Solution. There are simple closed curves on T^2 that do not separate the surface. The curve C in Figure 1.51 is such a curve.

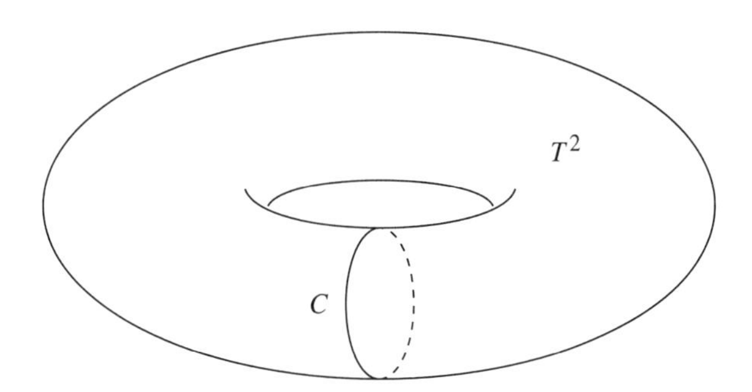

FIGURE 1.51
The curve C does not separate the torus T^2.

However, a homeomorphism $h : T^2 \to S^2$ would map C to a simple closed curve in S^2. By the Jordan Curve Theorem, $S^2 - h(C)$ would have two path components. By Corollary 1.48, it is impossible for the homeomorphism $h|_{T^2-C} : (T^2 - C) \to (S^2 - h(C))$ to map a space $T^2 - C$ with one path component onto a space $S^2 - h(C)$ with two path components. Hence, T^2 and S^2 are not topologically equivalent. ❄

Exercises 1.5

1. Prove that the relation defined in Definition 1.40 is an equivalence relation.

2. **(a)** Prove that there is no path $\alpha : [0, 1] \to [1, 3] \cup [4, 7]$ with $\alpha(0) \in [1, 3]$ and $\alpha(1) \in [4, 7]$.

   (b) Show that $[1, 3]$ and $[4, 7]$ are the path components of $[1, 3] \cup [4, 7]$.

3. Consider the x-axis $A = \{(x, 0) \in \mathbb{R}^2 \mid x \in \mathbb{R}\}$ and the hyperbola $B = \{(x, y) \in \mathbb{R}^2 \mid xy = 1\}$.

   (a) Prove that there is no path $\alpha : [0, 1] \to A \cup B$ with $\alpha(0) \in A$ and $\alpha(1) \in B$.

   (b) Determine the path components of $A \cup B$.

4. Show that $A = \{(x, 0) \in \mathbb{R}^2 \mid 0 < x \leq 1\}$, $B = \{(0, y) \in \mathbb{R}^2 \mid 0 < y \leq 1\}$ $C = \{(x, 0) \in \mathbb{R}^2 \mid -1 \leq x < 0\}$, and $D = \{(0, y) \in \mathbb{R}^2 \mid -1 \leq y < 0\}$ are the four path components of $A \cup B \cup C \cup D$.

5. Consider four real numbers a, b, c, and d with $a < b \leq c < d$. Show that the two open intervals (a, b) and (c, d) are the path components of the union $(a, b) \cup (c, d)$ of these disjoint intervals.

6. Let $A_0 = \{(0, y) \in \mathbb{R}^2 \mid 0 \leq y \leq 1\}$, and for $n = 1, 2, \ldots$ let $A_n = \left\{(\frac{1}{n}, y) \in \mathbb{R}^2 \mid 0 \leq y \leq 1\right\}$. Determine the path components of $A_0 \cup A_1 \cup A_2 \cup \cdots$.

7. Determine the path components of the set $\mathbb{Q}$ of rational numbers.

8. **(a)** Show that if any point is removed from $\mathbb{R}$, the resulting space has two path components.

(b) Show that if any point is removed from $\mathbb{R}^2$ the resulting space has one path component.

(c) Conclude that $\mathbb{R}$ and $\mathbb{R}^2$ are not homeomorphic.

9. Generalize Example 1.47 to find the path components of $R^{n+1} - S^n$. Be careful of the situation when $n = 0$. It is a little different from the cases when $n > 0$.

10. Write the upper-case letters of the alphabet using line segments and arcs of circles. Which of the resulting topological spaces are homeomorphic?

11. Show that the intervals $[0, 1)$ and $[0, 1]$ are not homeomorphic.

12. Show that the interval $[0, 1)$ is not homeomorphic to the square region $[0, 1] \times [0, 1]$.

13. Show that the cone $\{(x, y, z) \in \mathbb{R}^3 \mid x^2 + y^2 = z^2\}$ of two nappes is not homeomorphic to $\mathbb{R}^2$.

14. Suppose c is the least upper bound for a set $A \subseteq \mathbb{R}$. Show that there is a sequence $a_1, a_2, \ldots$ of points in A that converges to c. Suggestion: Use the fact that $c - \frac{1}{n}$ is not an upper bound of A to show that there is $a_n \in A$ with $c - \frac{1}{n} < a_n \leq c$.

1.6 Isotopy

The topological spaces we study are frequently subsets of Euclidean spaces. In this situation it is often possible to show that two subsets are homeomorphic by deforming the entire Euclidean space so that one subset is transformed into the other. If you envision this deformation as a motion picture, you can see the effects of the deformation on the original subset. As it changes through a family of homeomorphisms parameterized by time, you get the strong sense that the final subset really is the same as the original subset. Although the images throughout the course of the isotopy are different sets, we tend to think of the transformed images as if they were a single object moving around like an actor on a stage.

A deformation as described in the previous paragraph is known as an **ambient isotopy**. The term isotopy refers to the fact that we have a continuously parameterized family of homeomorphisms starting with the identity function. The adjective ambient signifies that the isotopy transforms the surrounding space as well as the subset of interest.

Example 1.52 Show that there is an ambient isotopy of $\mathbb{R}^2$ that deforms any triangle into any other triangle.

Solution. We proceed by describing a sequence of motions of $\mathbb{R}^2$ that will carry one of the triangles along with it. Fix the other triangle in its original position to serve as a guide for the motions we will use.

Begin by sliding $\mathbb{R}^2$ so that one of the vertices of one triangle is moved to the original position of a vertex of the fixed triangle. Next rotate $\mathbb{R}^2$ about this vertex so that one of the edges of the first triangle lines up with an edge of the fixed triangle. Choose the edge so that both triangles end up on the same side of this edge. With radial dilation or contraction of $\mathbb{R}^2$ centered at the matching vertex, stretch or shrink the triangle so the lined-up edge agrees in length with the edge of the other triangle. Now stretch or squeeze $\mathbb{R}^2$ in the direction perpendicular to this edge so the altitude to this edge has the same length as the corresponding altitude of the other triangle. Finally, shear $\mathbb{R}^2$ in the direction of this edge so the third vertex matches the third vertex of the fixed triangle. At this stage, the first triangle has been deformed to lie on top of the original position of the other triangle.

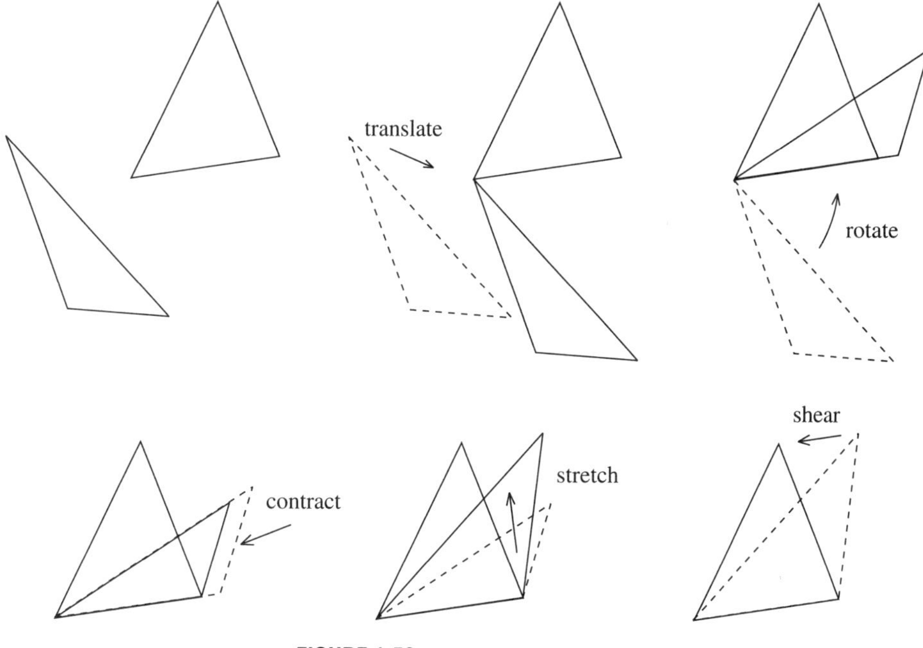

FIGURE 1.53
Any two triangles in $\mathbb{R}^2$ are isotopic.

Since each of the motions was a deformation of all of $\mathbb{R}^2$, this sequence of motions can be spliced together to yield an isotopy of $\mathbb{R}^2$ that takes the first triangle onto the second. ❄

Topologists have devised an ingeniously elegant way to make precise the intuitive idea an ambient isotopy. Although an intuitive understanding of this concept as a parameterized family of deformations will suffice for most purposes, the following definition puts the concept of ambient isotopy on a firm foundation.

Definition 1.54 *Suppose A and B are two subsets of a space X. An* ***ambient isotopy*** *from A to B in X is a continuous function $h : X \times [0, 1] \to X$ that satisfies the following three conditions. We denote $h(x, t)$ by $h_t(x)$.*

1. *$h_t : X \to X$ is a homeomorphism for every $t \in [0, 1]$.*
2. *h_0 is the identity function on X.*
3. *$h_1(A) = B$.*

The first condition captures the essence of an isotopy as a parameterized family of deformations of the ambient space. The second condition says that $h_0(x) = x$ for all $x \in X$; in particular, $h_0(A) = A$. The third condition says that h_1 restricted to A is a homeomorphism from A to B. Thus, having an ambient isotopy from A to B requires A and B to be topologically equivalent. But an ambient isotopy says more: from a topological point of view, the way A is embedded in X is equivalent to the way B is embedded in X. Figure 1.55 illustrates some intermediate stages of a simple isotopy that deforms a star A into a circle B in $\mathbb{R}^2$.

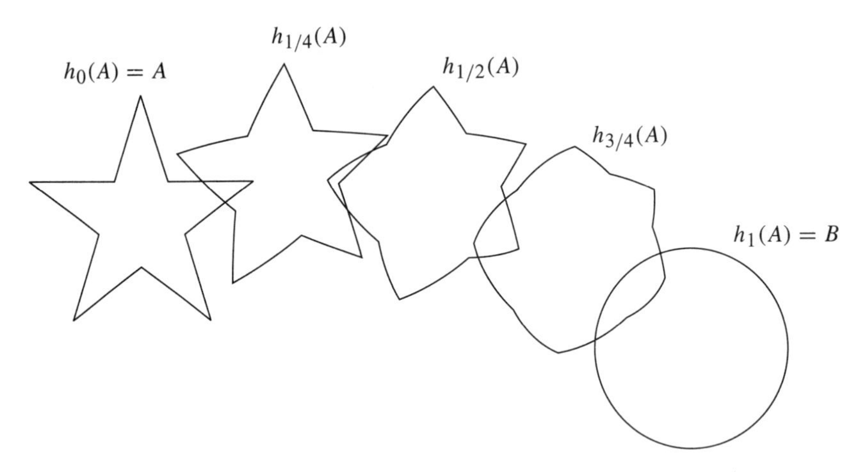

FIGURE 1.55
An ambient isotopy from a star to a circle in $\mathbb{R}^2$.

The next example illustrates the distinctions between homeomorphism and ambient isotopy. It also shows the important role of the ambient space in which the isotopy takes place.

Recall from Example 1.33 that the circle S^1 is homeomorphic to the knot K given in that example. It seems unlikely that K can be deformed (without cuts or self-intersections) to a circle lying in the xy-plane. One of our goals in Chapter 2 is to provide a mathematical basis for your intuition that there is no ambient isotopy of $\mathbb{R}^3$ that will take the circle $\{(x, y, z) \in \mathbb{R}^3 \mid x^2 + y^2 = 1,\ z = 0\}$ onto the knot K.

Example 1.56 Think of S^1 and K in Example 1.33 as subsets of $\mathbb{R}^3$ embedded as the three-dimensional slice of $\mathbb{R}^4$ consisting of all points whose fourth coordinates equal zero. Show there is an ambient isotopy of $\mathbb{R}^4$ that deforms S^1 into K.

Solution. Let us deform K into S^1. This isotopy can be played backwards to deform S^1 into K. Figure 1.57 shows a shaded fin attached to K along an arc in its boundary and intersecting another part of K in an interior point.

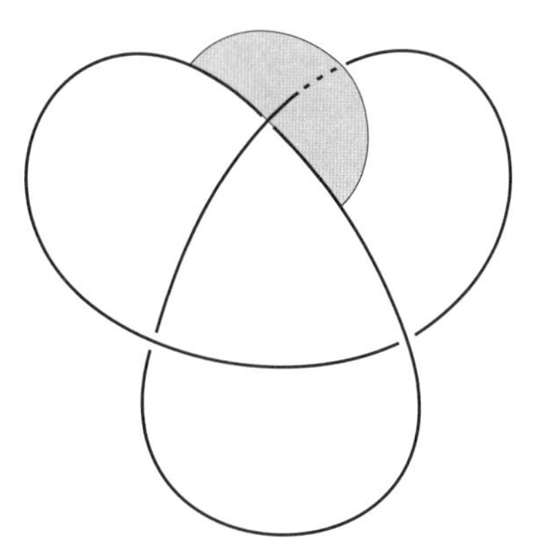

FIGURE 1.57
A fin attached to the knot K.

Without moving the boundary of this disk, gently push its interior into the portion of $\mathbb{R}^4$ with negative fourth coordinates. The modified disk now intersects K only along an arc in its boundary. Slide this arc of K across the modified disk to the other arc in the boundary of the disk. This deformation of K will be a simple closed curve in $\mathbb{R}^3 \times \{0\}$ that can easily be deformed to the unit circle in the xy-plane. ❄

With a time machine the Count of Monte Cristo could have used this same idea to escape from the Château d'If. He simply travels back to a time before the castle was built. He moves over in space to a position beyond the castle walls. When he returns to the present, he will find himself outside the dungeon.

Example 1.58 Here are two solid objects embedded in $\mathbb{R}^3$. Describe an ambient isotopy that deforms the first to the second.

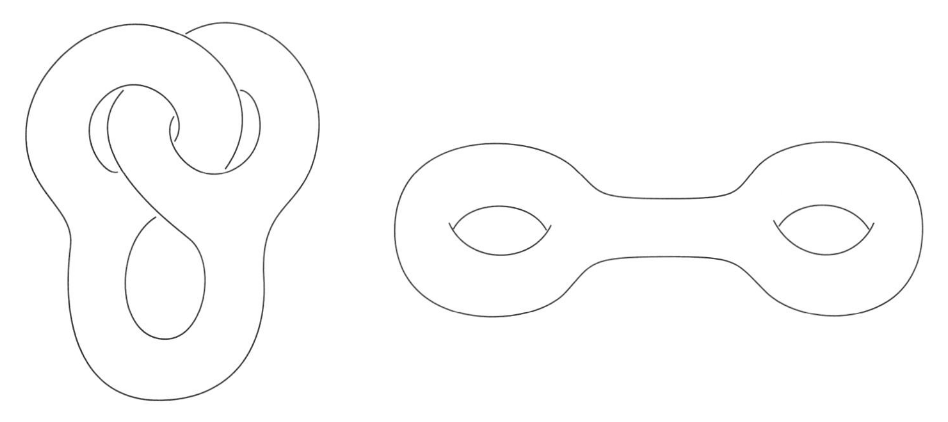

FIGURE 1.59
Two solid objects in $\mathbb{R}^3$.

Solution. Think of the objects as being made of soft clay or very flexible and malleable rubber. As we stretch the object and slide part of it along another part, we can extend the deformation to an ambient isotopy of $\mathbb{R}^3$. Begin by sliding the inner branch of the right loop down the connecting arm between the two loops. Pull this through the base of the left loop and down around the back of the object. Now shrink the right loop and slide the connecting arm around so the object looks like the desired finished product. ❄

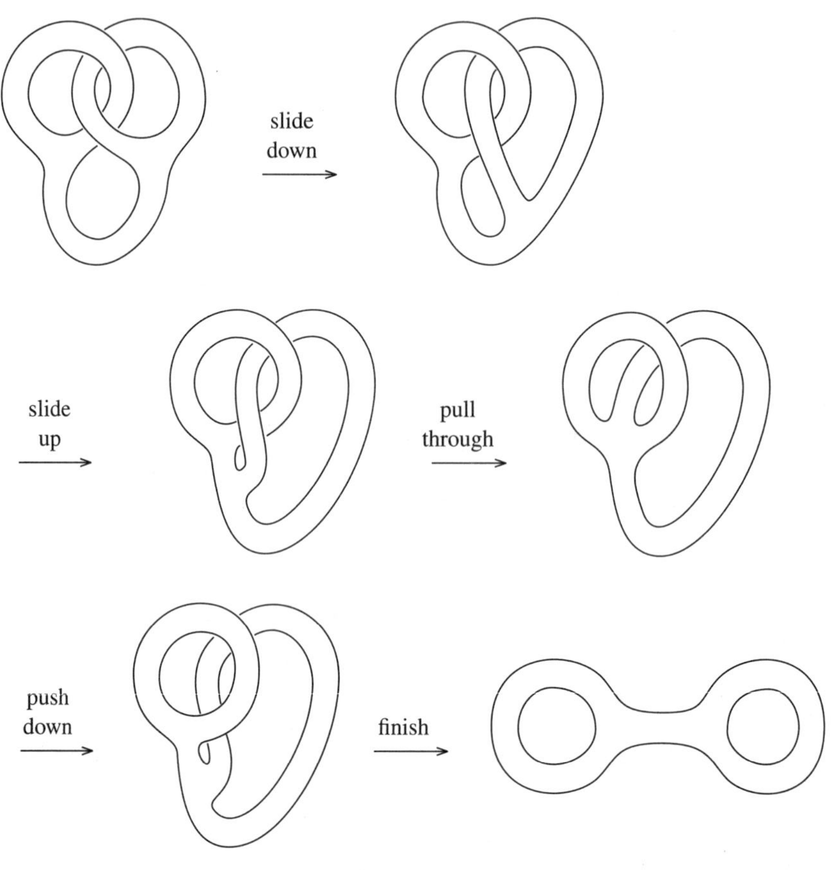

FIGURE 1.60
Unlinking the dogbone.

Example 1.58 shows how tricky it can be to find an isotopy between two objects. It can be even more challenging to show that no isotopy exists. Figure 1.61 illustrates a dogbone toy that cannot be deformed by an ambient isotopy of $\mathbb{R}^3$ to an untangled dogbone. In the article "A Topological Puzzle" [1], Inta Bertuccioni applies some techniques of algebraic topology to handle this difficult problem.

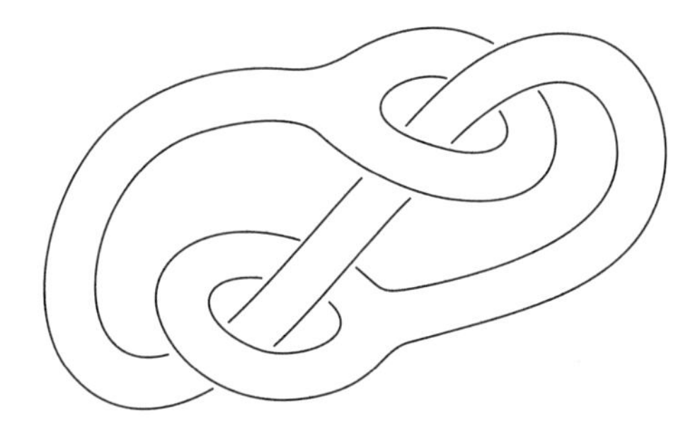

FIGURE 1.61
Another dogbone toy.

Exercises 1.6

1. Show that the relation among subsets of a given set X of being ambient isotopic is an equivalence relation on the subsets of X.

2. Show that there is an ambient isotopy of $\mathbb{R}^2$ that deforms any quadrilateral into any other quadrilateral.

3. Show that there is an ambient isotopy of $\mathbb{R}^2$ that deforms any quadrilateral into any triangle.

4. **(a)** Which of the objects in Figure 1.62 are ambient isotopic in $\mathbb{R}^2$?
   (b) Which of the objects in Figure 1.62 are ambient isotopic in $\mathbb{R}^3$?
   (c) Which of the objects in Figure 1.62 are homeomorphic?

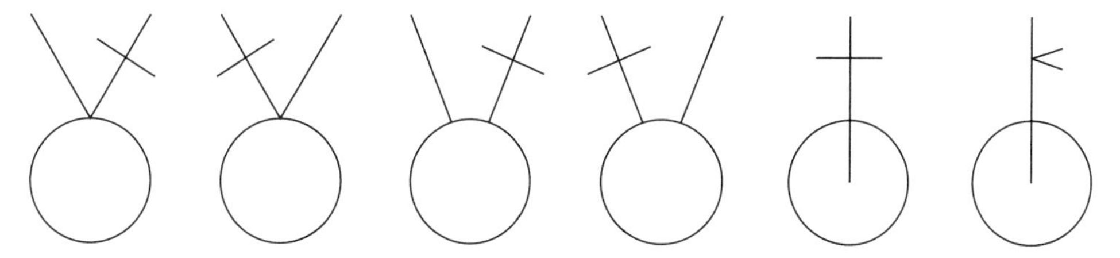

FIGURE 1.62
Which pairs are isotopic in $\mathbb{R}^2$? Which in $\mathbb{R}^3$?

5. **(a)** Describe a homeomorphism between the two objects in Figure 1.63.
   (b) Describe a homeomorphism between the complements in $\mathbb{R}^3$ of the two objects in Figure 1.63. Suggestion: Observe that the complements of these two figures are topologically the same as the complements of the dogbones in Figure 1.59.

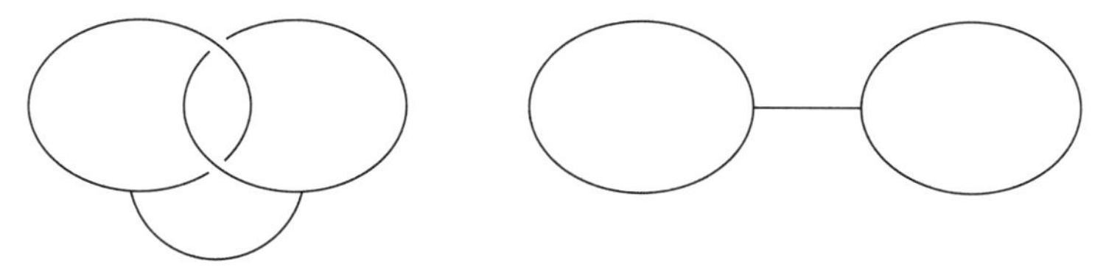

FIGURE 1.63
Can you unlink the eyeglass frames?

(c) Explain the difficulty in deforming the first object into the second by an ambient isotopy of $\mathbb{R}^3$.

(d) What makes this situation different from that of Figure 1.59?

6. Cut the toe off an old sock. Bring the two circular ends around and sew them together along matching arcs approximately three-quarters of the way around the circles. You will have a model of a torus with a hole. Reach your fingers through the hole and turn the surface inside out. What happens to the curves that define the coordinate system for the torus as in Example 1.12?

7. Figure 1.64 shows a knotted arc spanning the region between two concentric spheres. Explain how to construct an ambient isotopy of the region between the spheres so that the endpoints of the arc do not move during the isotopy and so that the arc is deformed to a straight line segment between the spheres. Suggestion: What would you do if the inner sphere were a light bulb hanging from a knotted cord?

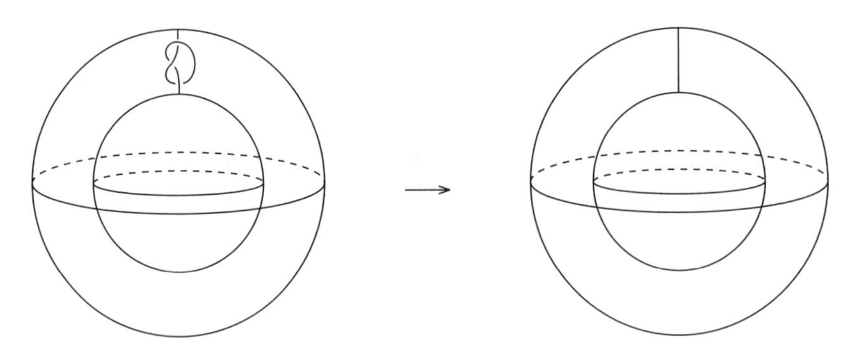

FIGURE 1.64
Can you untie the knot spanning between the two spheres?

8. Draw a sequence of pictures to illustrate an ambient isotopy of $\mathbb{R}^3$ that deforms the knotted dogbone on the left to the unknotted dogbone on the right in Figure 1.65.

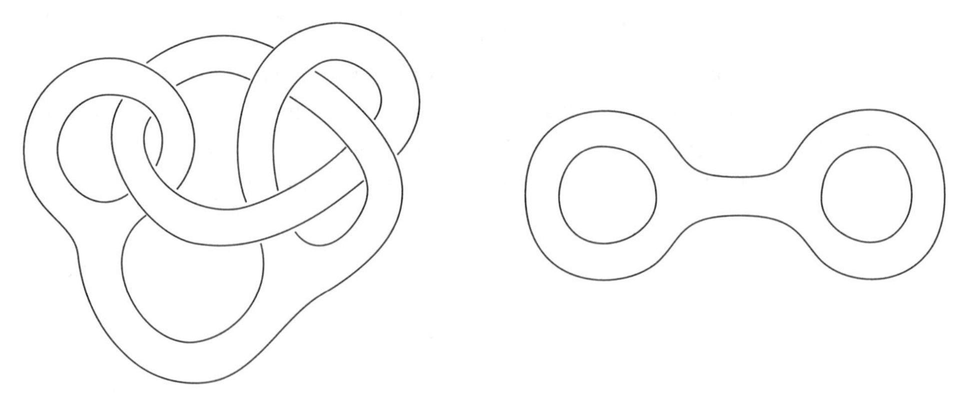

FIGURE 1.65
Two dogbone toys.

9. Draw a sequence of pictures to illustrate an ambient isotopy of $\mathbb{R}^3$ that deforms a figure-eight knot to its mirror image (see Figure 1.66).

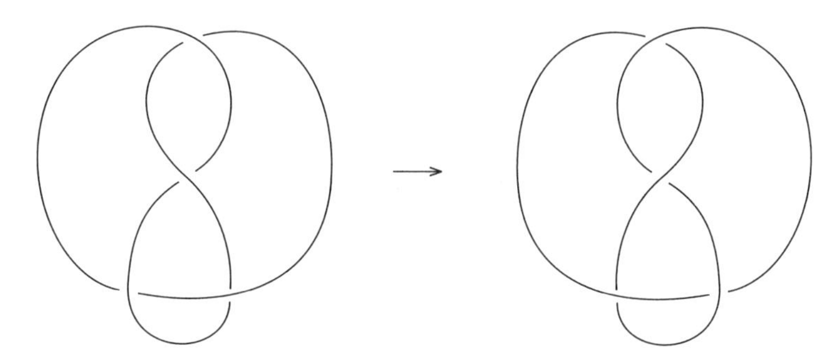

FIGURE 1.66
A figure-eight knot and its mirror image.

10. Let S^2 be the standard 2-sphere in $\mathbb{R}^3$. Let $A = S^2 \cup \{(0,0,z) \in \mathbb{R}^3 \mid 1 \leq z \leq 2\}$ be the sphere with a whisker, and let $B = S^2 \cup \{(0,0,z) \in \mathbb{R}^3 \mid 0 \leq z \leq 1\}$ be the sphere with an ingrown whisker.

(a) In what topological sense are A and B the same?

(b) In what topological sense are A and B different?

(c) Consider A and B as embedded in the three-dimensional slice of $\mathbb{R}^4$ consisting of all points whose fourth coordinates equal zero. Describe an ambient isotopy of $\mathbb{R}^4$ that deforms A into B.

11. Consider a disk with a polygonal boundary in the plane. A **triangulation** of the disk is a subdivision of the disk into triangular regions such that the intersection of any pair of distinct triangles is either empty, a vertex of both triangles, or an edge of both triangles.

(a) Draw a picture of a triangulated disk showing one of the triangles whose intersection with the boundary of the disk is a side of the triangle. We want to remove this triangle from the disk, leaving the two interior sides as part of the boundary of the remaining region. Describe an ambient isotopy of the disk in the plane that accomplishes the removal of this triangle.

(b) Draw a picture of a triangulated disk showing one of the triangles whose intersection with the boundary of the disk is two sides of the triangle. We want to remove this triangle from the disk, leaving the interior side as part of the boundary of the remaining region. Describe an ambient isotopy of the disk in the plane that accomplishes the removal of this triangle.

(c) Draw a picture of a triangulated disk showing one of the triangles whose intersection with the boundary of the disk is an edge and the opposite vertex of the triangle. What happens when we remove this triangle?

(d) **Shelling** the disk is the process of removing triangles one at a time until only one triangle remains and so that at each stage the figure is a disk. Prove that it is possible to shell any triangulated disk in the plane. Suggestion: Apply mathematical induction to the number of triangles in the triangulation; incorporate into the induction hypothesis the stipulation that any triangle can be designated as the final triangle.

(e) Read the paper "An Unshellable Triangulation of a Tetrahedron" [8] by Mary Ellen Rudin.

12. Prove that any polygonal disk in the plane can be triangulated.

References and Suggested Readings for Chapter 1

1. Inta Bertuccioni, "A topological puzzle," *American Mathematical Monthly*, 110 (2003), 937–939.
   The fundamental group of the complement of the dogbone in Figure 1.61 is used to analyze this object.
2. Stephan Carlson, *Topology of Surfaces, Knots, and Manifolds*, Wiley, New York, 2001.
   A visual approach to knots, surfaces, and other topological spaces.
3. H. Graham Flegg, *From Geometry to Topology*, Dover, New York, 2001.
   An intuitive approach to topology of geometric objects in Euclidean spaces. This leads to a more formal presentation of continuity, sets, functions, metric spaces, and topological spaces.
4. Sue Goodman, *Beginning Topology*, Brooks/Cole, Pacific Grove, CA 2005.
   Encourages readers to try their hands at proving basic topological results and to explore the interactions of topology with algebra, combinatorics, geometry, calculus, and differential equations.
5. Michael Henle, *A Combinatorial Introduction to Topology*, W. H. Freeman, San Francisco, 1979.
   Includes a geometric approach to homology theory leading to a proof of the Jordan curve theorem.
6. Stephen Huggett and David Jordan, *A Topological Aperitif*, Springer-Verlag, New York, 2001.
   Arouses curiosity about topology and provides a firm geometrical foundation for further study.
7. V. V. Prasolov, *Intuitive Topology*, American Mathematical Society, Providence, RI, 1995.
   The intuitive idea of deformations leads to results about knots, surfaces, and vector fields with lots of pictures and problems.
8. Mary Ellen Rudin, "An unshellable triangulation of a tetrahedron," *Bulletin of the American Mathematical Society*, 64 (1958), 90–91.
   An explicit triangulation with 14 *vertices and* 41 *tetrahedra shows that shellability does not extend to* 3*-dimensional cells.*
9. Nathan Salmon, *Frege's Puzzle*, MIT Press, Cambridge, MA, 1986.
   An investigation into Frege's problem of equality with many examples and approaches for resolving this paradox.
10. Saul Stahl, *Introduction to Topology and Geometry*, Wiley Interscience, Hoboken, NJ, 2005.
    Includes an introduction to differential geometry of surfaces.

2

Knots and Links

Everyone is familiar with knots as tangles in lamp cords and garden hoses. The first order of business in this chapter is to provide a mathematical abstraction of this everyday concept of a knot. Along with a geometric way to describe knots, we will investigate an equivalence relation among knots. Number theory, colorings, determinants, and polynomials come into play in defining invariants that will help us distinguish different types of knots.

2.1 Knots, Links, and Equivalences

We begin the study of knots in this section by considering difficulties with some definitions of a knot that might first come to mind. Mathematicians have devised ingenious remedies to overcome these problems.

First of all, most people will readily agree that the mathematical model of a knot should be a one-dimensional geometrical figure. Although the thickness of a string and the method of laying the strands in a rope are important for the strength and durability of the knot, such considerations are distractions to a topological description of the knot.

Just as we abstract an idealized concept of a geometric line from a line drawn with a pencil or chalk, we should think of a knot as a curve in $\mathbb{R}^3$. The curve follows the centerline of the string or rope in which the knot is tied. Therefore we do not want the curve to intersect itself. This leads us to the possibility of defining a knot as the image of an injection from an interval $[0, 1]$ to $\mathbb{R}^3$.

With this proposed definition, we can describe tying a knot in a piece of string in terms of an ambient isotopy of a line segment in $\mathbb{R}^3$. The isotopy deforms the line segment into a curve that follows the knotted string. But since we want to take the topological point of view and ignore angles and distances, the straight string is equivalent to the knotted string. In fact, all knots would be equivalent to a straight line segment. This would not yield a very interesting or useful theory.

We could cure this problem by devising a theory of knot equivalence in which endpoints are not moved during the isotopies and portions of the curve are not allowed to pass over the ends of the curve. However, a simpler solution is to consider a knot as a closed loop rather than as a curve with two loose ends.

We are getting close to an official definition, but one tricky problem remains. We want to rule out infinitely tangled curves. They are certainly beyond the realm of the everyday concept of a knot. Even mathematicians who delight in studying such monsters build their research in this area on the simpler theory of finitely tangled knots.

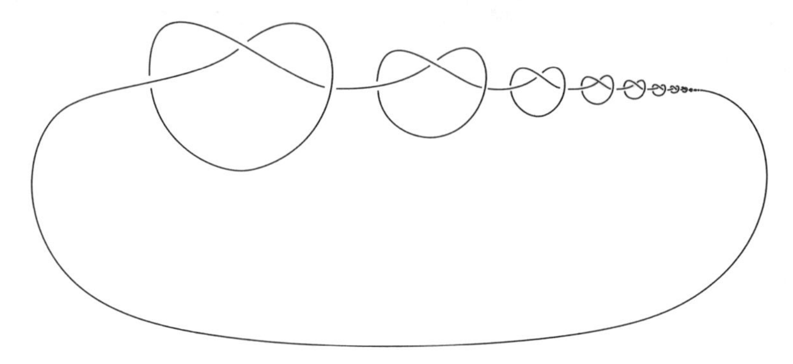

FIGURE 2.1
An infinitely tangled curve.

One remedy for this problem is to require a knot to be a differentiable simple closed curve in $\mathbb{R}^3$. This rules out the infinite amount of twisting necessary to follow along the curve such as that in Figure 2.1. We will circumvent the problem of infinitely tangled curves with an equivalent approach that is even simpler. We will require a knot to be made up of a finite number of straight line segments. This not only prevents the infinite amount of twisting seen in Figure 2.1, but it also relates to the combinatorial techniques we will study to classify knots. The book *Introduction to Knot Theory* [14] by Richard Crowell and Ralph Fox contains a concise discussion of the relation between these two approaches.

Here at last is our official definition of a knot.

Definition 2.2 *A **knot** K is a simple closed curve in $\mathbb{R}^3$ that can be broken into a finite number of straight line segments $e_1, e_2, \ldots, e_n$ such that the intersection of any segment e_k with the other segments is exactly one endpoint of e_k intersecting an endpoint of e_{k-1} (or e_n if $k = 1$) and the other endpoint of e_k intersecting an endpoint of e_{k+1} (or e_1 if $k = n$).*

Even though the definition requires a knot to be made of line segments, we will continue the tradition of drawing knots as smooth curves. If you are bothered by this artistic license in smoothing out the corners, you can imagine that the number of segments is so large that you cannot see the corners at the resolution of the drawings.

Figure 2.3 illustrates some typical knots.

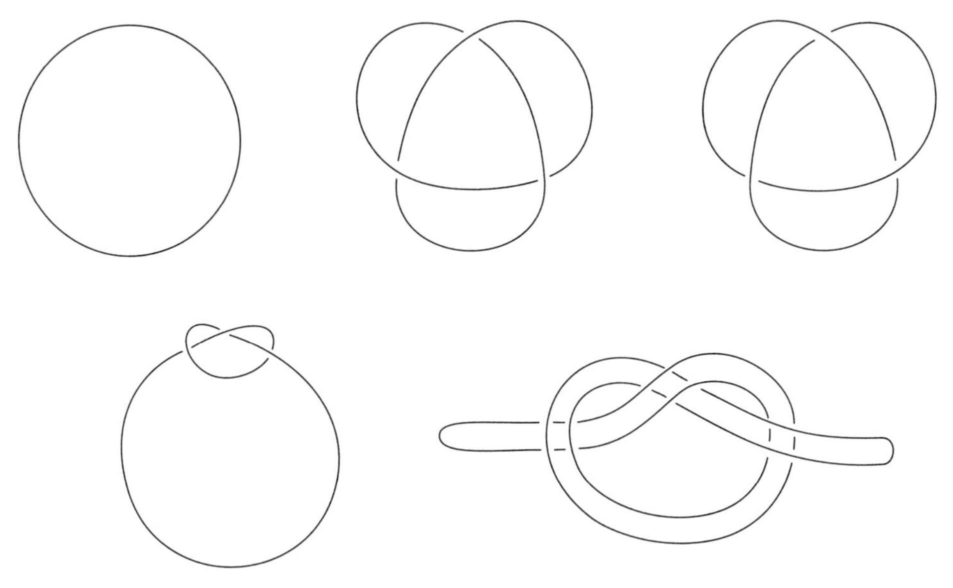

FIGURE 2.3
Which of these knots are equivalent?

We want to define an equivalence relation on knots that corresponds to the intuitive idea of deforming a loop of rope in three-dimensional space from one configuration to another. The definition should rule out cutting the rope or magically passing one portion of the rope through another portion. We saw in Example 1.33 that the unit circle in the plane is homeomorphic to a knot we would regard as truly knotted. In fact, since we defined a knot to be a simple closed curve, we realize that this automatically makes any knot the homeomorphic image of a circle. Thus, we must impose a stronger sense of equivalence on knots than merely being homeomorphic sets.

You also know from experience that you cannot straighten out a tangled string by brute force. The tighter you pull the ends, the tighter the tangle becomes. Thus, we must rule out deformations that shrink the tangled portion of a knot until it vanishes as a single point of a circle.

The moral of these observations is that the concept of knottedness involves the way a curve is embedded in $\mathbb{R}^3$. We need to involve the ambient space in the definition of equivalence between knots. One possibility is to require the deformation to be differentiable. As we did with the definition of a knot, we will instead adopt what is known as the piecewise-linear approach to knot equivalence.

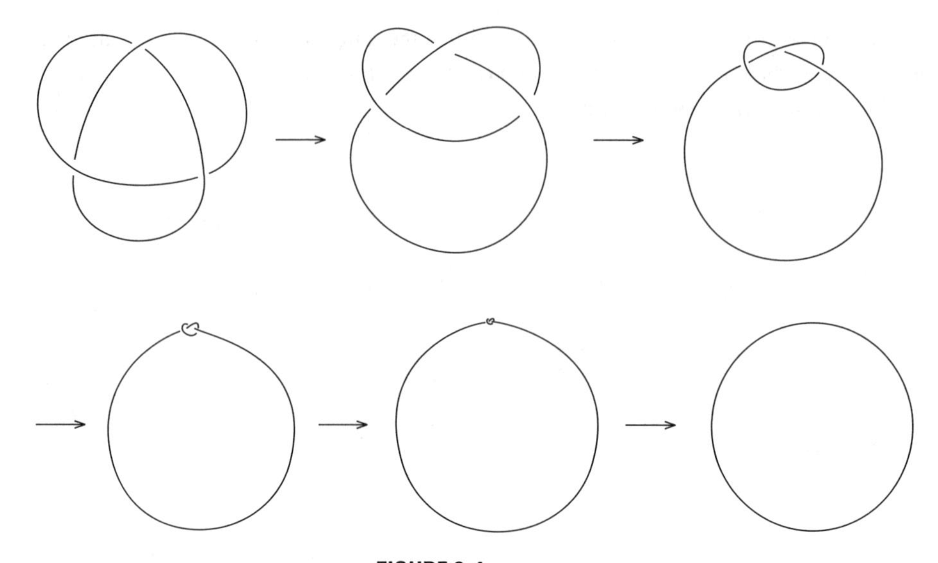

FIGURE 2.4
Illegal knot popping.

Definition 2.5 *Consider a triangle ABC with side AC matching one of the line segments of a knot K. In the plane determined by the triangle, we require that the region bounded by ABC intersects K only in the edge AC. A* ***triangular detour*** *involves replacing the edge AC of knot K with the two edges AB and BC to produce a new knot L. With the same notation, a* ***triangular shortcut*** *involves replacing the two edges AB and BC of knot L with the single edge AC to produce knot K. A* ***triangular move*** *is either a triangular detour or a triangular shortcut. Two knots are* ***equivalent*** *if and only if there is a finite sequence of triangular moves that changes the first knot into the second.*

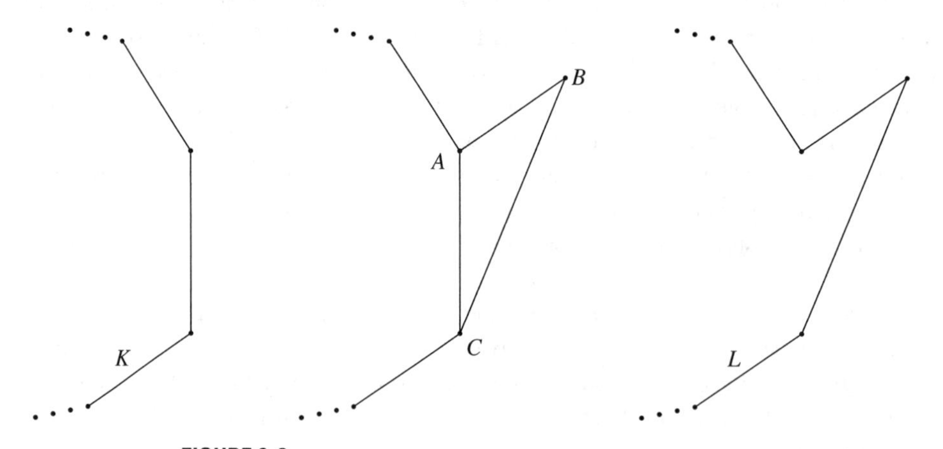

FIGURE 2.6
A triangular detour on knot K; a triangular shortcut on knot L.

Exercise 1 at the end of this section asks you to verify that equivalence between knots does indeed form an equivalence relation. The resulting equivalence classes are the **knot types**. Saying that two knots are equivalent or that they are of the same type are simply alternative ways of saying that they are in the same equivalence class.

See Exercise 6 at the end of this section to convince yourself that a triangular move can be implemented as an ambient isotopy of $\mathbb{R}^3$. In fact the movement of points can be confined to lie within an arbitrarily small distance of the triangular region involved. Now if knot K is equivalent to knot L, we can implement the sequence of triangular moves as a sequence of ambient isotopies. Performing these isotopies in sequence will yield an ambient isotopy that deforms K onto L.

The real impact of the definition of knot equivalence is its ability to reproduce any reasonable deformation of a knot in $\mathbb{R}^3$ as a sequence of triangular moves. Of course, an isotopy might move more than one vertex at a time. So the triangular moves may not reproduce the details of the isotopy. We must be content with triangular moves that ultimately transform the knot at the beginning of the isotopy into its image at the final stage of the isotopy. Exercise 7 at the end of this section gives some hints as to how you might do this.

The compatibility of a sequence of triangular moves with an ambient isotopy of a knot in $\mathbb{R}^3$ means we can use whichever approach is more convenient. When we analyze the effect of a deformation on an invariant of knot type, we need only consider triangular moves. When we want to experiment with loops of string or drawings of knots, we can envision the deformations as ambient isotopies.

Frequently in our study of knots we will be gluing together the ends of arcs or looking at the simple closed boundary curves of a surface in $\mathbb{R}^3$. In these situations we need to consider several knots at once. This motivates the following definition.

Definition 2.7 *A* ***link*** *is the nonempty union of a finite number of disjoint knots.*

Since the knots in a link are disjoint, each knot is a connected component of the link. In particular, a knot is the special case of a link with one component.

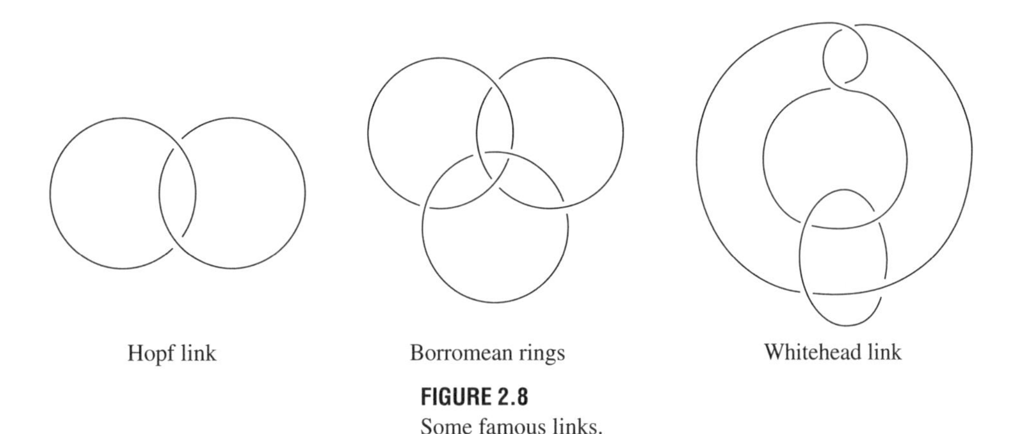

FIGURE 2.8
Some famous links.

A definition of equivalence for links is entirely similar to the definition for knots. In Exercise 8 you are asked to formulate this definition and discuss its compatibility with ambient isotopies.

Exercises 2.1

1. Show that the definition of equivalence between knots satisfies the three conditions of an equivalence relation.

2. Show that for any two triangles in $\mathbb{R}^2$ there is a sequence of triangular moves in $\mathbb{R}^2$ that transforms one triangle to the other.

3. **(a)** Show that any two triangles in parallel planes of $\mathbb{R}^3$ are equivalent knots.

   (b) Show that any two triangles in $\mathbb{R}^3$ are equivalent knots.

4. **(a)** Show that any two quadrilaterals in $\mathbb{R}^3$ are equivalent knots.

   (b) Show that any triangle in $\mathbb{R}^3$ and any quadrilateral in $\mathbb{R}^3$ are equivalent knots.

   (c) What appears to be the fewest number of edges necessary to create a knot that is not equivalent to a triangle?

5. **(a)** Prove that if a knot is equivalent to a triangle, then it is the boundary of a disk in $\mathbb{R}^3$.

   (b) Use the idea of shelling a disk (Exercise 11 of Section 1.6) to prove that the boundary of any disk in $\mathbb{R}^3$ is equivalent to a triangle.

6. **(a)** Draw a typical triangular detour on a knot.

   (b) Draw some intermediate stages of the knot as it undergoes an isotopy to implement this detour.

   (c) Show how the isotopy might stretch a small circle around the edge of the knot that is being modified.

   (d) Show how the isotopy might stretch segments perpendicular to the triangular region that defines the detour.

7. Consider a vertex of a knot and the two edges that share this vertex. Suppose the vertex is moved a small distance, and the edges incident with that vertex are adjusted to run to the new vertex. The other vertices and edges of the knot should not move.

   (a) Draw a picture to show that the new knot is equivalent to the old knot.

   (b) Draw pictures to show the kinds of problems that can arise if we do not restrict the distance the vertex is moved.

   (c) Describe how all vertices can be moved a suitably small amount, so that when the edges are redrawn between corresponding vertices, the resulting knot is equivalent to the original one.

   (d) Consider an ambient isotopy of a knot that moves no point in $\mathbb{R}^3$ more than a suitably small amount. Assume that all stages of the isotopy preserve the knot as a polygonal simple closed curve. Show that the final image of the knot is equivalent to the original knot.

(e) Show that any translation of a knot does not change the knot type.

(f) Show that any rotation of a knot does not change the knot type.

8. **(a)** Emulate Definition 2.5 to formulate similar definitions for equivalence between links.

(b) Show that your definition results in an equivalence relation.

(c) Discuss how a link can be deformed to an equivalent link by an ambient isotopy.

(d) Discuss the extent to which an ambient isotopy of a link can be reproduced as a sequence of triangular moves.

9. **(a)** If any one of the three simple closed curves is removed from the Borromean rings depicted in Figure 2.8, show that the remaining components can be separated.

(b) Construct a set of four linked simple closed curves so that if any one is removed, the remaining three can be separated.

(c) Read the chapter on Borromean rings in the book *Perspectives in Mathematics* [23] by David Penney.

(d) Look at the figure for Exercise 15 in Section 3F of the book *Knots and Links* [25] by Dale Rolfsen.

10. Draw a sequence of pictures or use loops of string to show that the Whitehead link depicted in Figure 2.8 can be deformed by an isotopy so as to interchange the two components of the link.

2.2 Knot Diagrams

Because knots are curves embedded in 3-dimensional Euclidean space, they are often difficult to envision directly. On the other hand, the shadow that a knot casts in a plane captures the basic geometric shape of the knot. In this section we will see how to handle a few problems that arise with knot projections and in particular how to keep track of the stratification when several points of the knot project to the same point in the plane.

If you randomly choose two lines in the plane, you would expect the lines to intersect in a single point. When the lines are perturbed slightly, the point of intersection may shift a short distance, but the nature of the intersection will not change. While the lines might conceivably be parallel or coincident, these possibilities are very special cases. Furthermore, any rotation (no matter how minute) of one of the lines will resolve such configuration into the more general situation. This is an example of the phenomenon of **general position**. Most likely two lines will already be in general position (and will remain so under small perturbations). But even if we aren't so lucky, we can gently nudge the lines into general position.

The bending and stretching permitted in topology usually allow us to assume that geometric objects intersect in general position. This is particularly useful when the objects are composed of points, line segments, planar polygonal regions, and higher-dimensional polyhedra. We can push such objects into general position by rotations and translations that are small enough not to knock other intersections out of general position. Of course, the moves will preserve the piecewise-linear structure of the objects.

Here is a simple rule to help determine how two objects intersect when they are in general position.

General Position Rule of Thumb 2.9 *Suppose two piecewise-linear objects are embedded in general position in $\mathbb{R}^n$. Suppose A is a vertex, edge, face, or analogous higher-dimensional part of one object and B is a vertex, edge, face, or analogous higher-dimensional part of the other object. If the intersection $A \cap B$ is nonempty, then*

$$\dim(A \cap B) = \dim(A) + \dim(B) - n.$$

Example 2.10 Describe the intersection of two polyhedral surfaces in general position in $\mathbb{R}^3$.

Solution. The surfaces are 2-dimensional piecewise-linear objects. By the General Position Rule of Thumb, if the surfaces intersect, the intersection will be of dimension $2+2-3=1$. The edges of either surface will intersect the other surface only at interior points of its faces (since $1+2-3=0$ and $1+1-3=-1$). No vertices of either surface will intersect the other surface (since $0+2-3=-1$).

Figure 2.11 illustrates how the 1-dimensional intersections join together. The intersection of a face of one surface with a face of the other is a line segment that misses the vertices of the faces. Adjacent faces contribute additional line segments of the intersection

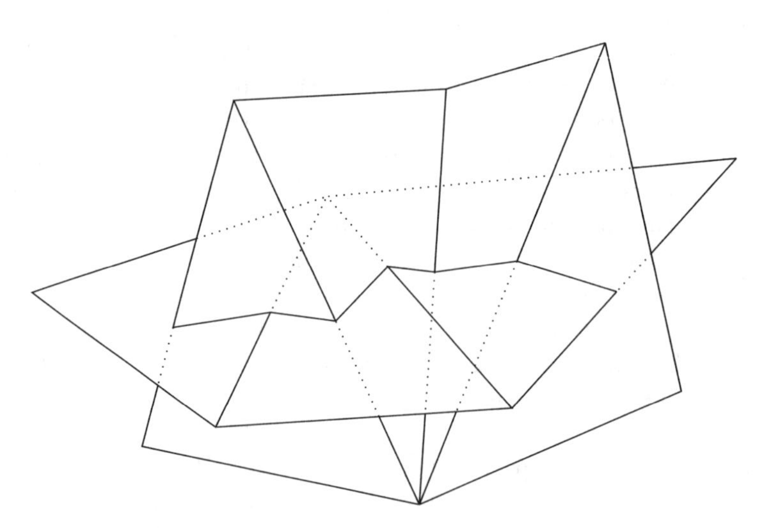

FIGURE 2.11
The intersection of two polyhedral surfaces in general position.

between the surfaces. These attach to the endpoints of the first segment. We can follow this curve until it either meets up with itself to form a simple closed curve, ends at a boundary curve of one of the surfaces, or wanders off to infinity (a possibility if the surfaces are unbounded). Thus, the intersection of the two surfaces is a disjoint union of arcs and simple closed curves. ❄

Consider the orthogonal projection of a knot K into a plane in $\mathbb{R}^3$. We might use something as simple as the projection that maps points (x, y, z) to $(x, y, 0)$. Ideally, the projection will cast a clear shadow of K: adjacent edges of K will project to line segments that meet at a common endpoint, disjoint edges of K will project to line segments that are disjoint or meet at a point interior to both, and we will not find three points of K that project to the same point in the plane. Such a point with three elements in its inverse image is known as a **triple point**.

If we are not so lucky, we can still adjust the direction of the projection so that the image of the knot is in general position with itself. As indicated in the following definition, there are really only two problems to take care of.

Definition 2.12 *The orthogonal projection of a knot onto a plane is a* ***regular projection*** *if and only if no vertex projects to the image of another point of the knot and there are no triple points.*

Suppose a vertex and another point of the knot project to the same point. For example, we might have an edge that projects to a point, a pair of edges that project to overlapping segments, or an edge whose projection touches the projection of another edge without crossing it. In this situation, we are looking down on the knot from a distance and see the vertex directly above or below the other point of the knot. We simply adjust our point of view by moving at an angle away from the edge (or edges) containing this other point. We can make the adjustment as small as we please and at the same time ensure that the vertex is not aligned with any other point of the knot.

Once we no longer have vertices projecting to the image of other points of the knot, we can deal with triple points. As we look down on the knot, a triple point results from an alignment of interior points of three edges of the knot. Although it is a bit difficult to specify exactly what direction to move, Exercise 4 should convince you that it is easy to adjust our point of view by an arbitrarily small amount to eliminate the triple point.

Figure 2.13 illustrates some typical adjustments being made to portions of the projection of a knot.

As we fix up one problem, we must of course be careful not to create another. But if a vertex or edge is in general position, a suitably small adjustment to the projection plane will maintain general position. Furthermore, each problem can be eliminated by arbitrarily small adjustments of our point of view. Thus, it is always possible to proceed without destroying previous work.

Instead of adjusting our point of view to obtain a regular projection of a knot, we can obtain the same result by rotating the knot in the opposite direction. According to Exercise 7 of Section 2.1, the rotated knot is equivalent to the original one. Since we are

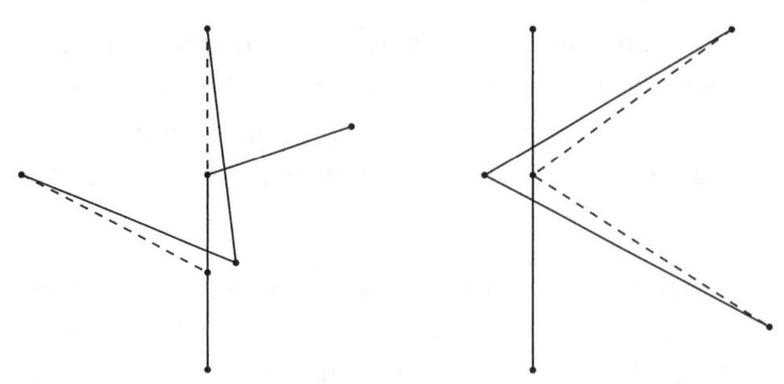
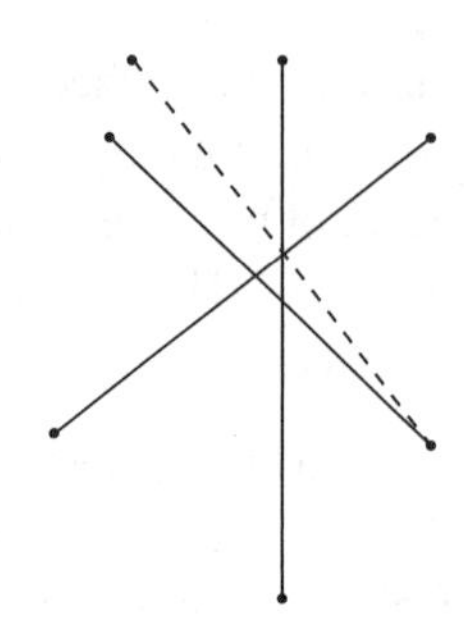

FIGURE 2.13
Adjusting a knot projection.

more interested in studying knot types than in the geometric knots themselves, we will freely assume from now on that any knot is depicted by a regular projection.

Now that we see how to simplify the projection of a knot, our next task is to find a way to reconstruct the knot (or at least the knot type) from a regular projection. The main problem occurs at crossing points of the projections of two edges. We cannot tell from the projection which edge in the knot goes over and which goes under the other. This ambiguity is easily resolved by drawing a small break in the projection of the lower edge to indicate that it crosses under the other edge. When this bit of artistic license is applied to each crossing point in a regular projection, the result is known as a **knot diagram**. Notice that the diagram of a knot consists of disjoint arcs each of which begins at one side of a crossing as the lower strand, possibly passes through some other crossing points as the upper strand, and ends at one side of a crossing as the lower strand.

Of course, many knots will have the same diagram. For example, slide a vertex of a knot up or down in the direction of the projection to change the knot without changing its projection. Or introduce a new vertex along an edge. Or slide such a vertex along the edge. Or remove such a vertex. But these kinds of vertex shifting are the only differences between knots with identical diagrams. As long as the shifts do not result in edges crossing, we can reproduce this modification of the knot by triangular moves. It is reassuring to know that any two knots with the same diagram are equivalent. Exercise 10 at the end of this section will guide you in supplying some details of the proof of this result.

Definition 2.14 *A* ***knot invariant*** *is a mathematical property or quantity associated with a knot that does not change as we perform triangular moves on the knot.*

Since an invariant of two equivalent knots will have the same value, this opens the possibility for us to distinguish knots of different types. With any invariant we face two questions: Why does the quantity in fact have the same value on all equivalent knots? And how can we compute the values of the invariant for various knots? In knot theory, the simplicity in answering one of these questions is often proportional to the difficulty in answering the other. The following definition uses knot diagrams to give a simple invariant.

However, computing the value of this invariant is quite challenging, and knot theorists are still struggling to prove basic properties of this invariant.

Definition 2.15 *The* ***crossing number*** *of a knot K is the minimum number of crossing points that occur in the knot diagrams for all knots equivalent to K.*

Example 2.16 Show that there is only one knot type with crossing number 0.

Solution. A knot with crossing number 0 is equivalent to a knot whose projection is a simple closed curve in a plane. As in Exercise 10 at the end of this section, we can perform a sequence of triangular moves to slide the vertices of this knot into the plane. By the Schönflies Theorem, this polygon bounds a disk. Exercises 11 and 12 in Section 1.6 then give a sequence of triangular moves that simplify this polygon to a triangle. Finally, Exercise 3 of Section 2.1 gives a few additional triangular moves that will transform this triangle to your favorite standard triangle. ❄

The preceding example leads to the following definition of the simplest types of knots and links.

Definition 2.17 *A* ***trivial knot*** *is a knot that is equivalent to a triangle. A* ***trivial link*** *is a link that is equivalent to the union of disjoint triangles lying in a plane.*

Here is another invariant property defined in terms of knot diagrams.

Definition 2.18 *A knot is* ***alternating*** *if and only if it is equivalent to a knot with a diagram in which underpasses alternate with overpasses as you travel around the knot.*

This is a nice property that divides knots into two types, alternating and nonalternating. You can easily draw knot diagrams with alternating underpasses and overpasses (see Exercise 7 at the end of this section) and thus exhibit some alternating knots. You can also draw a knot diagram in which the underpasses and overpasses do not alternate. But that doesn't guarantee a nonalternating knot. There might be an equivalent knot with an alternating projection. In 1877, Peter Guthrie Tait published the first article [29] on the mathematical theory of knots. In this paper he listed all knots with crossing number up to seven, and all of these were alternating knots. Although he suspected that all knots were alternating, he later found nonalternating examples. But it was not until 1930 that C. Bankwitz proved that these did not have alternating diagrams.

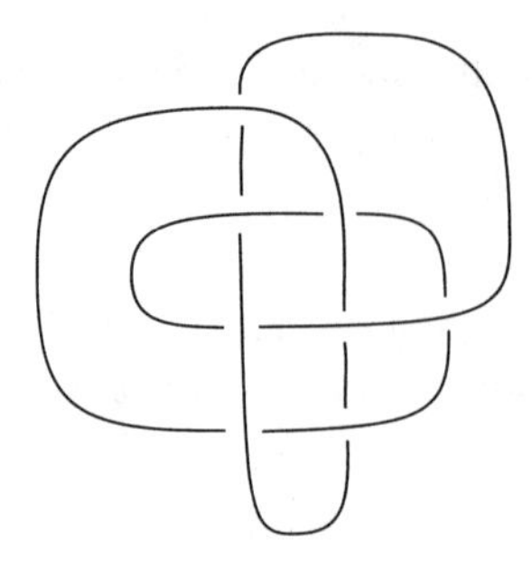

FIGURE 2.19
A nonalternating knot with crossing number 8.

Although these two invariants are computationally difficult to handle, a 1986 result by Lou Kauffman, Kunio Murasugi, and Morwen Thistlethwaite makes it easy to determine the crossing number of a knot given by a diagram with alternating underpasses and overpasses. Suppose we have an alternating diagram of a knot for which the number of crossings cannot be reduced by flipping over a portion of the diagram. Such a diagram is called a **reduced alternating diagram**. Kauffman, Murasugi, and Thistlethwaite proved that the number of crossings in a reduced alternating diagram for a knot is the crossing number of the knot.

Example 2.20 Use the Kauffman-Murasugi-Thistlethwaite result to determine the crossing number of the knot diagram on the left in Figure 2.21.

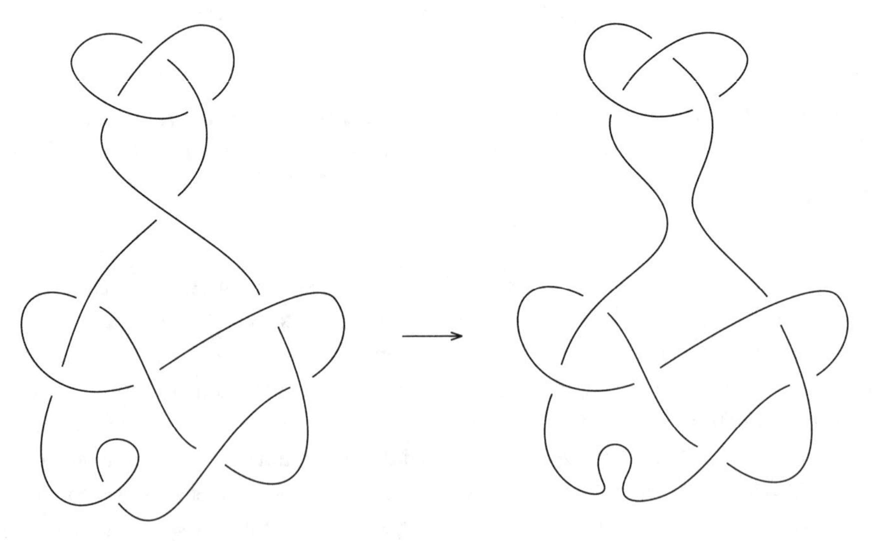

FIGURE 2.21
Determining the crossing number of an alternating knot.

Solution. A couple of simple flips of portions of the knot diagram produces a reduced alternating diagram for the knot. Hence, the knot has crossing number 9 as indicated in the diagram on the right of Figure 2.21. ❄

Exercises 2.2

1. **(a)** Describe the intersection of two polyhedral surfaces in general position in $\mathbb{R}^4$.
   (b) Describe the intersection of two polyhedral surfaces in general position in $\mathbb{R}^5$.
2. Develop a rule of thumb for the dimension of the intersection of three piecewise-linear objects embedded in general position in $\mathbb{R}^n$.
3. Find a condition on the dimensions of two piecewise-linear objects so that embedding them in general position in $\mathbb{R}^n$ will guarantee that they are disjoint.
4. Suspend three straight pieces of string across the room. Find a direction from which you see the strings line up in a triple point. Move your line of sight slightly in different directions. How easy is it to eliminate the triple point? Is it possible to move your line of sight so that a triple point remains in view?
5. Show that a knot with a diagram having one or two crossings is equivalent to the unknotted circle.
6. How many knot types have diagrams with three crossings?
7. Consider a regular projection of any knot. Show that we can modify the crossings to produce a knot diagram in which underpasses alternate with overpasses as we go around the new knot. Here are some suggestions from a paper "Some Elementary Properties of Closed Plane Curves" [28] published in 1877 by P. G. Tait. Tait is dealing with polygonal curves in general position in the plane.
   (a) A closed curve intersects a straight line in an even number of crossings.
   (b) A closed curve intersects a curve whose ends go off to infinity in an even number of crossings. This situation can be simplified to the previous case by "opening up" the unbounded curve whenever it crosses itself. That is, replace each crossing by two curves that approach each other but do not cross. There are two ways to open up each crossing. Be sure to retain all the crossings with the closed curve.
   (c) Any two closed curves intersect in an even number of crossings.
   (d) The portion of a regular knot projection from one crossing back to the same crossing passes through an even number of other crossings.
8. Prove that if a regular projection of a knot has n crossings, then the projection partitions the plane into $n + 2$ regions.
9. **(a)** Slide a vertex of a knot along a straight line, moving the two edges incident with the vertex along with it. Show that if the edges do not cross other parts of the knot during this motion, then the new knot is equivalent to the old one.
   (b) Introduce a new vertex along an edge of a knot. Show that the new knot is equivalent to the old one.

(c) Suppose three consecutive vertices of a knot lie on a line. Slide the inner vertex along the line and adjust the two edges incident with this vertex. Show that the new knot is equivalent to the old one.

10. Consider two knots with the same knot diagram.

(a) Draw a typical pair of adjacent edges of one of the knots. Slide the vertex common to the two edges in a direction perpendicular to the plane of the projection. Keep the other vertices of the edges fixed and add the displaced edges to your drawing. Indicate triangular moves that give an equivalence between the original and modified knot. Illustrate the possible obstructions to sliding the vertex an arbitrarily large distance.

(b) Introduce enough new vertices in the knots so that the vertices and edges of both knots match the vertices and edges of the diagram and so that there are at least three vertices between any two crossing points in the knot diagram. Consider the vertices of the first knot that are not endpoints of edges involved in any of the crossings in the diagram. Slide these vertices to coincide with the corresponding vertices of the other knot. Show that this preserves the knot type.

(c) Draw a picture of a typical crossing in the knot diagram and the corresponding edges of the two knots. Extend the picture to include the edges leading into this crossing. Remember that you have modified the first knot so that the outer vertices coincide with the corresponding vertices of the other knot. Illustrate triangular moves that will slide the overcrossing edge of the first knot to the overcrossing edge of the other knot. Do the same for the undercrossing edges.

11. Fill in the details of the following steps to show that any knot can be transformed to the trivial knot by changing overcrossings to undercrossings in a regular projection of the knot.

(a) Find a vertex whose image is a maximum in some direction in the plane of the projection.

(b) Start from that vertex and travel around the knot. You will encounter each crossing twice. The first time you encounter a crossing, change it if necessary so you are traveling on an overcrossing.

(c) Picture the resulting knot from the side with the knot above the plane of the projection and the initial vertex on the far right. Show that the knot can be modified by sliding vertices vertically so that your travel around the knot is decreasing except for one vertical segment connecting the initial vertex with the terminal point on the journey along the knot.

(d) Project the knot in the direction of your view. Show that it has crossing number 0. The result of Example 2.16 shows that this knot is the trivial knot.

12. Consider a knot K as subsets of $\mathbb{R}^3$ embedded as the three-dimensional slice of $\mathbb{R}^4$ consisting of all points whose fourth coordinates equal zero. Show that K can be unknotted in $\mathbb{R}^4$. Suggestion: Combine the results of Example 1.56 and Exercise 11.

13. Show that any link can be transformed to a trivial link by changing overcrossings to undercrossings in a regular projection of the link.

14. Use the Kauffman-Murasugi-Thistlethwaite result to determine the crossing numbers of the knots illustrated in Figure 2.22.

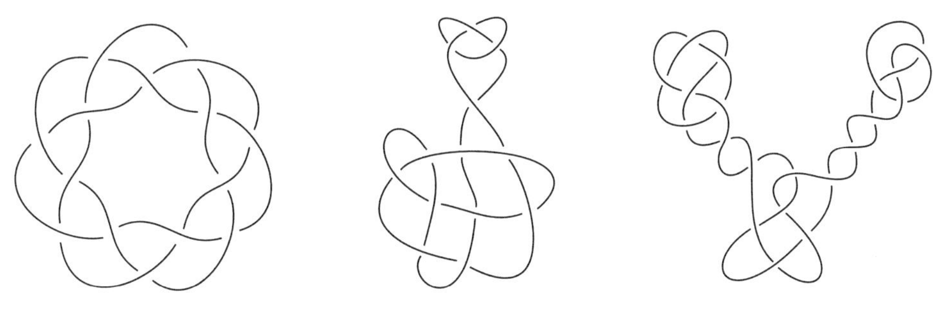

FIGURE 2.22
Determine the crossing numbers.

15. Use the Kauffman-Murasugi-Thistlethwaite result to determine the crossing numbers of the knots illustrated in Figure 2.23.

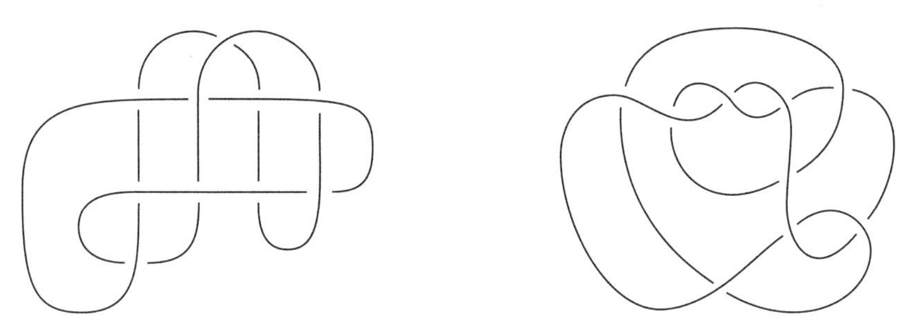

FIGURE 2.23
Determine the crossing numbers.

2.3 Reidemeister Moves

Although a triangular move is a simple enough step in transforming one knot into an equivalent knot, such a change in a knot can significantly alter the diagram of the knot. We would like to consider triangular moves that only affect the knot diagram one vertex or one edge at a time. This will enable us to analyze the effect of the move on some quantity we are trying to establish as a knot invariant.

In the 1920s, Kurt Reidemeister and other knot theorists realized that triangular moves can be broken down so the effect on the knot diagram is one of three simple modifications or their inverses. Figure 2.24 depicts the three types of Reidemeister moves. These schematic drawings illustrate the change in the combinatorial relations among the arcs, crossings, and regions in the knot diagram. Of course, the geometric aspects of distance and angles are completely arbitrary. So the Reidemeister moves can be combined with isotopies of the plane that do not change these combinatorial relations in the knot diagram.

Also included among the Reidemeister moves are the inverses of those illustrated in Figure 2.24 and the reflections through the plane of the knot projection. Exercise 1 at the end of this section asks you to investigate these variations of the basic three types of Reidemeister moves.

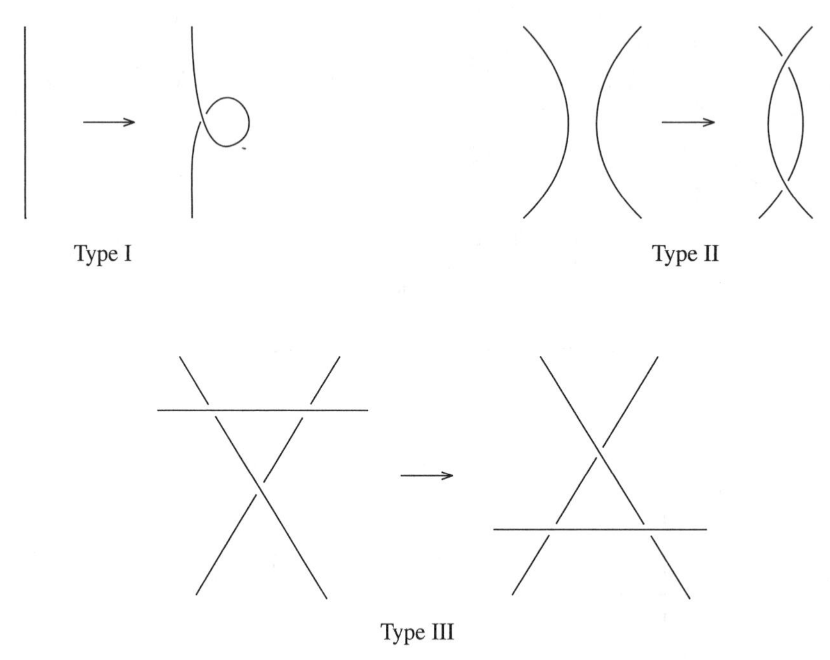

FIGURE 2.24
The three types of Reidemeister moves.

At first, you may be amazed that the three types of Reidemeister moves suffice to transform a knot diagram into a diagram of any equivalent knot. The proof of this fact involves some technical details, but it is not hard to understand the basic ideas involved.

We begin with diagrams of two equivalent knots. Rotate the knots so that the diagrams of the two knots are based on projections into the same plane. Now adjust the projection so that both knots, as well as the triangles involved in all triangular moves, project in general position. Make this adjustment so that it does not disturb the combinatorics of the knot diagrams.

Next, break down each triangular move into moves involving smaller triangles. Choose these triangles small enough so that the projection of each of them to the plane of the knot diagram contains at most one interesting feature of the rest of the knot. The interesting feature may be another edge of the knot, parts of two adjacent edges including their common vertex, or portions of two nonadjacent edges including an overcrossing. See Figure 2.25.

Finally, use the ideas illustrated in Exercise 2 at the end of this section to realize each of these simpler triangular moves as one or more Reidemeister moves.

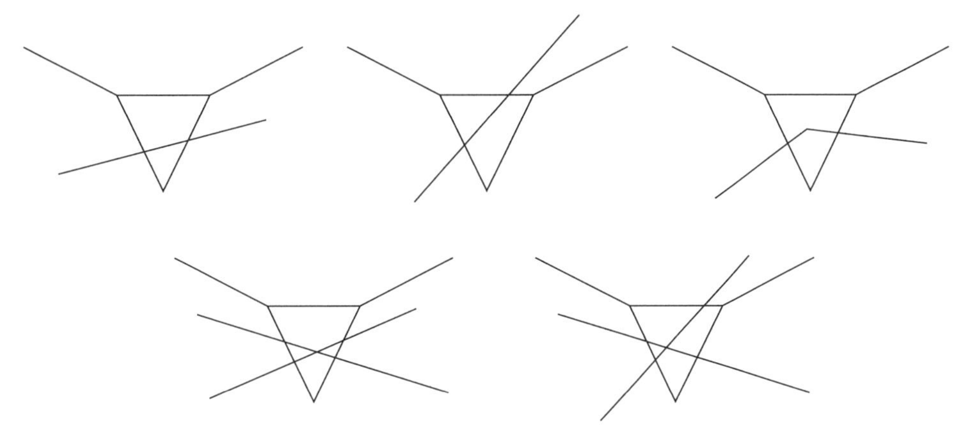

FIGURE 2.25
Typical cases of some simpler triangular moves.

We are now ready to summarize this important conclusion. First, however, look back over the steps we have taken: triangular moves, equivalence, regular projections, diagrams, Reidemeister moves. They pertain equally as well to links of more than one component. Hence we state our result in the more general context of links.

Theorem 2.26 *If two links are equivalent, then their diagrams, subject to ambient isotopies of the plane, are related by a sequence of Reidemeister moves.*

Example 2.29 illustrates the typical way we can use Reidemeister moves to establish a quantity as an invariant. We first define the quantity in terms of a regular projection.

Definition 2.27 *An* ***orientation*** *of a link is a choice of direction to travel around each component of the link. Consider a crossing in a regular projection of an oriented link. Stand on the overpass and face in the direction of the orientation. The crossing is* ***right-handed*** *if and only if traffic on the underpass goes from right to left; the crossing is* ***left-handed*** *if and only if traffic on the underpass goes from left to right. In a regular projection of an oriented link of two components, assign* $+1$ *to right-handed crossings and* -1 *to left-handed crossings. Add up the numbers assigned to crossings involving both components. One half this sum is the* ***linking number*** *of the two oriented components of the link.*

Figure 2.28 illustrates the computation of the linking number of the two components of the given link. Notice that we use only the crossings involving both of the components of the link.

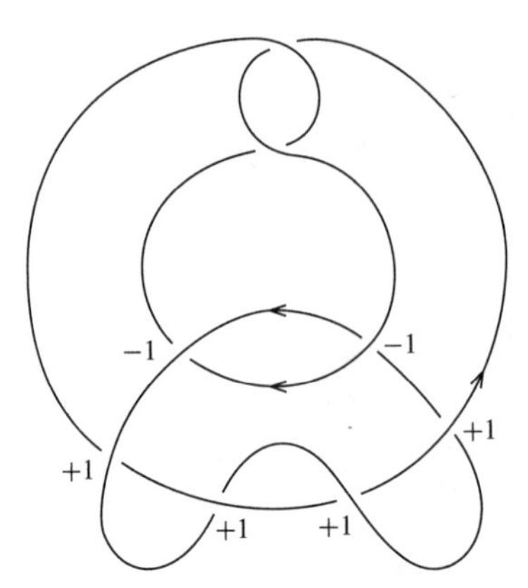

FIGURE 2.28
Linking number $\frac{1}{2}(-2+4)=1$.

Example 2.29 Show that the linking number is an invariant of an oriented link of two components.

Solution. Let us check how Reidemeister moves affect the linking number.

A move of Type I generates a new crossing of a component with itself. These are ignored in computing the linking number. Thus, there is no change to this quantity.

A move of Type II generates two crossings, both involving the same component or components of the link. One crossing will be assigned $+1$, and the other will be assigned -1. Thus, even if the crossings involve both components of the link, the sum of the numbers will not change.

A move of Type III merely alters the relative positions of three crossings without changing the components or orientations involved. Hence, there is no change to the linking number.

Because the diagrams of two equivalent links are related by a sequence of Reidemeister moves, it follows that the linking number is an invariant of the oriented link. ❄

Since the linking number between two components of a trivial link is 0, we can conclude that Figure 2.28 depicts a nontrivial link.

Exercises 2.3

1. As noted in the text, the inverses and mirror images of the Reidemeister moves illustrated in Figure 2.24 are also Reidemeister moves.
   - **(a)** Draw pictures to illustrate the inverses of the three types of Reidemeister moves. Show that the inverse of a Type III move is also a Type III move.
   - **(b)** Draw pictures to illustrate the mirror images of the three types of Reidemeister moves. Assume the mirror is parallel to the plane of the projection, so the reflection merely interchanges overcrossings and undercrossings. Show that the mirror image of a Type II move is also a Type II move.

2. Suppose a triangular move on a knot involves a triangle that projects to a triangle in the diagram of the knot. By subdividing the triangle involved in a triangular move, we can arrange that the edges in the knot diagram not involved in the triangular move passes through the triangle only with a portion of one edge, portions of two adjacent edges including the common vertex, or portions of two nonadjacent edges including an overcrossing.
   (a) Draw pictures to illustrate the possible ways the triangular move can affect a knot diagram. Don't forget the cases as illustrated in Figure 2.30 in which one or both of the additional edges are adjacent to edges of the triangle.

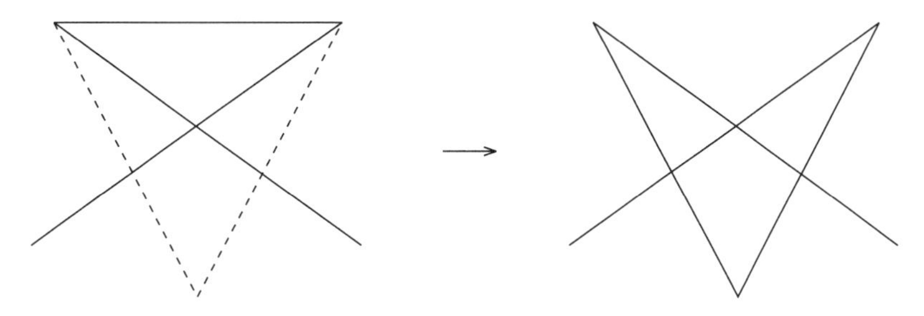

FIGURE 2.30
A tricky triangular move.

   (b) In each case, show how to subdivide the triangle so that the triangular move can be realized as a sequence of moves across smaller triangles each of which yields a Reidemeister move in the knot diagram.

3. Compute the linking numbers of the oriented links in Figure 2.31. Which of these links can you conclude are nontrivial?

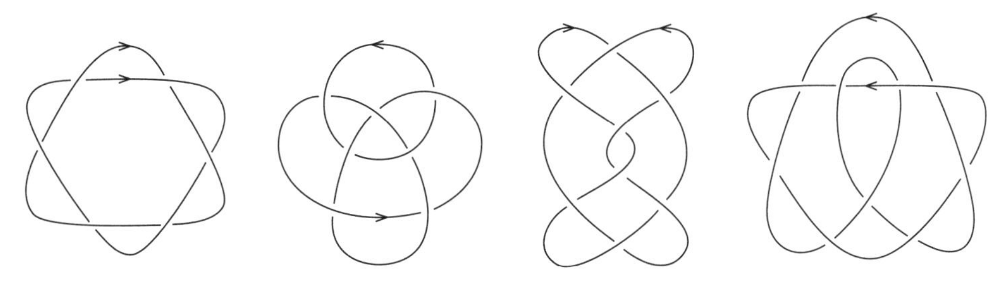

FIGURE 2.31
Compute the linking numbers.

4. Orient the components of the Hopf link, the Borromean rings, and the Whitehead link in Figure 2.8. Compute the linking numbers between pairs of components. Which links can you conclude are nontrivial?

5. Develop a right-hand rule for determining whether a crossing is right-handed or left-handed.

6. In an oriented link of two components, show that changing the orientation of one of the components will change the sign of the linking number. Conclude that the absolute value of the linking number is an invariant of a link (without orientation) of two components.

7. How does the linking number of an oriented link of two components compare with the linking number of its mirror image?

8. Give a sequence of examples to show that any integer is a possible value of the linking number of an oriented link of two components.

9. Consider an oriented link of two components.

(a) Show that changing an overcrossing to an undercrossing changes the sign of the number assigned to that crossing.

(b) What effect does this have on the linking number?

(c) Use the result of Exercise 13 in Section 2.2 to conclude that the linking number of any oriented link of two components is an integer.

10. (a) Compute the linking number between the boundary curve of the Möbius band illustrated in Figure 1.19 and its centerline.

(b) Show that there is no ambient isotopy of $\mathbb{R}^3$ that will deform a left-handed Möbius band onto a right-handed Möbius band.

(c) Show that a Möbius band with three half-twists cannot be deformed to a Möbius band constructed with a single half-twist.

11. Find a sequence of Reidemeister moves that will reduce the number of crossings in the knot diagram in Figure 2.32. Notice that you must first introduce additional crossing into the knot diagram before you can reduce the number of crossings.

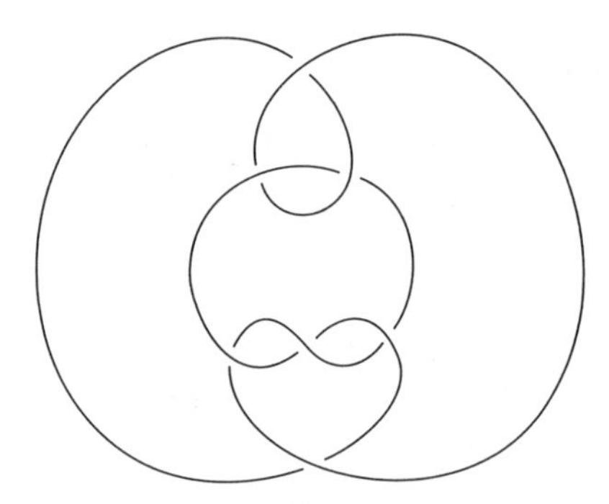

FIGURE 2.32
Simplify this knot diagram.

12. Find sequences of Reidemeister moves that will reduce the number of crossings in the knot diagrams in Figure 2.23.

2.4 Colorings

At this stage in the development of knot theory, we have to face the realization that we have no conclusive way to distinguish among different types of knots. We have not even seen a proof of the existence of knots that are not equivalent to the trivial knot. In this section we will take the first steps toward remedying this situation. We will describe rules for coloring the arcs that comprise a knot diagram. These rules can be satisfied in some knot diagrams, but not in others. We will show that the possibility of coloring the knot diagram according to the rules does not change when we perform Reidemeister moves. Hence, the colorability of a diagram is an invariant property of the knot. In particular, knots with colorable diagrams are distinct from knots with diagrams that are not colorable.

The basic concept of coloring a knot diagram is extremely simple. Here is the definition.

Definition 2.33 *The diagram of a knot is* ***colorable*** *if and only if each arc can be assigned one of three colors subject to two conditions:*

1. *At least two colors appear.*
2. *At any crossing where two colors appear, all three colors appear.*

Example 2.34 Determine which of the knot diagrams in Figure 2.35 can be colored.

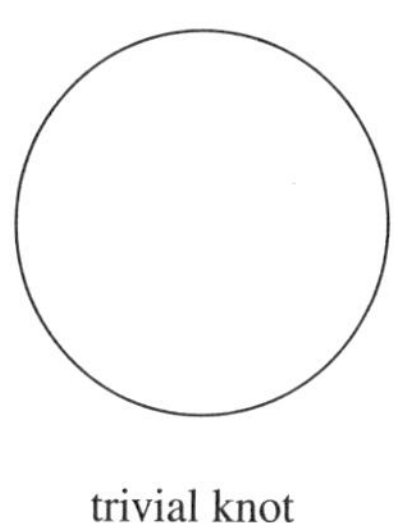
trivial knot

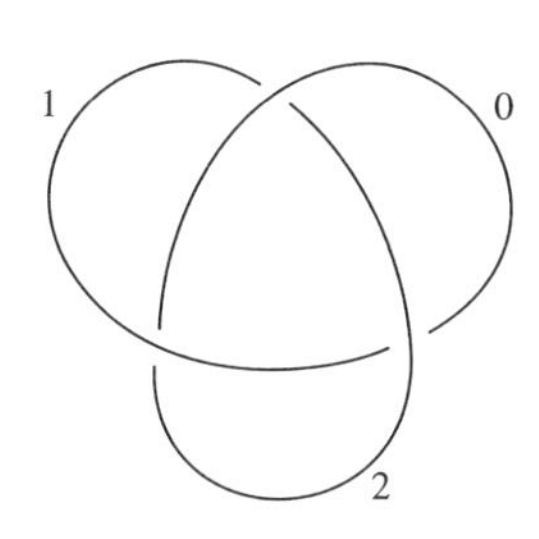

trefoil knot

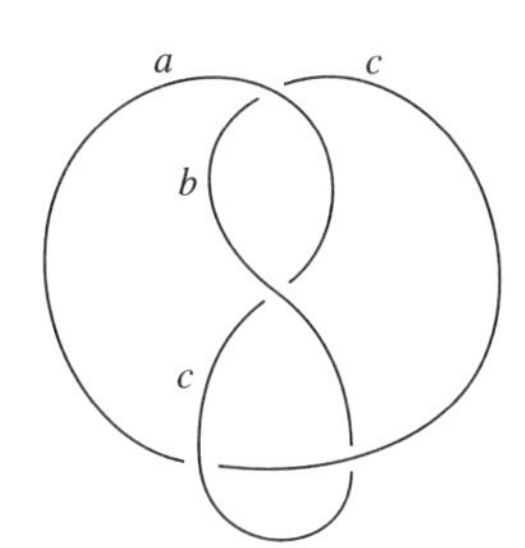

figure-eight knot

FIGURE 2.35
Coloring knot diagrams.

Solution. The diagram of the trivial knot has only one arc (actually a simple closed curve). Since that does not allow for more than one color, the first condition of colorability fails.

The three arcs in the diagram of the trefoil knot can be colored with the three different colors. All three of the colors appear at each of the three crossings. Thus, we have a coloring of this knot diagram.

In the diagram of the figure-eight knot, consider the portions of the arcs involved in the crossing at the top of the diagram. These three strands can be assigned colors a, b, and c as indicated. One possibility is that these are three different colors. The strands labeled a and b meet again in the middle of the diagram. So in this case, the fourth arc of the diagram must be colored c. But then both of the bottom two crossings have two colors. This violates the second condition of colorability. The other possibility is that the three strands at the top of the diagram are all the same color. But then the fourth strand must also be of that same color. This violates the first condition. We conclude that the diagram cannot be colored. ❄

The following result is the key to being able to distinguish between knots with colorable diagrams and knots whose diagrams cannot be colored.

Theorem 2.36 *The colorability of a knot diagram is an invariant property of the knot type.*

Proof. This proof involves checking a number of cases, all fairly straightforward. We simply examine the different possibilities for coloring the arcs affected by each type of Reidemeister move and its inverse. The key is that any changes in the color scheme will occur on the arcs that are modified by the move. We will not have to change the color of any arcs leading out of the picture of a Reidemeister move.

A Reidemeister move of Type I introduces a new crossing that breaks an arc into two arcs. The overpass arc loops around to become one of the strands of the underpass. Since these two strands are the same color, the third strand at that crossing must also be the same color. Thus, a Type I move or its inverse will not result in any change in the coloring of the diagram.

A Reidemeister move of Type II passes a portion of one arc under another arc. If the two arcs are the same color, the new arc must also be that same color. If the two arcs are different colors, the new arc must be assigned the third color. The same situations occur with the inverse of this type of Reidemeister move. If all arcs are the same color, they stay that way as the strands are pulled apart. Otherwise, one arc enters with one color, and as it crosses under the arc of another color, it changes to the third color. Then it changes back to the original color at the second undercrossing. When it is pulled out from under, it can be colored with the original color. Thus, a Type II move or its inverse will not alter the colorability of a knot diagram.

A Reidemeister move of Type III involves three strands: a top strand passes over the other two, a bottom strand passes under the other two, and a middle strand passes under the top strand and over the bottom strand. If the top and middle strands are the same color, the bottom strand can be entirely that same color, or it can change from a second color to a third and back to the second color as it passes under the other strands. If the middle strand changes color as it passes under the top strand, there are three possible color schemes for the lower strand. Exercise 4 at the end of this section asks you to verify that colorability is preserved in all five of these cases. Since the inverse of a move of Type III is also a move of Type III, this completes the verification that colorability is an invariant of the knot type. ❄

The invariant property of colorability is a good start at distinguishing knot types. In particular, the trefoil knot can be colored and the trivial knot cannot. Thus, we now have a proof of the existence of nontrivial knots. However, the figure eight knot and many knots besides the trivial knot cannot be colored. We would like to be able to prove these are in fact nontrivial. We would also like to distinguish among the knots that can be colored.

Colorability is merely the simplest of a sequence of invariants involving coloring the arcs of knot diagrams.

Definition 2.37 *Let p be an odd prime number. Consider a circle with p radii of different colors equally spaced around the circle. The diagram of a knot has a* ***representation on a wheel with p colors*** *(or the diagram is* ***p-colorable****) if and only if each arc of the diagram can be assigned one of the colors subject to two conditions:*

1. *At least two colors appear.*
2. *At any crossing, the colors of the two undercrossing arcs are from radii that are symmetric with respect to the radius with the color of the overpass.*

Since three colors will automatically satisfy the symmetry condition of Definition 2.37, we see that colorability is the particular case of 3-colorability.

It will be obvious whether a colored knot diagram uses more than one color. It is nearly as easy to check that the diagram satisfies the second condition. Simply travel around the knot and at each underpass make sure the color switches to the color of the reflected radius.

If we use numbers $0, 1, \ldots, p-1$ rather than colors, we can give an algebraic formulation of the second requirement for colorability. Suppose c is assigned to an overcrossing and a, b are assigned to the undercrossing strands. By the symmetry condition, the angle on the wheel from a to c must equal the angle from c to b. That is, $a - c$ is congruent modulo p to $c - b$. Or more concisely, $a + b = 2c$ modulo p. Thus, at each crossing the sum of the two numbers assigned to the arcs of the underpass must be congruent modulo p to twice the number assigned to the overpass.

Example 2.38 Show that the figure-eight knot has a representation on a wheel with five colors.

Solution. See Figure 2.39. ❄

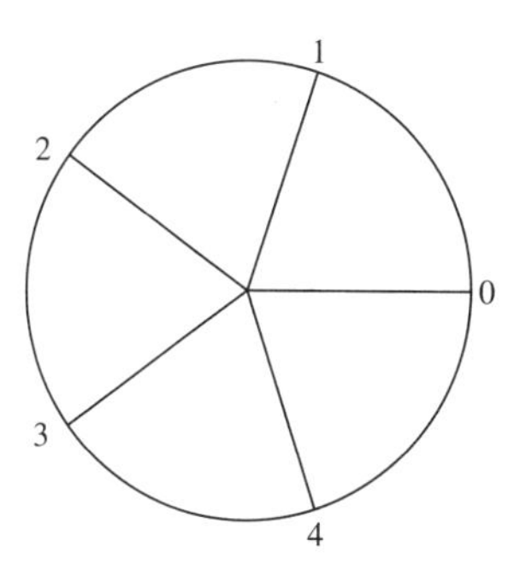

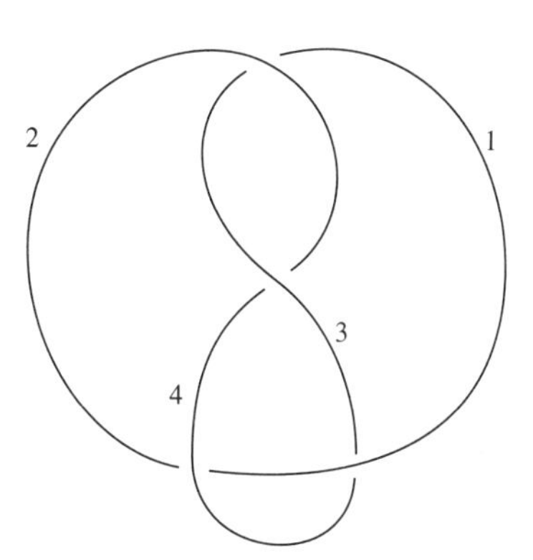

FIGURE 2.39
Coloring the figure-eight knot with five colors.

Theorem 2.40 *The representation of a knot diagram on a wheel with p colors is an invariant property of the knot type.*

The proof of this theorem is very similar to the proof of the invariance of colorability with three colors. Exercise 9 asks you to supply the details.

Exercises 2.4

1. There are two knot types of crossing number 5. Figure 2.41 gives knot diagrams for these two types. Show that neither of these is colorable with three colors.

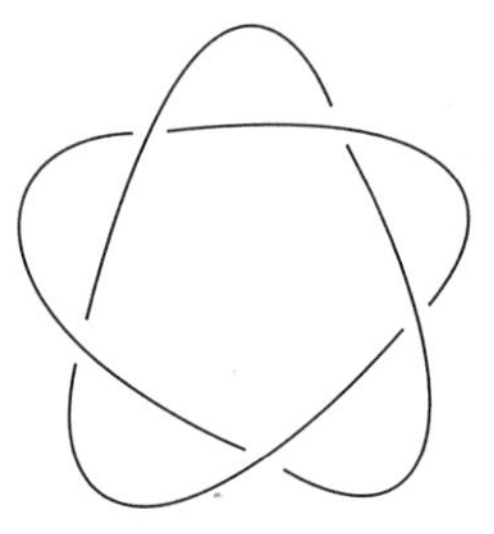
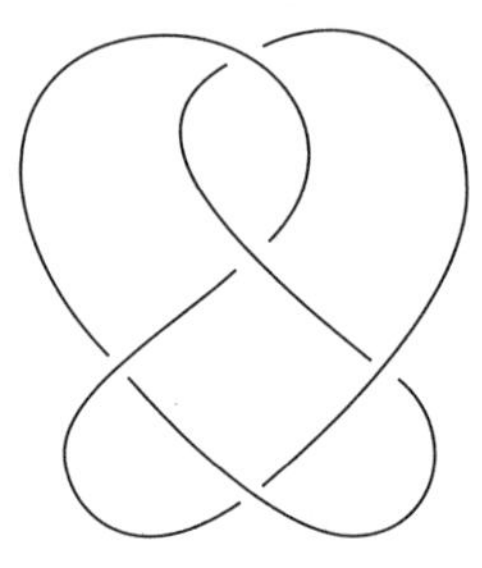

FIGURE 2.41
The two knot types of crossing number 5.

2. There are three basic knot types of crossing number 6. Figure 2.42 gives knot diagrams for these three types. Determine which of these are colorable with three colors.

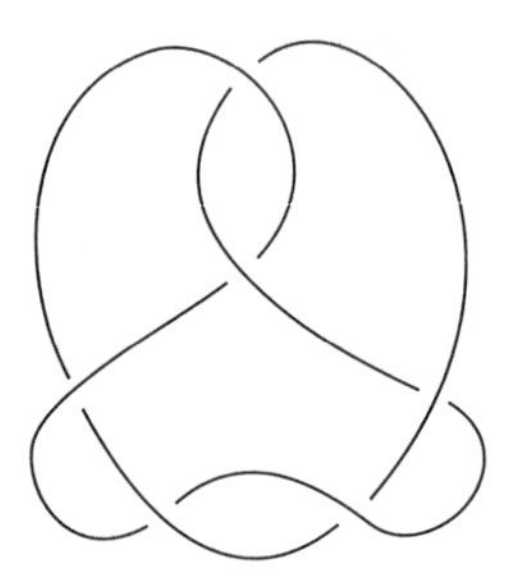
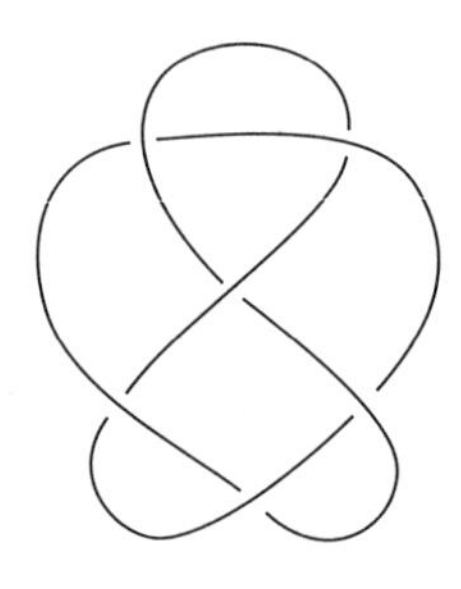
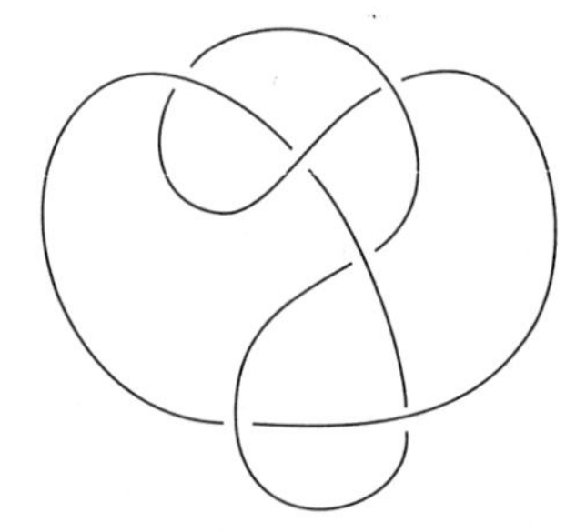

FIGURE 2.42
The three basic knot types of crossing number 6.

3. **(a)** Draw diagrams to illustrate the possible ways of coloring with three colors the arcs involved in a Reidemeister move of Type II.
   (b) In each case, draw the diagram after the Reidemeister move and give the color scheme that preserves the colors of the strands leading out of your pictures.
4. **(a)** Draw diagrams to illustrate the possible ways of coloring with three colors the arcs involved in a Reidemeister move of Type III.

(b) In each case, draw the diagram after the Reidemeister move and give the color scheme that preserves the colors of the strands leading out of your pictures.

5. (a) Show that there are six ways to use three colors to color the diagram of a trefoil knot.

(b) A granny knot is made by tying two trefoil knots as indicated in Figure 2.43. Show that there are 24 ways to color this diagram of the granny knot with three colors.

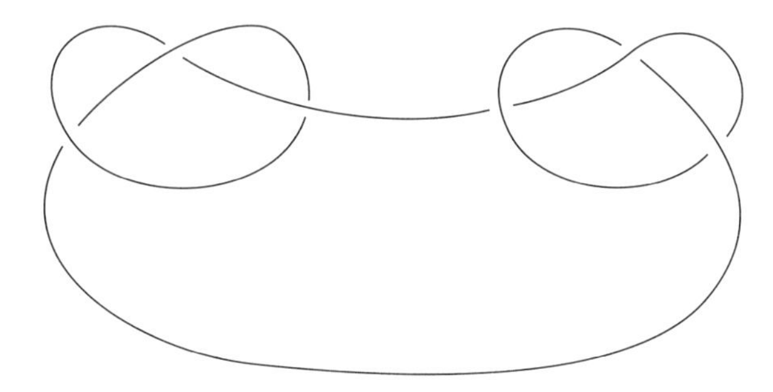

FIGURE 2.43
A granny knot.

(c) Show that the number of different ways of coloring the diagram of a knot with three colors is an invariant of the knot type.

(d) Find a formula for the number of ways to color a string of n trefoil knots with three colors.

(e) Conclude that there is an infinite number of distinct knot types.

6. Show that the trefoil knot cannot be represented on a wheel with five colors.

7. Which of the knots of crossing number 5 (see Figure 2.41) are 5-colorable?

8. Which of the basic knots of crossing number 6 (see Figure 2.42) are 5-colorable?

9. Prove that p-colorability is an invariant property of the knot type.

10. The proof that p-colorability is an invariant property extends to links of more than one component.

(a) Prove that the Whitehead link (see Figure 2.8) is not 3-colorable. Conclude that the Whitehead link is nontrivial.

(b) Prove that the Borromean rings (see Figure 2.8) are not 3-colorable. Conclude that the Borromean rings are a nontrivial link.

2.5 The Alexander Polynomial

In 1928, J. W. Alexander introduced a polynomial invariant that is straightforward to compute, has a moderately easy proof that it is an invariant, and yet is fairly good at distinguishing among different knots. You may want to track down the original paper "Topological Invariants of Knots and Links" [12], and compare Alexander's approach to the slightly simpler presentation in this section. In both approaches, the knot diagram gives the entries of a matrix whose determinant is the Alexander polynomial. Alexander used arcs and regions formed by the diagram of a knot; we will determine the entries of a matrix from the crossings and arcs of the diagram.

At each crossing of the knot diagram, the breaks in the lower strand divide the curve into arcs that begin and end at crossings. Since each arc has two ends and each crossing involves the ends of two arcs, the number of arcs is equal to the number of crossings. Index the crossings $x_1, x_2, \ldots, x_n$, and index the arcs $a_1, a_2, \ldots, a_n$. Exercise 7 at the end of this section explains how to deal with the trivial knot, which has a knot diagram with no crossings.

Next choose an orientation of the knot. As you travel around the knot, stand on the overpass of each crossing and face in the direction of the orientation. Label the three strands of the knot diagram that form this crossing: put $1-t$ on the overpass itself, label the end of the underpass arc on your left with t, and label the end of the underpass arc on your right with -1. Figure 2.44 illustrates this notation in a simple case.

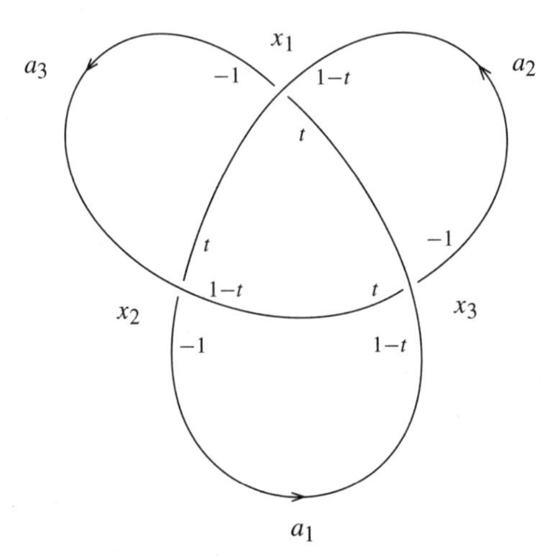

FIGURE 2.44
Labeling a diagram of the trefoil knot.

Form an $n \times n$ matrix by writing the label on arc a_j at crossing x_i as the ij-entry of the matrix. If an arc has more than one of its labels at the crossing, put the sum of the labels as the entry of the matrix. If the arc is not involved in forming the crossing, put a 0 as the entry. This matrix is the **crossing/arc** matrix of the knot diagram. The crossing/arc matrix of the trefoil knot in Figure 2.44 is

$$\begin{array}{c} \\ x_1 \\ x_2 \\ x_3 \end{array} \begin{array}{c} \begin{array}{ccc} a_1 & a_2 & a_3 \end{array} \\ \begin{bmatrix} t & 1-t & -1 \\ -1 & t & 1-t \\ 1-t & -1 & t \end{bmatrix} \end{array}.$$

Next we delete one column and one row of the matrix and compute the determinant of the resulting $(n-1) \times (n-1)$ matrix. This will be a polynomial in the variable t. In our example, if we delete the third row and third column, the determinant of the resulting 2×2 matrix is

$$\det \begin{bmatrix} t & 1-t \\ -1 & t \end{bmatrix} = t^2 - (-1)(1-t) = t^2 - t + 1.$$

This polynomial depends on quite a few choices: the index system for the crossings and arcs, the orientation of the knot, the selection of a row and a column to eliminate, to say nothing of the choice of the knot diagram. Alexander's amazing discovery is that although these choices may produce different polynomials, any one of them will be plus or minus a power of t times any other. Thus, if we normalize the polynomial to have a positive constant term, the resulting **Alexander polynomial** will be a knot invariant.

Before we investigate the invariance of the Alexander polynomial, let us try another example.

Example 2.45 Compute the Alexander polynomial of the knot in Figure 2.46.

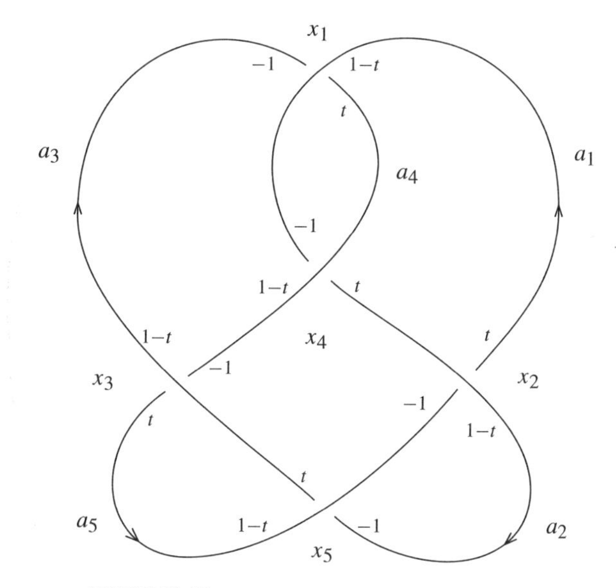

FIGURE 2.46
Compute the Alexander polynomial of this knot.

Solution. Based on the orientation and labeling in Figure 2.46, the matrix is

$$\begin{array}{c} \\ x_1 \\ x_2 \\ x_3 \\ x_4 \\ x_5 \end{array} \begin{array}{c} \begin{array}{ccccc} a_1 & a_2 & a_3 & a_4 & a_5 \end{array} \\ \begin{bmatrix} 1-t & 0 & -1 & t & 0 \\ t & 1-t & 0 & 0 & -1 \\ 0 & 0 & 1-t & -1 & t \\ -1 & t & 0 & 1-t & 0 \\ 0 & -1 & t & 0 & 1-t \end{bmatrix} \end{array}.$$

Eliminate the first row and the last column, and compute the determinant of the resulting 4×4 matrix by expansion along the top row.

$$\det \begin{bmatrix} t & 1-t & 0 & 0 \\ 0 & 0 & 1-t & -1 \\ -1 & t & 0 & 1-t \\ 0 & -1 & t & 0 \end{bmatrix}$$

$$= t \det \begin{bmatrix} 0 & 1-t & -1 \\ t & 0 & 1-t \\ -1 & t & 0 \end{bmatrix} - (1-t) \det \begin{bmatrix} 0 & 1-t & -1 \\ -1 & 0 & 1-t \\ 0 & t & 0 \end{bmatrix}$$

$$= t(-(1-t)(1-t) - t^2) - (1-t)(-(1-t) \cdot 0 - (-t))$$

$$= t(-1 + 2t - 2t^2) - t + t^2$$

$$= -2t^3 + 3t^2 - 2t.$$

Finally we multiply by $-t^{-1}$ to normalize this polynomial to have a positive constant term. Thus, $2t^2 - 3t + 2$ is the Alexander polynomial of this knot. ❄

The proof that the Alexander polynomial is a knot invariant relies on some basic properties of determinants. In particular, we need the facts that interchanging two rows or two columns of the matrix changes the sign of the determinant and that multiplying a row or column by some value multiplies the determinant by that value. We also need the fact that adding a multiple of one row to another row does not change the determinant and neither does adding a multiple of one column to another column.

Now we are ready to consider some of the steps in proving that the Alexander polynomial does not depend on the choices involved in its definition. This material is fairly technical, involving careful bookkeeping and attention to detail. You may want to check a few cases to get the flavor of the proofs even if you choose not to work through all the details.

Theorem 2.47 *The Alexander polynomial does not depend on the indexing of the crossings or the arcs of the knot diagram.*

Proof. A change in the indexing of the crossings will permute the rows of the matrix. By a sequence of row interchanges, we can restore the rows to their original order. A change in the indexing of the arcs will have a similar effect on the columns of the matrix. Thus, a change in the indexing will change the determinant by a factor of ± 1. Since the Alexander polynomial is normalized by a factor of $\pm t^k$ to have a positive constant term, both matrices will yield the same Alexander polynomial. ❄

Theorem 2.48 *The Alexander polynomial does not depend on which column is eliminated from the crossing/arc matrix.*

Proof. Let $\hat{a}_1, \hat{a}_2, \ldots, \hat{a}_n$ denote the columns of the matrix with one row eliminated. The sum of the entries in each row of the matrix is $t + (1 - t) + (-1) = 0$. Thus, $\hat{a}_1 + \hat{a}_2 + \cdots + \hat{a}_n$ is the zero vector. Suppose column $\hat{a}_p$ has been eliminated and we want to eliminate column $\hat{a}_q$ instead. Multiply $\hat{a}_q$ by -1 and add -1 times all the other remaining columns. This will yield $-\hat{a}_1 - \cdots - \hat{a}_{p-1} - \hat{a}_{p+1} - \cdots - \hat{a}_n = \hat{a}_p$ in the place of $\hat{a}_q$. A few column interchanges will restore the matrix with $\hat{a}_p$ in place at the expense of eliminating $\hat{a}_q$. These operations will affect the determinant by a factor of ± 1. Thus, the Alexander polynomial does not depend on which column is eliminated. ❄

The same kind of algebra is involved in showing that the Alexander polynomial does not depend on which row is eliminated. However, we have to look a little harder at the knot diagram to find a suitable linear combination of the rows that adds up to zero. We first need a couple of definitions. These definitions and their properties will rely on your geometric intuition until you have a chance to work out the details of Exercise 11 at the end of this section to add a bit of rigor to these concepts.

Definition 2.49 *The projection of an oriented knot divides the plane into a number of regions. The* **index** *of one of these regions is the net number of times the projection winds counterclockwise around any point in the region.*

As you travel around in the plane, the index of the region will increase by one if you cross the projection with the curve going from left to right, and it will decrease by one if you cross the projection with the curve going from right to left. Thus, as you travel around the knot diagram in the direction of the orientation, the index of the region on your left will always be one more than the index of the region on your right.

Near a crossing in a regular knot projection, the arcs that cross divide the region of the plane into four sections. Through the eyes of a topologist, we can view these arcs as a pair of coordinate axes dividing the region into four quadrants. The first and third quadrants will have the same index, the second quadrant will have an index one higher than this common value, and the fourth quadrant will have an index one lower.

Definition 2.50 *The* **index** *of a crossing of a knot diagram is the common value of the index of two of the regions near the crossing.*

Example 2.51 Determine the indices of the regions in the complement of the oriented knot diagram in Figure 2.52. Also determine the indices of each of the crossings.

Solution. Starting from the outer region (with index 0) we can cross the curve of the knot projection to reach the other regions. If the knot passes from left to right as we cross it, the index of the region we are entering is one greater than the index of the region we are leaving. If the knot passes from right to left, the index of the new region is one less than

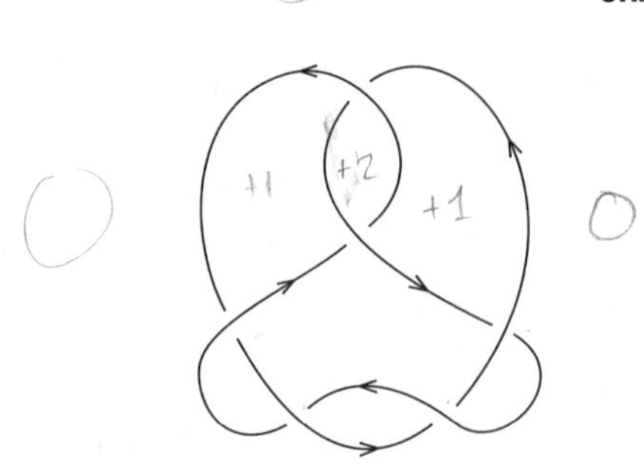

FIGURE 2.52
Determine the indices of the regions and crossings.

the index of the old. Thus, the three regions across the top of the diagram have indices $+1$, $+2$, and $+1$. The index of the central region is 0. The two small regions on the sides along the bottom both have index -1. And the small region at the center along the bottom has index $+1$.

The vertex at the top and the vertex in the center of the diagram have index $+1$. All four vertices along the bottom have index 0. ❄

Theorem 2.53 *The Alexander polynomial does not depend on which row is eliminated from the crossing/arc matrix.*

Proof. Let $x_1, \ldots, x_n$ denote the rows of the crossing/arc matrix. The entries of x_i are determined by the labels on the three arcs that meet at the ith crossing. Form a multiplier $M_i = \pm t^p$ where p is the index of the crossing and the sign is positive for a right-handed crossing, negative for a left-handed crossing.

Let us verify that $M_1 x_1 + \cdots + M_n x_n = 0$. We do this by traveling along an arc in the direction of its orientation and multiplying the labels we encounter at each vertex by the multiplier for that vertex. As we add up these products we will find that the cumulative total is always equal to t^p where p is the index of the region on the left side of the arc.

Indeed, let $p_0, \ldots, p_m$ be the indices of the crossing we encounter as we travel along an arc. The arc begins at an undercrossing of index p_0. If the crossing is left-handed, we pick up a label -1 times a multiplier $-t^{p_0}$ and begin with a total of t^{p_0}. If the crossing is right-handed, we pick up a label t times a multiplier $+t^{p_0}$ and begin with a total of t^{p_0+1}. As you can see from the labels on the regions in Figure 2.54, the total in either case is a power of t with exponent equal to the index of the region on the left side of the arc.

We now proceed along the arc and apply mathematical induction to the number of crossings. Let us assume that the cumulative total of the products from the initial crossing up to the kth crossing is t raised to the power that is the index of the region to the left of the arc. At the kth overcrossing we again face two possibilities. At a left-handed crossing, the label $1 - t$ is multiplied by $-t^{p_k}$. We add the product $-t^{p_k} + t^{p_k+1}$ to the cumulative total t^{p_k} prior to the crossing to obtain an updated cumulative total of t^{p_k+1}. At a right-handed crossing, the label $1 - t$ is multiplied by $+t^{p_k}$. We add the product $t^{p_k} - t^{p_k+1}$ to the

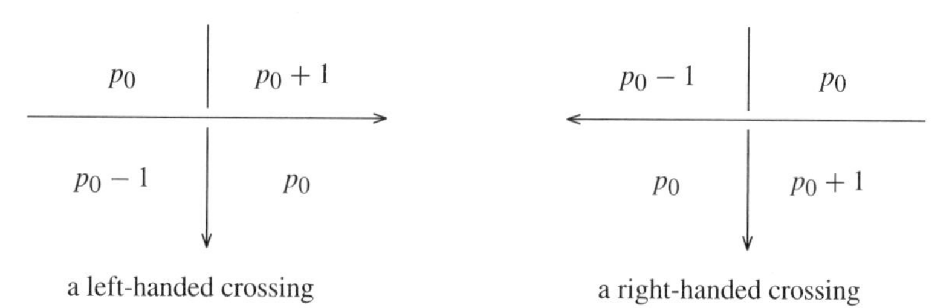

FIGURE 2.54
Indices at the beginning of an arc.

cumulative total t^{p_k+1} prior to the crossing to obtain an updated cumulative total of t^{p_k}. As you can see from the labels on the regions in Figure 2.55, the total in either case retains the form of a power of t with exponent equal to the index of the region on the left side of the arc.

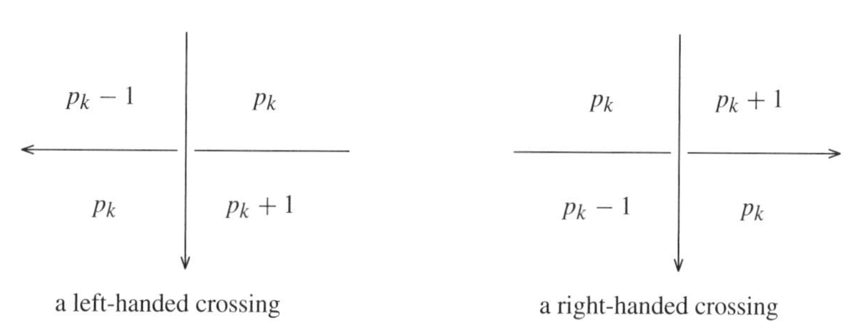

FIGURE 2.55
Indices at the kth overcrossing of an arc.

The arc ends at an undercrossing of index p_m. If this is a left-handed crossing, the label t is multiplied by $-t^{p_m}$. We add the product $-t^{p_m+1}$ to the cumulative total t^{p_m+1} prior to the crossing to obtain a grand total of 0. At a right-handed crossing, the label -1 is multiplied by $+t^{p_m}$. We add the product $-t^{p_m}$ to the cumulative total t^{p_m} prior to the crossing to obtain a grand total again of 0.

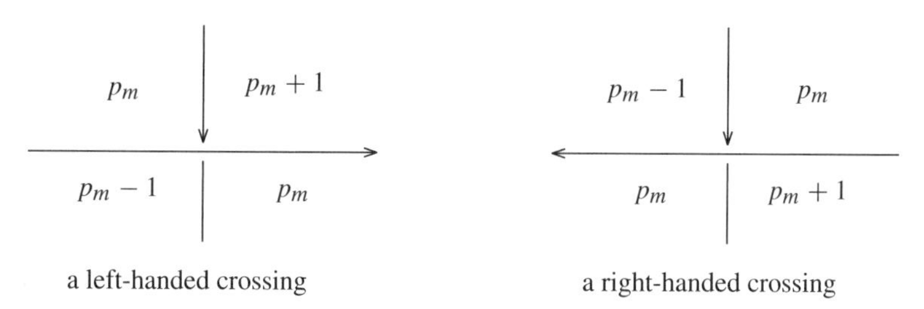

FIGURE 2.56
Indices at the end of an arc.

If we let $\hat{x}_1, \ldots, \hat{x}_n$ denote the rows of the matrix with one column deleted, we have that $M_1\hat{x}_1 + \cdots + M_n\hat{x}_n = 0$. Suppose row $\hat{x}_i$ has been eliminated and we want to eliminate row $\hat{x}_k$ instead. Multiply $\hat{x}_k$ by $-M_k$ and add $-M_j\hat{x}_j$ for each other row. This will yield $-M_1\hat{x}_1 - \cdots - M_{i-1}\hat{x}_{i-1} - M_{i+1}\hat{x}_{i+1} - \cdots - M_n\hat{x}_n = M_i\hat{x}_i$. A few row interchanges will restore the matrix with $M_i\hat{x}_i$ in place at the expense of eliminating $\hat{x}_k$. Since the multipliers M_i and M_k are each ± 1 times some power of t, it follows that eliminating a different row of the crossing/arc matrix will not affect the Alexander polynomial. ❄

Now we know that the Alexander polynomial is well-defined for a given diagram of an oriented knot. The final step is to show that it is an invariant of the oriented knot.

Theorem 2.57 *The Alexander polynomial of an oriented knot is an invariant under Reidemeister moves.*

Proof. When we take into account the two mirror images and the two orientations, the basic Reidemeister move of Type I leads to the four possibilities illustrated in Figure 2.58.

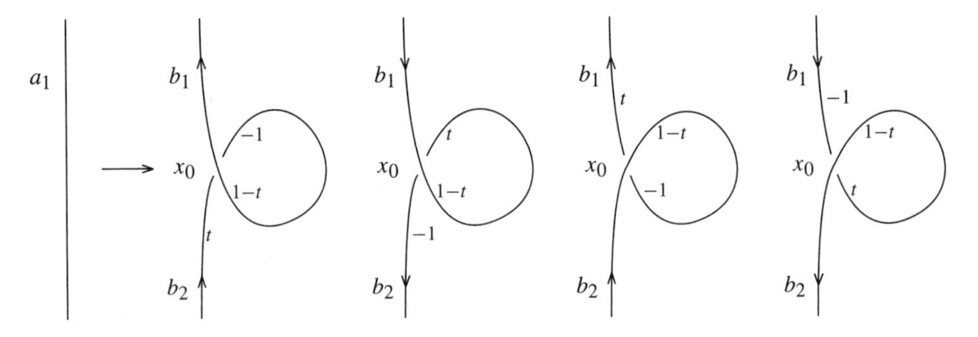

FIGURE 2.58
The effect of a Reidemeister move of Type I.

In each case an arc a_1 breaks into two arcs b_1 and b_2, and a new crossing x_0 appears. The crossing/arc matrix has a new row (corresponding to crossing x_0) and the entries of the column corresponding to arc a_1 are split between the two columns (corresponding to b_1 and b_2) that replace this column. For example, the first case produces the following change in the crossing/arc matrix:

$$\begin{array}{c} \\ x_1 \\ \\ \\ \\ \end{array}\overset{\displaystyle a_1\qquad\qquad}{\begin{bmatrix} a_{11} & * & \cdots & * \\ * & & & \\ \vdots & & Q & \\ * & & & \end{bmatrix}} \longrightarrow \begin{array}{c} x_0 \\ x_1 \\ \\ \\ \\ \end{array}\overset{\displaystyle \quad b_1 \qquad\quad b_2 \qquad\qquad\qquad}{\begin{bmatrix} (1-t)-1 & t & 0 & \cdots & 0 \\ * & * & * & \cdots & * \\ * & * & & & \\ \vdots & \vdots & & Q & \\ * & * & & & \end{bmatrix}}.$$

From the original crossing/arc matrix we can delete the row corresponding to x_1 and the column corresponding to a_1, and compute the determinant of the remaining matrix Q. From the crossing/arc matrix of the resulting diagram, we can delete the row corresponding to x_1 and the column corresponding to b_2. By expanding along the row corresponding to x_0, we see that the determinant of the new matrix is $-t \det Q$. The factor of $-t$ will wash out when we normalize the polynomial to have a positive constant term. Thus, the Reidemeister move does not change the Alexander polynomial. In Exercise 14 at the end of this section, you are asked to verify that the other three cases behave similarly.

Since a Reidemeister move of Type II is its own mirror image, we need only consider the two orientations of the overcrossing arc as illustrated in Figure 2.59.

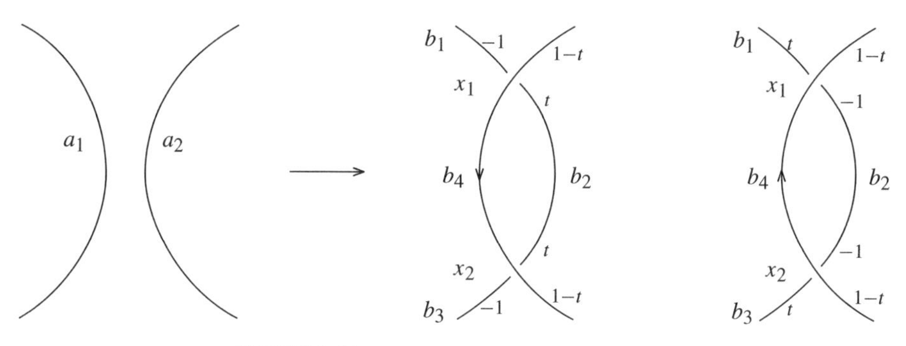

FIGURE 2.59
The effect of a Reidemeister move of Type II.

In either case, one arc a_1 breaks into three arcs b_1, b_2, and b_3, and two new crossings x_1 and x_2 appear. The other arc a_2 becomes the new overcrossing arc, which we relabel b_4. Notice that the full extent of arc b_2 appears in Figure 2.59. The crossing/arc matrix has two new rows (corresponding to the new crossings) and the entries of the column corresponding to the arc a_1 are split between the two columns corresponding to b_1 and b_3. For example, the first case produces the following change in the crossing/arc matrix:

$$\begin{array}{cc} a_1 & a_2 & \\ \end{array} \left[\begin{array}{ccc} a_{11} & a_{12} & \\ \vdots & \vdots & Q \\ a_{m1} & a_{m2} & \end{array}\right] \longrightarrow \begin{array}{c} \\ x_1 \\ x_2 \\ \\ \\ \\ \end{array} \begin{array}{c} \begin{array}{cccc} b_1 & b_2 & b_3 & b_4 \end{array} \\ \left[\begin{array}{ccccccc} -1 & t & 0 & 1-t & 0 & \cdots & 0 \\ 0 & t & -1 & 1-t & 0 & \cdots & 0 \\ b_{11} & 0 & b_{13} & a_{12} & & & \\ \vdots & \vdots & \vdots & \vdots & & Q & \\ b_{m1} & 0 & b_{m3} & a_{m2} & & & \end{array}\right] \end{array}.$$

We can reconstitute the column corresponding to a_1 by adding the column for b_1 to the column for b_3. If we also add in t^{-1} times the column for b_2, we can even arrange to have zeros as the first two entries. Likewise we add $1 - t^{-1}$ times the column for b_2 to the column for b_4 to top off this column with zeros. The resulting matrix is

$$\begin{array}{cccccc} b_1 & b_2 & b_3+b_1+t^{-1}b_2 & b_4+(1-t^{-1})b_2 & & \\ \end{array}$$
$$\begin{bmatrix} -1 & t & 0 & 0 & 0 & \cdots & 0 \\ 0 & t & 0 & 0 & 0 & \cdots & 0 \\ b_{11} & 0 & a_{11} & a_{12} & & & \\ \vdots & \vdots & \vdots & \vdots & & Q & \\ b_{m1} & 0 & a_{m1} & a_{m2} & & & \end{bmatrix}.$$

The row and the column that were deleted from the original matrix have counterparts that can be deleted from the new matrix. By Exercise 6, the determinant of the new matrix is

$$\det\begin{bmatrix} -1 & t \\ 0 & t \end{bmatrix} = -t$$

times the determinant of the original matrix. This factor will not affect the Alexander polynomial.

Finally, we examine the effect of a Reidemeister move of Type III. Let us use the variables $a_{12}, a_{13}, a_{21}, a_{22}, a_{34}$, and a_{35} to denote the labels on the underpass arcs of crossings x_1, x_2, and x_3. In this way we incorporate all possible orientations and reflections into one argument.

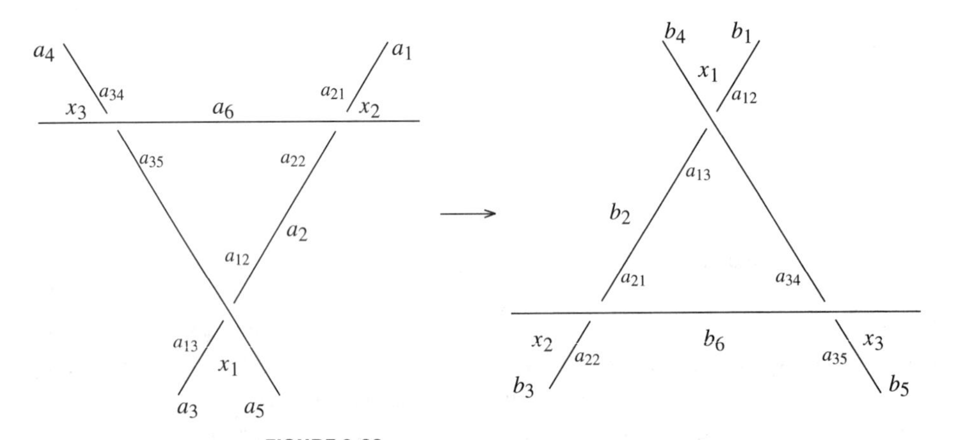

FIGURE 2.60
The effect of a Reidemeister move of Type III.

Here is what happens to the column/arc matrix:

$$\begin{array}{c} \\ x_1 \\ x_2 \\ x_3 \\ \\ \\ \\ \end{array}
\begin{array}{c} \begin{array}{cccccccccc} a_1 & a_2 & a_3 & a_4 & a_5 & a_6 & & & \end{array} \\
\begin{bmatrix} 0 & a_{12} & a_{13} & 0 & 1-t & 0 & 0 & \cdots & 0 \\ a_{21} & a_{22} & 0 & 0 & 0 & 1-t & 0 & \cdots & 0 \\ 0 & 0 & 0 & a_{34} & a_{35} & 1-t & 0 & \cdots & 0 \\ & 0 & & & & & & & \\ & \vdots & & & & & & & \\ & 0 & & & & & & & \end{bmatrix} \end{array}$$

$$
\longrightarrow \quad
\begin{array}{c|ccccccccc|}
 & b_1 & b_2 & b_3 & b_4 & b_5 & b_6 & & & \\
x_1 & a_{12} & a_{13} & 0 & 1-t & 0 & 0 & 0 & \cdots & 0 \\
x_2 & 0 & a_{21} & a_{22} & 0 & 0 & 1-t & 0 & \cdots & 0 \\
x_3 & 0 & 0 & 0 & a_{34} & a_{35} & 1-t & 0 & \cdots & 0 \\
 & & 0 & & & & & & & \\
 & & \vdots & & & & & & & \\
 & & 0 & & & & & & &
\end{array} .
$$

Delete the first row and first column of both matrices. Then it is easy to reconstitute the original matrix by subtracting a_{22}/a_{21} times the column for b_2 from the column for b_3 and multiplying the column for b_2 by a_{22}/a_{21}. This changes the determinant by a factor of a_{22}/a_{21}. But since this fraction is $-t$ or $-t^{-1}$, this has no effect on the Alexander polynomial. ❄

Exercises 2.5

1. Compute the Alexander polynomial of the figure-eight knot.
2. For the knot in Example 2.45, choose a different row and column to delete from the crossing/arc matrix. Verify that you get the same Alexander polynomial.
3. Give the crossing/arc matrix for the first knot in Figure 2.41. Compute the Alexander polynomial of this knot. Take advantage of any available computer algebra system for computing the determinant.
4. Use the determinant function of a computer algebra system to assist you in computing the Alexander polynomials of the knots in Figure 2.42. Experiment with deleting different rows and columns of the crossing/arc matrix to show that this choice does not affect the Alexander polynomial.
5. What is the determinant of the crossing/arc matrix of any knot diagram if we do not delete a row and a column? Why?
6. Let $Q = [q_{ij}]$ be an $n \times n$ matrix.
   (a) Show that
$$
\det \begin{bmatrix} a & 0 & \cdots & 0 \\ & q_{11} & \cdots & q_{1n} \\ \vdots & \vdots & & \vdots \\ & q_{n1} & \cdots & q_{nn} \end{bmatrix} = a \det Q
$$
   no matter what entries appear in the first column below the entry a.
   (b) Show that
$$
\det \begin{bmatrix} a & b & 0 & \cdots & 0 \\ c & d & 0 & \cdots & 0 \\ & * & q_{11} & \cdots & q_{1n} \\ \vdots & \vdots & \vdots & & \vdots \\ & * & q_{n1} & \cdots & q_{nn} \end{bmatrix} = \left(\det \begin{bmatrix} a & b \\ c & d \end{bmatrix} \right) (\det Q)
$$

no matter what entries appear in the first and second columns below the entries a, b, c, d.

7. Our definition of the Alexander polynomial will not work for the standard projection of the trivial knot, since there are no crossings and no arcs. So, to compute the Alexander polynomial of the trivial knot, we must use a different projection. Here are two ways to do it.

(a) Draw a knot diagram with one crossing and use it to compute the Alexander polynomial of the trivial knot. Notice that upon deleting a row and column, you will get a 0×0 matrix. It is a useful convention that the determinant of a 0×0 matrix is the multiplicative identity 1. That way, formulas such as those in Exercise 6 will hold for 0×0 matrices.

(b) Since you still might object to taking the determinant of a 0×0 matrix, draw a knot diagram with two crossings and use it to compute the Alexander polynomial of the trivial knot.

8. There is a wonderful relation between the Alexander polynomial and knot coloring. The **determinant** of a knot is defined to be its Alexander polynomial evaluated at -1. It turns out that for any prime p, a knot can be represented on a wheel of p colors if and only if p is a divisor of the determinant of the knot! For a proof of this fact, see Chapter 3 of the book *Knot Theory* [20] by Charles Livingston. Use this result and your work from Exercises 3 and 4 to find values of p for which the knots in Figures 2.41 and 2.42 can be represented on a wheel of p colors. Then find such representations.

9. Determine the indices of the regions and the indices of the crossings for the second and third of the knots in Figure 2.42.

10. Determine the indices of the regions and the indices of the crossings for the oriented knot diagram in Figure 2.61.

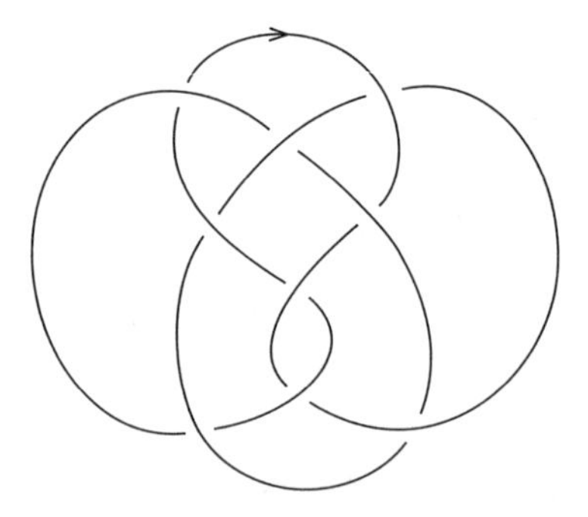

FIGURE 2.61
Determine the indices of the regions and crossings.

11. Let $v_1, \ldots, v_n$ be the consecutive vertices of a polygonal closed curve in $\mathbb{R}^2$. Let P be a point of $\mathbb{R}^2$ not on the curve.

(a) For any pair of consecutive vertices v_k and v_{k+1} (or v_1 if $k = n$) show that the directed angle $v_k P v_{k+1}$ has a well-defined measure in the interval $(-\pi, \pi)$.

(b) Show that the sum of these angles is an integer multiple of 2π. This integer is the **winding number** of the curve around the point P.

(c) Show that the measure of the angle $v_k P v_{k+1}$ is continuous as a function of P.

(d) Conclude that the winding number is constant throughout any of the complementary regions determined by the curve. This is the **index** of the region.

(e) Suppose P is on the curve. Move a short portion of the curve slightly to the left or slightly to the right to miss P. Compare the winding numbers of these two curves around P.

(f) Prove that if the curve is oriented from left to right as you cross it, the index of the region you are entering is one more than the index of the region you are leaving.

12. Show that the knot winds zero times around any point that is sufficiently far away from the knot. Thus, the unbounded region determined by a knot diagram has index zero.

13. An arc of a knot diagram might begin or end in a loop. In such a case the arc meets the crossing more than once, and the entry in the arc/crossing matrix is the sum of the labels on the arc at this crossing.

(a) Confirm that the proof of Theorem 2.53 is valid in these special cases.

(b) Confirm that the proof of Theorem 2.57 is valid in these special cases.

14. (a) Check that the remaining three possibilities of a Reidemeister move of Type I yield determinants that are ± 1 or $\pm t$ times the original determinant.

(b) Check that the other possible orientation of the overcrossing arc in a Reidemeister move of Type II yields a determinant that is $-t$ times the original determinant.

15. Extend the definition of the Alexander polynomial to oriented links. Extend to oriented links the proofs that the Alexander polynomial is an invariant.

16. (a) Verify that the Alexander polynomial of the oriented link in Figure 2.62 is $-2t + 2$.

(b) Reverse the orientation of one of the curves in Figure 2.62. Verify that the Alexander polynomial of the resulting link with four left-handed crossings is $-t^3 + t^2 - t + 1$.

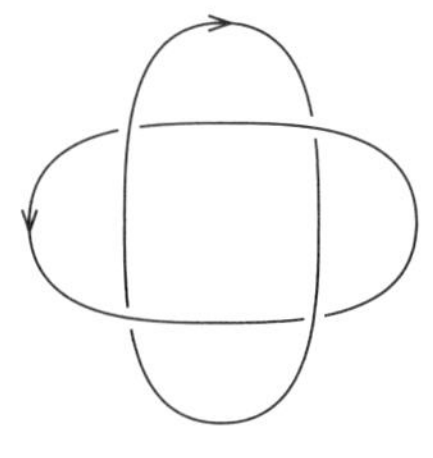

FIGURE 2.62
An oriented link with four right-handed crossings.

2.6 Skein Relations

Since its discovery in 1928, the Alexander polynomial has been a useful invariant of knots and the subject of much research. However, it was not until 1969 that John Conway noticed a relation among the Alexander polynomials of links whose diagrams differ at a single overcrossing.

In Section 2.5, we normalized the Alexander polynomial with a factor of $\pm t^p$ so that it has a positive constant term. Conway's relation modifies this choice to introduce terms with negative powers of t so that the highest positive degree is equal to the absolute value of its lowest negative degree. For an Alexander polynomial of the form $a_n t^n + a_{n-1}t^{n-1} + \cdots + a_1 t + a_0$, with $a_n \neq 0$ and $a_0 \neq 0$, we can multiply by $\pm t^{-n/2}$ to produce an equivalent polynomial Δ with this property.

Of course, Δ need not be a polynomial in the strict sense since it is likely to have terms involving negative powers of t. For a knot, it turns out that the Alexander polynomial has even degree. So the terms of Δ at least have integer powers of t. For a link, however, Δ may involve fractional exponents.

Before we set up the assembly line to mass produce Δ polynomials using what has become known as a skein relation, here are a couple of facts that can be used for quality control. First, the coefficients of the positive powers of t match up with the coefficients of the corresponding negative powers of t. For a knot, the correspondence is exact; that is, $\Delta(t^{-1}) = \Delta(t)$. For a link of more than one component, the corresponding coefficients are negatives; that is, $\Delta(t^{-1}) = -\Delta(t)$.

Second, for a knot $\Delta(1) = 1$ (see Exercise 9 at the end of this section). This will pin down the sign on the factor of $\pm t^p$ used to renormalize the Alexander polynomial to form Δ. For a link of more than one component, $\Delta(1) = 0$ (see Exercise 10). So we will have to be careful to keep track of the signs when we compute Δ for links.

Example 2.63 Determine Δ for the Hopf link with left-handed crossings.

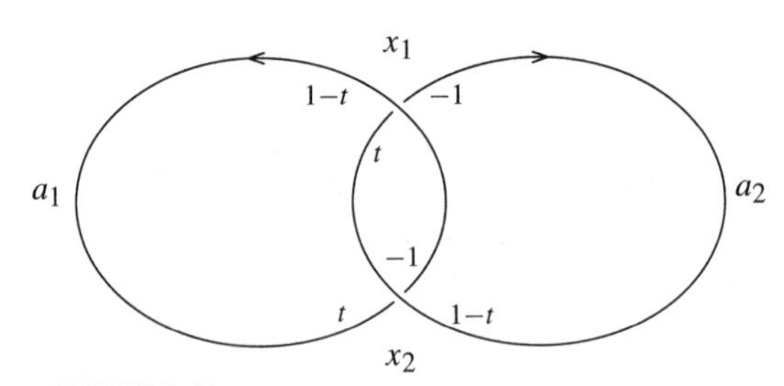

FIGURE 2.64
Compute the Alexander polynomial of the Hopf link.

Solution. Based on Figure 2.64, the crossing/arc matrix of the Hopf link is

$$\begin{array}{c} \\ x_1 \\ x_2 \end{array}\begin{array}{c} \begin{array}{cc} a_1 & \quad a_2 \end{array} \\ \begin{bmatrix} 1-t & t-1 \\ t-1 & 1-t \end{bmatrix} \end{array}.$$

Notice that at crossing x_1 the two labels on arc a_2 are added; likewise at crossing x_2 the two labels on arc a_1 are added. Eliminating the first row and second column leaves a 1×1 matrix with determinant $t - 1$. Since this polynomial is of degree 1, we multiply by $t^{-1/2}$ to get $\Delta(t) = t^{1/2} - t^{-1/2}$. Of course, we do not know at this point whether we have to reverse the signs of these two terms. We will see in the next example how Conway's skein relation settles this issue. ❄

Here is Conway's skein relation. Let L_+, L_-, and L_0 be three diagrams of an oriented link that are identical except at one crossing where L_+ has a right-handed crossing, L_- has a left-handed crossing, and L_0 has the strands joined to eliminate the crossing. The skein relation for the Alexander polynomial is the identity

$$\Delta(L_+) - \Delta(L_-) + (t^{1/2} - t^{-1/2})\Delta(L_0) = 0.$$

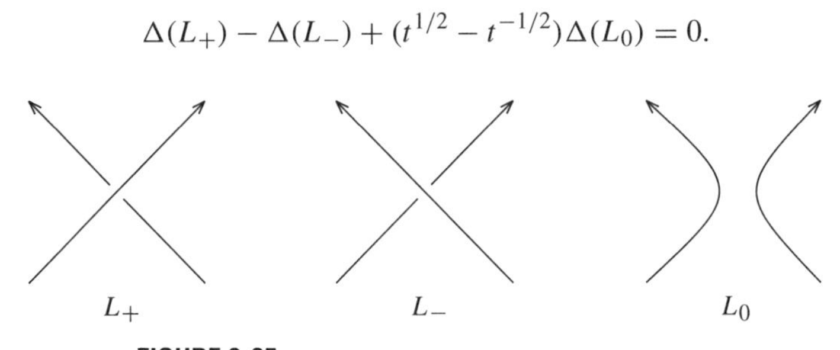

FIGURE 2.65
The crossing at which L_+, L_-, and L_0 are different.

Together with the fact that the trivial knot has Δ equal to 1, this skein relation can be used to compute Δ for any knot or link.

Example 2.66 Use the skein relation to compute Δ for the Hopf link with left-handed crossings.

Solution. We can regard the Hopf link in Figure 2.64 as L_- with x_1 as the left-handed crossing. Then L_+ is a trivial link with two components, and L_0 is a trivial knot. By Exercise 2, $\Delta(L_+) = 0$. And we know that $\Delta(L_0) = 1$. Thus,

$$\begin{aligned}\Delta(L_-) &= \Delta(L_+) + (t^{1/2} - t^{-1/2})\Delta(L_0) \\ &= 0 + (t^{1/2} - t^{-1/2})1 \\ &= t^{1/2} - t^{-1/2}.\end{aligned}$$ ❄

Skein relations are often written more graphically by drawing the link projections in place of symbols for the links. The next example illustrates this convention in computing Δ for the trefoil knot.

Example 2.67 Compute $\Delta\left(\quad\right)$.

Solution. We apply the skein relation to the marked crossing and use the result of Example 2.66 together with the fact that Δ for the trivial link is 1:

$$\Delta\left(\quad\right) = \Delta\left(\quad\right) + (t^{1/2} - t^{-1/2})\Delta\left(\quad\right)$$

$$= 1 + (t^{1/2} - t^{-1/2})(t^{1/2} - t^{-1/2})$$

$$= 1 + (t - 2 + t^{-1})$$

$$= t - 1 + t^{-1}.$$

The correspondence between the coefficients of the positive and negative powers of t and the value of this polynomial at $t = 1$ provide a degree of assurance that our calculations are correct. ❄

Here is another illustration of how we use the skein relation to compute Δ for a knot in terms of the corresponding polynomials of simpler links.

Example 2.68 Compute $\Delta\left(\quad\right)$.

Solution. We apply the skein relation to the marked crossing to write Δ in terms of the polynomials of the trefoil knot (which appears as L_+) and the Hopf link with left-handed crossings (which appears as L_0):

$$\Delta\left(\quad\right) = \Delta\left(\quad\right) + (t^{1/2} - t^{-1/2})\,\Delta\left(\quad\right)$$

$$= (t - 1 + t^{-1}) + (t^{1/2} - t^{-1/2})(t^{1/2} - t^{-1/2})$$

$$= (t - 1 + t^{-1}) + (t - 2 + t^{-1})$$

$$= 2t - 3 + 2t^{-1}.$$

Again it is reassuring that the coefficients of the positive and negative powers of t correspond and that this polynomial has a value of 1 at $t = 1$. ❄

In simple examples it is easy to choose a crossing so that modifications as required by the skein relation result in links that are simpler than the original. But how do we measure this simplicity, and what guarantees that we can find a crossing so the modified links are simpler than the original? Well, for any crossing, L_0 will clearly have fewer crossings than either L_+ or L_-. Also, recall from Exercises 11 and 13 of Section 2.2, that we can always find a sequence of crossing changes that will trivialize a knot or link. Thus, by choosing this sequence of crossing changes, we can ensure that the link with the changed crossing will also be simpler than the original. Exercise 11 at the end of this section asks you to fill in some of the details of this argument.

Exercises 2.6

1. Consider a link with n components. Break this link apart at a pair of points that map to a crossing in a regular projection of the link. Reattach the strands to eliminate the crossing in the projection. Show that the number of components of the resulting link is either $n-1$ or $n+1$.
2. Suppose L_1 and L_2 are links such that in some projection the diagrams for L_1 and L_2 are disjoint. Show that $\Delta(L_1 \cup L_2) = 0$. Suggestion: Apply the skein relation with $L_1 \cup L_2$ as the L_0 crossing and take advantage of some cancellation.
3. Confirm that the skein relation gives $\Delta = t^{1/2} - t^{-1/2}$ for the Hopf link when applied to the left-handed crossing x_2 in Figure 2.64.
4. Reverse the orientation of one of the curves of the Hopf link in Figure 2.64 so the two crossings are both right-handed. Show that the skein relation gives $\Delta = -t^{1/2} + t^{-1/2}$.
5. Use the skein relation to compute Δ for the trefoil knot with three right-handed crossings. Verify that the result is compatible with the Alexander polynomial $t^2 - t + 1$ for this knot.
6. **(a)** Compute Δ for the figure-eight knot by applying the skein relation to one of the right-handed crossings.
   (b) Compute Δ for the figure-eight knot by applying the skein relation to one of the left-handed crossings.
   (c) Verify that the results are compatible with the Alexander polynomial $t^2 - 3t + 1$ for this knot.
7. **(a)** Use the skein relation to compute Δ( ). Verify that the result is compatible with the Alexander polynomial $-2t + 2$ of this link.
   (b) Reverse the orientation of one of the curves of this link and compute Δ( ) for the resulting link. Verify that the result is compatible with the Alexander polynomial $-t^3 + t^2 - t + 1$ of this link.
8. Compute Δ( ) by applying the skein relation to the left-handed crossing at the top of the diagram. Verify that the result is compatible with the result obtained in Example 2.68.
9. Use the skein relation to show for any knot that $\Delta(1) = 1$.
10. Use the skein relation to show for any link with more than one component that $\Delta(1) = 0$.
11. Define the complexity of a regular projection of a link to be the ordered pair (c, u) where c is the number of crossings in the projection and u is the minimum number of crossings that need to be changed to trivialize the link (Exercise 13 of Section 2.2 guarantees that it is always possible to find such a number). Order the complexities using lexicographic order: $(c_1, u_1) < (c_2, u_2)$ if and only if $c_1 < c_2$, or $c_1 = c_2$ and $u_1 < u_2$.

(a) Show that any nonempty set of complexities has a minimum element.

(b) Show that no infinite decreasing sequence of complexities can exist.

(c) For a regular projection of a link, show that the skein relation can be applied to some crossing so that both of the modified links have smaller complexity.

(d) Conclude that the skein relation can be applied to compute Δ for any link.

2.7 The Jones Polynomial

In 1985, the New Zealand mathematician Vaughan Jones announced a new polynomial invariant for knots and links [16]. He had noticed some relations in the field of operator theory that appeared similar to relations among knots. This new polynomial distinguishes between knots that have the same Alexander polynomial. In fact, the Jones polynomial distinguishes every knot of 10 or fewer crossings. Although there are distinct knots with the same Jones polynomial, no one knows of a nontrivial knot with the same Jones polynomial as that of the trivial knot.

Four months after the announcement, four independent groups of mathematicians developed a polynomial in two variables that generalizes both the Alexander polynomial and the Jones polynomial. The six knot theorists, Hoste, Ocneanu, Millett, Freyd, Lickorish, and Yetter published a joint paper [15] announcing their result and using their initials to christen their new polynomial HOMFLY. Another pair of mathematicians, Przytycki and Traczyk, had also developed this same polynomial, but several months' delay in the mail from Poland prevented them from adding two more letters to this acronym.

In this section, we will concentrate on the Jones polynomial as defined by Louis Kauffman [18] in terms of his bracket polynomial. Then a simple change of variable will yield the original Jones polynomial. The definition of Kauffman's bracket polynomial relies on a skein relation for a regular projection of a link.

Rules for the Bracket Polynomial 2.69 *The Kauffman* ***bracket polynomial*** *of a regular projection of a link is a polynomial in integer powers of the variable A defined by the following three rules:*

1. $\langle \bigcirc \rangle = 1,$
2. $\langle L \cup \bigcirc \rangle = (-A^2 - A^{-2})\langle L \rangle,$
3. $\langle \times \rangle = A\langle \asymp \rangle + A^{-1}\langle)(\rangle.$

The first rule gives the bracket polynomial of the trivial link of one component. The second rule gives the bracket polynomial of the disjoint union of the regular projection L of a link and an unknotted component in terms of the polynomial of L. The third rule is the skein relation for reducing a regular projection with a crossing to the two regular projections changed only by eliminating this crossing in the two ways illustrated.

For any regular projection of a link, we can repeatedly apply Rule 3 each time reducing the number of crossings. We can apply Rule 2 to eliminate disjoint unknotted components. Ultimately, we will be left with a single, unknotted component to which we can apply Rule 1. By Exercise 1, the order in which we apply these rules does not affect the final result. Thus, Kauffman's bracket polynomial is a well-defined quantity assigned to a regular projection of a link.

Notice that these rules do not require an orientation of the link. In Rule 3, simply treat the overcrossing as the x-axis and the undercrossing as the y-axis in a coordinate system for the plane. Join the strands so the first and third quadrants flow together and then join the strands the other way so the second and fourth quadrants flow together. Multiply the polynomial of the first by A, multiply the polynomial of the second by A^{-1}, and add the results.

Example 2.70 Compute $\langle\ \rangle$.

Solution. We repeatedly apply Rule 3 to eliminate the crossings; then we apply Rules 1 and 2.

$$\begin{aligned}\langle\ \rangle &= A\langle\ \rangle + A^{-1}\langle\ \rangle\\ &= A^2\langle\ \rangle + \langle\ \rangle + \langle\ \rangle + A^{-2}\langle\ \rangle\\ &= A^3\langle\ \rangle + A\langle\ \rangle + A\langle\ \rangle + A^{-1}\langle\ \rangle\\ &\quad + A\langle\ \rangle + A^{-1}\langle\ \rangle + A^{-1}\langle\ \rangle + A^{-3}\langle\ \rangle\\ &= A^3(-A^2 - A^{-2})^2 + A(-A^2 - A^{-2}) + A(-A^2 - A^{-2}) + A^{-1}\\ &\quad + A(-A^2 - A^{-2}) + A^{-1} + A^{-1} + A^{-3}(-A^2 - A^{-2})\\ &= A^7 - A^3 - A^{-5}.\end{aligned}$$

❄

The next order of business is to examine the effects of Reidemeister moves on the bracket polynomial. There is a problem with Type I moves, so we will save those for last.

Checking the invariance of the bracket polynomial under Type II moves is a straightforward calculation. A few applications of Rule 3 introduce factors of A and A^{-1} that cancel out the factor of $-A^2 - A^{-2}$ introduced by Rule 2:

$$\begin{aligned}\langle\ \rangle &= A\langle\ \rangle + A^{-1}\langle\ \rangle\\ &= A^2\langle\ \rangle + \langle\ \rangle + \langle\ \rangle + A^{-2}\langle\ \rangle\\ &= A^2\langle\ \rangle + (-A^2 - A^{-2})\langle\ \rangle + \langle\ \rangle + A^{-2}\langle\ \rangle\\ &= \langle\ \rangle.\end{aligned}$$

Checking the invariance under Type III moves is easy since we can use the invariance under Type II moves to slide some arcs around a little:

$$
\begin{aligned}
\langle\;\rangle &= A\langle\;\rangle + A^{-1}\langle\;\rangle \\
&= A\langle\;\rangle + A^{-1}\langle\;\rangle \\
&= A\langle\;\rangle + A^{-1}\langle\;\rangle \\
&= \langle\;\rangle.
\end{aligned}
$$

We must now face up to the effect on the bracket polynomial of a Type I Reidemeister move. Undoing a loop with a right-handed crossing gives

$$
\begin{aligned}
\langle\;\rangle &= A\langle\;\rangle + A^{-1}\langle\;\rangle \\
&= A(-A^2 - A^{-2})\langle\;\rangle + A^{-1}\langle\;\rangle \\
&= (-A^3 - A^{-1} + A^{-1})\langle\;\rangle \\
&= -A^3\langle\;\rangle.
\end{aligned}
$$

And applying a Type I move to a loop with a left-handed crossing gives

$$
\begin{aligned}
\langle\;\rangle &= A\langle\;\rangle + A^{-1}\langle\;\rangle \\
&= A\langle\;\rangle + A^{-1}(-A^2 - A^{-2})\langle\;\rangle \\
&= (A - A - A^{-3})\langle\;\rangle \\
&= -A^{-3}\langle\;\rangle.
\end{aligned}
$$

This looks like bad news. But we can take care of these pesky factors of $-A^3$ and $-A^{-3}$ that pop up when we perform Type I Reidemeister moves.

Definition 2.71 *The **writhe** $w(L)$ of the regular projection L of a link is the number of right-handed crossings minus the number of left-handed crossings.*

Example 2.72 Compute w ( ).

Solution. There are two right-handed crossings toward the top of the projection of this knot and four left-handed crossing toward the bottom. Thus, the writhe is $2 - 4 = -2$. ❄

As we have seen, a Type I move that eliminates a right-handed crossing will introduce a factor of $-A^3$ into the bracket polynomial, and a Type I move that eliminates a left-handed crossing will introduce a factor of $-A^{-3}$ into the bracket polynomial. Thus, multiplying the bracket polynomial by $(-A)^{-3w(L)}$ will cancel out the ill effects of Type I Reidemeister moves. Exercise 2 at the end of this section asks you to explain why the writhe is unaffected by Reidemeister moves of Type II and Type III.

The polynomial

$$X(L) = (-A)^{-3w(L)}\langle L\rangle$$

is therefore an invariant of the link type. For a knot, it does not even depend on a choice of orientation. However, for a link of more than one component, the writhe will depend on the choice of orientations of the components.

Definition 2.73 *The Jones polynomial $V(L)$ of a link L is obtained by making the substitution $A = t^{-1/4}$ in the polynomial $X(L)$.*

Example 2.74 Compute the Jones polynomial of the trefoil knot K with three left-handed crossings.

Solution. In Example 2.70, we determined that the bracket polynomial of K is $A^7 - A^3 - A^{-5}$. Since all three crossings are left-handed, the writhe is $w(K) = -3$. So

$$X(K) = (-A)^{-3w(K)}\langle K\rangle = (-A)^9(A^7 - A^3 - A^{-5}) = -A^{16} + A^{12} + A^4.$$

The substitution $A = t^{-1/4}$ gives the Jones polynomial $V(K) = -t^{-4} + t^{-3} + t^{-1}$. ❄

We can use the skein relation for the bracket polynomial to derive a skein relation for the Jones polynomial. This will simplify the computation of the Jones polynomial of a link by eliminating the need to compute the bracket polynomial and writhe. Since the Jones polynomial is an invariant of the knot type, we can also use different projections in the course of computing the Jones polynomial.

Theorem 2.75 *Let L_+, L_-, and L_0 be three diagrams of an oriented link that are identical except at one crossing where L_+ has a right-handed crossing, L_- has a left-handed crossing, and L_0 has the strands joined to eliminate the crossing (see Figure 2.65). The Jones polynomial of an oriented link satisfies the skein relation*

$$t^{-1}V(L_+) - tV(L_-) + (t^{-1/2} - t^{1/2})V(L_0) = 0.$$

Proof. Exercise 7 at the end of this section asks you to justify the steps in the following chain of equalities.

$$
\begin{aligned}
&t^{-1}V(L_+) - tV(L_-) + (t^{-1/2} - t^{1/2})V(L_0) \\
&\quad = A^4X(L_+) - A^{-4}X(L_-) + (A^2 - A^{-2})X(L_0) \\
&\quad = A^4(-A)^{-3w(L_+)}\langle L_+\rangle - A^{-4}(-A)^{-3w(L_-)}\langle L_-\rangle \\
&\qquad + (A^2 - A^{-2})(-A)^{-3w(L_0)}\langle L_0\rangle \\
&\quad = A^4(-A)^{-3(w(L_0)+1)}\langle L_+\rangle - A^{-4}(-A)^{-3(w(L_0)-1)}\langle L_-\rangle \\
&\qquad + (A^2 - A^{-2})(-A)^{-3w(L_0)}\langle L_0\rangle \\
&\quad = (-A)^{-3w(L_0)}\big(A^4(-A)^{-3}\langle L_+\rangle - A^{-4}(-A)^3\langle L_-\rangle + (A^2 - A^{-2})\langle L_0\rangle\big) \\
&\quad = (-A)^{-3w(L_0)}\big(-A\langle L_+\rangle + A^{-1}\langle L_-\rangle + (A^2 - A^{-2})\langle L_0\rangle\big) \\
&\quad = (-A)^{-3w(L_0)}\Big(-A\Big(A\langle\, \supset\subset \,\rangle + A^{-1}\langle\, \asymp \,\rangle\Big) \\
&\qquad + A^{-1}\Big(A\langle\, \asymp \,\rangle + A^{-1}\langle\, \supset\subset \,\rangle\Big) + (A^2 - A^{-2})\langle\, \supset\subset \,\rangle\Big) \\
&\quad = (-A)^{-3w(L_0)}\Big(-A^2\langle\, \supset\subset \,\rangle - \langle\, \asymp \,\rangle \\
&\qquad + \langle\, \asymp \,\rangle + A^{-2}\langle\, \supset\subset \,\rangle + (A^2 - A^{-2})\langle\, \supset\subset \,\rangle\Big) \\
&\quad = 0.
\end{aligned}
$$

❄

Example 2.76 Use Theorem 2.75 to compute the Jones polynomial of the trefoil knot with three left-handed crossings.

Solution. We first observe that the Jones polynomial of the trivial knot is 1:

$$V\big(\bigcirc\big) = X\big(\bigcirc\big) = \big\langle\bigcirc\big\rangle = 1.$$

Next we use Theorem 2.75 to compute the Jones polynomial of the trivial link of two components.

$$t^{-1}V\big(\infty_+\big) - tV\big(\infty_-\big) + (t^{-1/2} - t^{1/2})V\big(\bigcirc\bigcirc\big) = 0.$$

So

$$
\begin{aligned}
(t^{-1/2} - t^{1/2})V\big(\bigcirc\bigcirc\big) &= -t^{-1}\cdot 1 + t\cdot 1 \\
&= -(t^{-1} - t) \\
&= -(t^{-1/2} - t^{1/2})(t^{-1/2} + t^{1/2}).
\end{aligned}
$$

Thus, $V\big(\bigcirc\bigcirc\big) = -t^{-1/2} - t^{1/2}$.

We can solve for $V(L_-)$ in the skein relation for the Jones polynomial to obtain

$$V(L_-) = t^{-2}V(L_+) + (t^{-3/2} - t^{-1/2})V(L_0).$$

We now have all the ingredients for a straightforward computation of the Jones polynomial of the left-handed trefoil knot.

$$\begin{aligned}
V(\;) &= t^{-2}V(\;) + (t^{-3/2} - t^{-1/2})V(\;) \\
&= t^{-2}\cdot 1 + (t^{-3/2} - t^{-1/2})\left(t^{-2}V(\;) + (t^{-3/2} - t^{-1/2})V(\;)\right) \\
&= t^{-2} + (t^{-3/2} - t^{-1/2})(t^{-2}(-t^{-1/2} - t^{1/2}) + (t^{-3/2} - t^{-1/2})\cdot 1) \\
&= t^{-2} + (t^{-3/2} - t^{-1/2})(-t^{-5/2} - t^{-3/2} + t^{-3/2} - t^{-1/2}) \\
&= t^{-2} + (t^{-3/2} - t^{-1/2})(-t^{-5/2} - t^{-1/2}) \\
&= t^{-2} - t^{-4} + t^{-3} - t^{-2} + t^{-1} \\
&= -t^{-4} + t^{-3} + t^{-1}
\end{aligned}$$

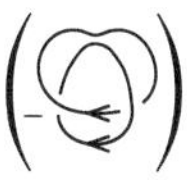

Exercises 2.7

1. **(a)** Show that when Rule 3 for the bracket polynomial is applied to two crossings of a regular projection, the order in which it is applied does not affect the result.
   (b) Show that when Rules 2 and 3 for the bracket polynomial are applied to a regular projection of a link with a disjoint unknotted component, the order in which they are applied does not affect the result.
2. Show that Reidemeister moves of Type II and Type III do not affect the writhe of the regular projection of a link.
3. Show that the Jones polynomial of the trefoil knot K with three right-handed crossings is $t + t^3 - t^4$. Conclude that the trefoil knot is not equivalent to its mirror image (a result that was difficult to prove prior to the discovery of the Jones polynomial).
4. Show that the Jones polynomials of the mirror image of a link is equal to the Jones polynomial of the link with the variable replaced by its reciprocal.
5. A knot is said to be **achiral** if and only if it can be deformed into its mirror image.
   (a) Use the result of Exercise 4 to give a condition on the Jones polynomial of an achiral knot.
   (b) By Exercise 9 of Section 1.6, the figure-eight knot is achiral. Compute the Jones polynomial of the figure-eight knot, and verify that it satisfies the condition.
6. Find values of A for which the bracket polynomial gives complex numbers that are invariant under all three Reidemeister moves.
7. Provide a justification for each of the steps in the proof of Theorem 2.75.

8. **(a)** Compute the Jones polynomial of the Hopf link with components oriented to have linking number $+1$.

(b) Compute the Jones polynomial of the Hopf link with components oriented to have linking number -1.

9. Compute the Jones polynomials of the two knots of crossing number 5 (see Figure 2.41).

10. Choose orientations for the components of the Whitehead link and compute the Jones polynomial. Use this result to conclude that the Whitehead link is not a trivial link.

11. Choose orientations for the components of the Borromean rings and compute the Jones polynomial. Use this result to conclude that the rings really are linked.

References and Suggested Readings for Chapter 2

11. Colin Adams, *The Knot Book: An Elementary Introduction to the Mathematical Theory of Knots*, W. H. Freeman, New York, 1994.

    A marvelous and lively introduction to knots with some applications to biology, chemistry, and physics.

12. J. W. Alexander, "Topological invariants of knots and links," *Transactions of the American Mathematical Society*, 30 (1928), 275–306.

    The original paper on the Alexander polynomial is fairly accessible to undergraduate math students.

13. Peter Cromwell, *Knots and Links*, Cambridge University Press, Cambridge, England, 2004.

    An introduction to modern knot theory.

14. Richard Crowell and Ralph Fox, *Introduction to Knot Theory*, Springer-Verlag, New York, 1963.

    A classical text in knot theory. It contains a proof that differentiable knots can be converted into polygonal knots.

15. P. Freyd, D. Yetter, J. Hoste, W. B. R. Lickorish, K. Millett, and A. Ocneanu, "A new polynomial invariant of knots and links," *Bulletin of the American Mathematical Society*, 12 (1985), 239–249.

    The presentation to the world of the HOMFLY polynomial.

16. V. F. R. Jones, "A polynomial invariant for knots via von Neumann algebras," *Bulletin of the American Mathematical Society*, 12 (1985), 103–111.

    This paper introduced the Jones polynomial and initiated a revolution in knot theory.

17. Louis Kauffman, *On Knots*, Princeton University Press, Princeton, NJ, 1987.

    Although this book uses some advanced concepts, it is full of fascinating ideas and visual arguments.

18. ———, "New invariants in the theory of knots," *American Mathematical Monthly*, 95 (1988), 195–242.

    A survey of Kauffman and other polynomials leading to a proof of the Kauffman-Murasugi-Thistlethwaite result on crossing numbers among other results.

19. W. B. R. Lickorish and K. C. Millett, "The new polynomial invariants of knots and links," *Mathematics Magazine*, 61 (1988), 3–23.

This article provides a concise definition of the HOMFLY polynomial and demonstrates how it is related to the Jones polynomial and the Alexander polynomial.

20. Charles Livingston, *Knot Theory*, Mathematical Association of America, Washington, DC, 1993.

A concise tour of the variety of techniques used in knot theory: combinatorial, geometric, algebraic.

21. Kunio Murasugi, *Knot Theory and Its Applications*, Birkhäuser, Boston, 1996.

The foundations of knot theory leading to application in statistical mechanics, molecular biology, and structural chemistry.

22. Lee Neuwirth, "The theory of knots," *Scientific American*, 240 (1979), 110–124.

A popular survey of classical knot theory.

23. David Penney, *Perspectives in Mathematics*, Benjamin, Mento Park, CA, 1972.

Chapter 2 discusses generalizations of the Borromean rings.

24. K. Reidemeister, *Knot Theory*, BCS Associates, Moscow, ID, 1983.

The source for Reidemeister moves.

25. Dale Rolfsen, *Knots and Links*, Publish or Perish, Berkeley, CA, 1976.

Somewhat advanced, but this book contains lots of examples, folklore, and techniques accessible to undergraduate students.

26. Alexei Sossinsky, *Knots: Mathematics with a Twist*, Harvard University Press, Cambridge, MA, 2002.

A lively exposition of selected topics in the theory and applications of knots.

27. Saul Stahl, *Geometry: From Euclid to Knots*, Prentice Hall, Upper Saddle River, NJ, 2003.

The chapter on knots and links provides an informal discussion of knot invariants including the Jones polynomial. The chapters on informal topology and surfaces may also be of interest.

28. P. G. Tait, "Some elementary properties of closed plane curves" in *Scientific Papers*, Cambridge University Press, Cambridge, England, 1898, 270–272.

Simple observations lead to an alternating knot for any knot projection.

29. ———, "On knots" in *Scientific Papers*, Cambridge University Press, Cambridge, England, 1898, 273–317.

The first published paper on the mathematical theory of knots.

3

Surfaces

Surfaces are natural generalizations of 2-dimensional Euclidean space: the region surrounding any point of a surface is topologically equivalent to $\mathbb{R}^2$. The 2-sphere and the torus are familiar examples of surfaces. Other examples such as the Klein bottle and the projective plane have challenged the imaginations of generations of mathematicians. One of the goals of this chapter is to fit these examples together in a comprehensive classification theorem. Surfaces provide further insights into knots, and they also play a key role in the study of spaces of higher dimensions.

3.1 Definitions and Examples

We normally picture a surface as a subset of Euclidean space, perhaps as the graph of a function from $\mathbb{R}^2$ to $\mathbb{R}$ or as the boundary of a solid object in $\mathbb{R}^3$. Although not all surfaces can be so nicely embedded in $\mathbb{R}^3$, we can avoid several technical difficulties by restricting our attention to spaces that can be embedded in some finite-dimensional Euclidean space. Definition 3.2 takes advantage of this, even though we often want to think of a surface as a topological space with its own intrinsic existence.

The characteristic feature of a surface is the simple topology of the region immediately surrounding any of its points. We need a preliminary definition to describe such regions.

Definition 3.1 *In a space with a way of measuring distances between points, a* ***neighborhood*** *of a point is a subset that contains all points within some positive distance of the point.*

The simplest neighborhoods of a point x_0 in a space X with a distance function d are usually the open balls $\{x \in X \mid d(x, x_0) < r\}$ of some radius $r > 0$. But in general, a neighborhood of a point is only required to contain such an open ball around the point.

Definition 3.2 *A* ***surface*** *(or* ***2-manifold****) is a space that is homeomorphic to a nonempty subset of a finite-dimensional Euclidean space and in which every point has a neighborhood homeomorphic to* $\mathbb{R}^2$. *We sometimes also wish to admit* ***boundary*** *points, which have neighborhoods homeomorphic to the half-plane* $\{(x, y) \in \mathbb{R}^2 \mid y \geq 0\}$.

Example 3.3 Show that the open disk $\{(x, y) \in \mathbb{R}^2 \mid x^2 + y^2 < 1\}$ is a surface.

Solution. Radial stretching of the radii of the disk onto corresponding rays from the origin of $\mathbb{R}^2$ is a homeomorphism of the open disk onto $\mathbb{R}^2$ (see Exercise 10 of Section 1.4). Thus, each point has the entire open disk as a neighborhood that is homeomorphic to $\mathbb{R}^2$. ❄

The homeomorphism of the previous example between the open disk and $\mathbb{R}^2$ means that we can use either of these sets as the model for the local structure of a surface.

Example 3.4 Show that the 2-sphere and the torus are surfaces.

Solution. Each point of the 2-sphere is the center of a hemisphere that is homeomorphic to a disk. Indeed, if we rotate the sphere so the point becomes the north pole, the orthogonal projection of the northern hemisphere onto the equatorial cross-section is such a homeomorphism. Thus, the interior of this hemisphere is a neighborhood of the point that is homeomorphic to an open disk.

Consider the torus as the subset of $\mathbb{R}^3$ obtained by rotating the circle $\{(x, y, z) \in \mathbb{R}^3 \mid (x - 2)^2 + z^2 = 1,\ y = 0\}$ about the z-axis. For any point of the torus consider the plane tangent to the torus at the given point. The orthogonal projection of the torus into the tangent plane gives a homeomorphism between a sufficiently small neighborhood of the point and an open disk in the tangent plane. ❄

Based on the principle that an arc (a set homeomorphic to a closed bounded interval) cannot separate $\mathbb{R}^2$, we can distinguish boundary points from nonboundary points of a surface. A point not on the boundary has a neighborhood homeomorphic to $\mathbb{R}^2$. Thus, no arc can separate such a neighborhood, let alone the surface. On the other hand, for any $r > 0$ the function $\alpha : [0, \pi] \to \{(x, y) \in \mathbb{R}^2 \mid y \geq 0\}$ defined by $\alpha(t) = (r \cos t, r \sin t)$ has as its image an arc that separates the half-plane $\{(x, y) \in \mathbb{R}^2 \mid y \geq 0\}$. Thus, any neighborhood of a boundary point contains arcs that separate the surface. Therefore, the property that any neighborhood can be separated by an arc characterizes the boundary points of a surface.

In dealing with a surface with boundary, we are usually interested in the number of boundary components. As you can see in Figure 3.5, the disk, for example, has one boundary component, the annulus $\{(x, y) \in \mathbb{R}^2 \mid 1 \leq x^2 + y^2 \leq 4\}$ has two boundary components, and the Möbius band has one boundary component. In all three of these examples, the boundary components are simple closed curves. Another possibility is a boundary component homeomorphic to the real line. This occurs as the boundary of the half-plane $\{(x, y) \in \mathbb{R}^2 \mid y \geq 0\}$ for example.

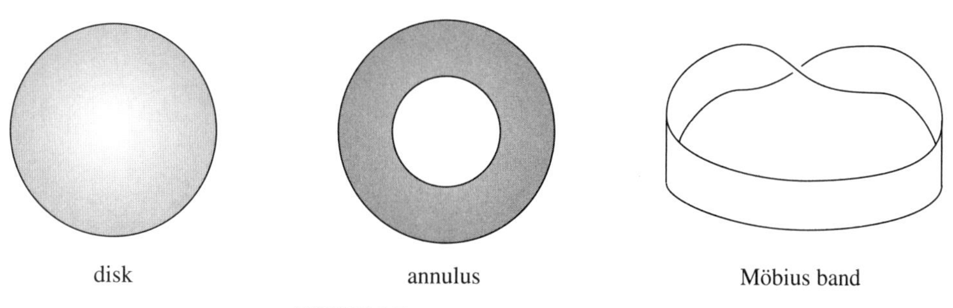

FIGURE 3.5
Simple surfaces with boundary points.

One method of constructing surfaces is to glue together portions of the boundary of a simpler surface. Typically, two simple-closed-curve boundary components are brought together so that each of the points in one curve can be identified with a corresponding point in the other curve. The new surface is formed by amalgamating the two curves into one curve. Portions of the original surface now extend on either side of the new curve. The seam along which we glued immediately heals over and no longer can be distinguished from other points of the new surface.

This gluing operation can be made precise by defining an equivalence relation on the original surface. In addition to each point being equivalent to itself, pairs of points to be identified are declared equivalent. The new surface is then the set of equivalence classes with a suitable definition of distance between points based on the corresponding distances in the original surface. Exercises 16 through 21 in Section 1.2 illustrate this technique. Section 7.5 provides a mathematically rigorous treatment of gluing.

Gluing can also be performed along a pair of arcs in the boundary of a surface. The only complication is that the endpoints of the arcs may still be boundary points. They will often be subject to further gluing. Hence, more than two endpoints of arcs on the original surface may ultimately be identified together in the final surface.

Figure 3.6 illustrates the construction of a torus formed by gluing together opposite edges of a rectangular disk. As we travel counterclockwise around the boundary of the disk, we can read off the word $aba^{-1}b^{-1}$ where the inverses indicate traveling in the direction opposite that of the orientation of the arrow. Notice that this word contains the essential gluing instructions for forming the torus. We will frequently use such words to indicate the gluing instructions for forming a surface from a polygonal disk.

The last step in this construction of a torus is to glue together the two simple closed curves that form the boundary of the cylinder. But wait! There is another way to glue together these two circles. If the orientation of one of the vertical sides of the original disk is reversed, the two ends of the cylinder cannot be so easily matched up. Perhaps the simplest way to remedy this is to create a hole in the side of the cylinder by pushing a disk slightly into the future (or with the surface embedded in the 3-dimensional slice of $\mathbb{R}^4$,

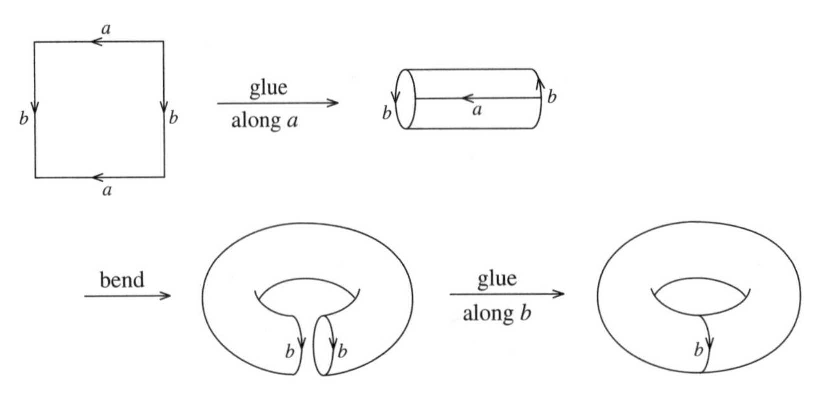

FIGURE 3.6
Gluing the edges of a disk to form a torus.

gradually increase the fourth coordinate from its original value of 0), passing one end of the cylinder through the hole, and aligning the two boundary components so they can be glued together with this alternative orientation. This surface was first described in 1882 by the great German mathematician Felix Klein. You can imagine the delight of some witty graduate students in transforming Klein's surface (*Fläche*) into Klein's bottle (*Flasche*).

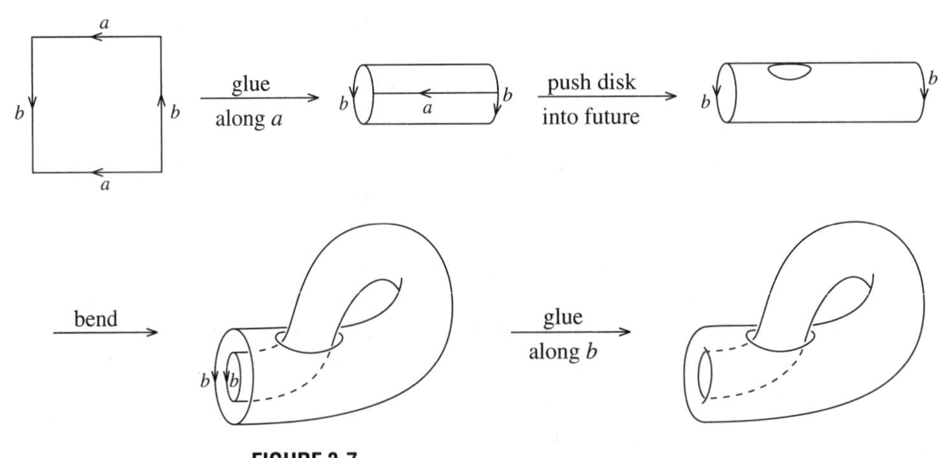

FIGURE 3.7
Gluing the edges of a disk to form a Klein bottle.

Another surface to include in your bag of basic examples is the projective plane. This surface can be constructed from a disk by gluing along two arcs that comprise the boundary of the disk. The arcs are oriented in the same direction so we read off the word aa as we travel around the boundary of the disk. Now it requires some ingenuity to picture the projective plane in $\mathbb{R}^3$. In fact, Exercise 10 at the end of this section demonstrates that the projective plane cannot be embedded in $\mathbb{R}^3$ as a surface made up of polyhedral faces. However, it is possible to picture the projective plane in $\mathbb{R}^3$ as a sphere with what is called a cross-cap. This cross-cap involves two portions of the surface intersecting transversely along an arc. The self-intersection can be removed if we allow ourselves the freedom of a fourth dimension so we can push one of the portions of the surface forming this intersection slightly into the future. This will eliminate the self-intersection and yield an embedding of the projective plane in $\mathbb{R}^4$.

We construct the sphere with a cross-cap by pulling down the interior of the disk and bringing together the arcs of the boundary. The initial set up looks like a purse that is about to be snapped shut. The problem is that as the arcs are brought together, only the midpoints are in position to be glued to the matching point on the other arc. We need to glue the initial half of one arc to the final half of the second arc. We also need to glue the final half of the first arc to the initial half of the second. Glue the midpoints together at the top of the cross-cap. From the midpoint, work down the four arcs, gluing together the opposite edges of the opening. The result is a single arc along which the two portions of the surface intersect transversely. The ends of the boundary curves come together at the bottom of this arc. Figure 3.8 may be of some help in picturing the projective plane.

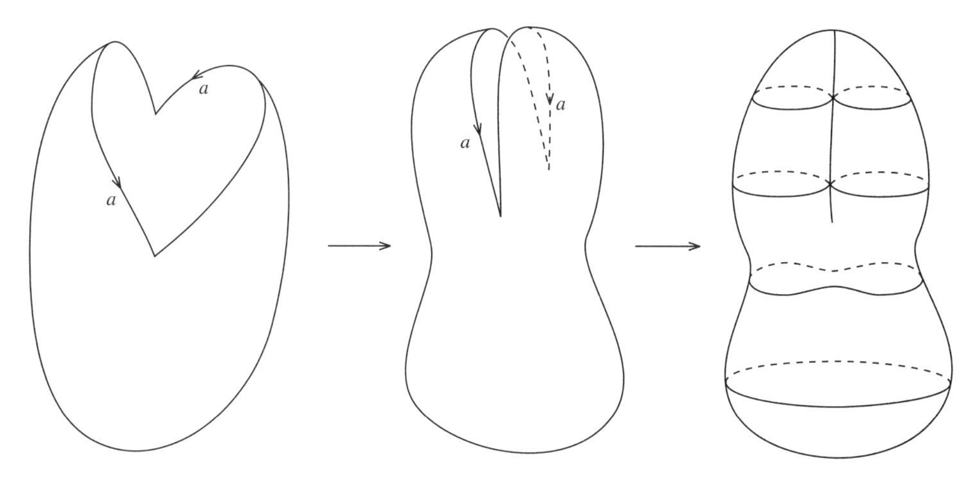

FIGURE 3.8
The projective plane as a sphere with a cross-cap.

In this section we have encountered some important examples of surfaces. In our future work we will use the following standard symbols to denote spaces homeomorphic to these basic examples: S^2 denotes a 2-sphere, T^2 denotes a torus, P^2 a projective plane, K^2 a Klein bottle, D^2 a disk, A^2 an annulus, and M^2 a Möbius band.

Exercises 3.1

1. Consider the boundary of a disk as two arcs with opposite orientation. Draw a picture to identify the surface that results from gluing the disk along these arcs.

2. Show that the Klein bottle can be cut into two Möbius bands that intersect along their boundaries. The following limerick uses poetic license to describe this property of the Klein bottle:

A mathematician named Klein
Thought the Möbius band was divine.
Said he,"If you glue
The edges of two,
You'll get a weird bottle like mine."

3. **(a)** Show that a projective plane contains a Möbius band.

(b) Show that a disk glued along the boundary of a Möbius band yields a projective plane.

4. **(a)** Consider a square disk with edges labeled as in Figure 3.6 so that we read off the word $aba^{-1}b^{-1}$ as we go around the boundary. Draw quarter circles near each of the four corners in such a way that each end of each quarter circle is identified with an end of another quarter circle. Trace around these quarter circles to confirm that when we glue the edges, the four corners are all glued together as a single point in the torus and this point has a neighborhood homeomorphic to a disk.

(b) Consider a hexagonal disk with edges glued according to the word $abca^{-1}b^{-1}c^{-1}$. How are the vertices of the hexagonal disk glued?

(c) Here is a technique to help identify the surface that results when we glue together the hexagonal disk. Cut the disk along a chord d that joins the end of one c edge to the corresponding end of the other c edge. Glue the two pieces along the b edges to get a new hexagonal disk with its edges glued together in pairs. Combine the adjacent a and c edges to form new edges. What surface do you now find as a result of the gluing?

5. **(a)** Two sides of a triangular disk are given opposing orientations. Draw a picture of the dunce cap that results when the disk is glued along these two arcs.

(b) The three sides of a triangular disk are given orientations according to the word $aa^{-1}a$. Draw a picture of the court-jester hat that results when the three sides are glued together.

6. Consider the homeomorphism described in Example 3.3 between the open disk $\{(x, y) \in \mathbb{R}^2 \mid x^2 + y^2 < 1\}$ and $\mathbb{R}^2$.

(a) Illustrate some typical curves in the open disk that correspond to parallel lines in $\mathbb{R}^2$.

(b) Each direction in $\mathbb{R}^2$ corresponds to a family of parallel lines. Adjoin a vanishing point at infinity for the lines in this family to converge to (from either direction). Show how this circle of ideal points can be realized when $\mathbb{R}^2$ is shrunk down to the size of the open disk.

(c) Show that the surface resulting from this disk with the circle of ideal points is the projective plane.

7. Show that the projective plane can be realized as the configuration space of lines in $\mathbb{R}^3$ passing through the origin. Suggestion: On the unit sphere in $\mathbb{R}^3$ choose one point in the northern hemisphere to represent each line that does not lie entirely in the xy-plane. What happens to lines that pass through the equator?

8. Show that the configuration space of chords and tangent lines to a circle can be realized as a Möbius band. Suggestion: Define a bijection so that each family of parallel chords and the two parallel tangent lines corresponds to a set of points spanning across a Möbius band.

9. Remove the boundary curve from a Möbius band to form what is known as an open Möbius band. Show that the set of straight lines in $\mathbb{R}^2$ can be realized as an open Möbius band. Suggestion: Define the bijection so that each family of parallel lines corresponds to a set of points spanning across an open Möbius band.

10. Fill in the details of this proof by contradiction that a projective plane cannot be embedded in $\mathbb{R}^3$ as a surface made up of polyhedral faces. This proof is known as the nonending arc trick.

(a) Suppose a projective plane is embedded in $\mathbb{R}^3$ as a surface P made up of polyhedral faces. By Exercise 3, P contains a Möbius band. Find a short arc that pokes through P at a point on the centerline of a Möbius band in P. Join the two ends of this arc by another arc that follows along the centerline of the Möbius band just to the side of P. The resulting simple closed curve C intersects P in only one point.

(b) From an arbitrary point in $\mathbb{R}^3$ draw arcs to each of the points of C. Define a function from a disk onto this set that is a homeomorphism between the boundary of the disk and the curve C.

(c) Assume the disk is triangulated so that the function maps triangles to triangles. Also assume that when the function is restricted to any of these triangles, it is linear in some coordinate system on these triangles. Adjust the image of the function so that it is in general position with respect to P. Show that the points of the disk that map to P form simple closed curves in the interior of the disk and arcs with endpoints on the boundary of the disk.

(d) Derive a contradiction from the fact that only one point of the boundary of the disk maps to P.

11. Use the nonending arc trick of Exercise 10 to prove that the Klein bottle cannot be embedded in $\mathbb{R}^3$ as a surface made up of polyhedral faces.

12. **(a)** Given any two points in a path-connected surface, draw pictures to illustrate an ambient isotopy of the surface that deforms the first point onto the second.

(b) Given any two points in a path-connected surface, draw pictures to illustrate an ambient isotopy of the surface that interchanges the two points.

(c) Given a permutation of any finite set of points in a path-connected surface, draw pictures to illustrate an ambient isotopy of the surface that yields the permutation on the set of points.

3.2 Cut-and-Paste Techniques

The previous section introduced the technique of gluing (or pasting) along a pair of arcs or pair of simple closed curves in the boundary of a surface. In particular, the basic examples of the 2-sphere, the torus, the projective plane, and the Klein bottle can be constructed by

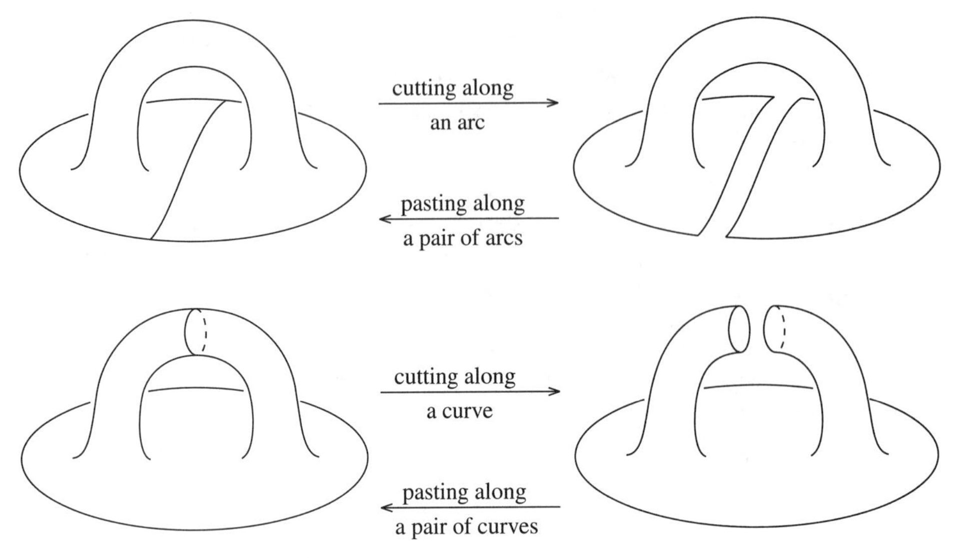

FIGURE 3.9
Cutting and pasting along arcs and simple closed curves.

gluing together pairs of edges of a polygonal disk. In Exercise 1 at the end of this section, you are asked to show that pasting pairs of edges of a polygonal disk always produces a surface.

Cutting along an arc or a simple closed curve is the reverse of the operation of pasting. The arc or simple closed curve we cut along must not intersect the boundary of the surface except at the endpoints of the arc. We cut along an arc by removing the arc to split the surface apart. Then we place copies of the arc on both portions of the surface that ran along the original arc. These arcs become part of the boundary of the resulting surface. We can likewise cut along a simple closed curve in a surface. The result is either two copies of the curve as new boundary components or one new boundary component that goes twice around the original curve. Figure 3.9 illustrates typical cases of these operations.

Be sure to convince yourself that cutting and pasting are inverse operations. Of course if you cut along a curve, you must keep track of the orientations of the duplicate curves so that you can paste them back together correctly to recover the original surface.

Example 3.10 Show that a rectangular strip of paper with ends pasted together after three half-twists is a Möbius band.

Solution. Cut the surface along the arc where it was pasted together. Untwist one of the ends one full revolution. This will return the end to its original alignment so the two ends can be pasted back together exactly as they were. The net result of this cutting, untwisting, and pasting gives a homeomorphism between the original surface and the traditional Möbius band with a single half-twist as illustrated in Figure 1.19.

By Exercise 10 of Section 2.3, there is no ambient isotopy of $\mathbb{R}^3$ that deforms one of these Möbius bands onto the other. Nevertheless, they are topologically the same intrinsic surface. ❄

As we saw in the preceding example, the arcs we cut along typically span across the surface so as to intersect the boundary of the surface only in their endpoints. However, this need not always be the case. For example, the disk with boundary arcs identified according to the word aa^{-1} is obtained by cutting a 2-sphere along an arc. In such a case, the endpoints of the arc are not duplicated; they become the endpoints for both copies of the new boundary arcs.

The next example illustrates the usefulness of cutting along a simple closed curve.

Example 3.11 Use cut-and-paste techniques to show that the Cartesian product of two circles $S^1 \times S^1 = \{((a,b),(c,d)) \in \mathbb{R}^2 \times \mathbb{R}^2 \mid a^2 + b^2 = 1 \text{ and } c^2 + d^2 = 1\}$ is a torus.

Solution. Cutting $S^1 \times S^1$ along the circle $S^1 \times \{(1,0)\}$ opens up the second factor into a closed interval I. The result is a cylinder $S^1 \times I$. Notice that an orientation of the first factor gives a consistent orientation to both boundary components of $S^1 \times I$. Thus, we recognize from Figure 3.6 that a torus results when we reattach the two ends of the cylinder to reconstitute $S^1 \times S^1$. This is a lot easier than setting up the toroidal coordinate system as we did in Example 1.12. ❄

Cut-and-paste techniques are also useful in forming new surfaces from old ones. The following definition introduces an important algebraic operation for surfaces.

Definition 3.12 *Let S and T be path-connected surfaces. Remove the interior of a disk from each surface by cutting along the boundaries of the disks. Glue the remaining surfaces together along the newly formed boundary components. The resulting surface is the* **connected sum** *of S and T. It is denoted S#T.*

Example 3.13 Draw a sequence of diagrams to illustrate the steps in forming the connected sum of two tori.

Solution.

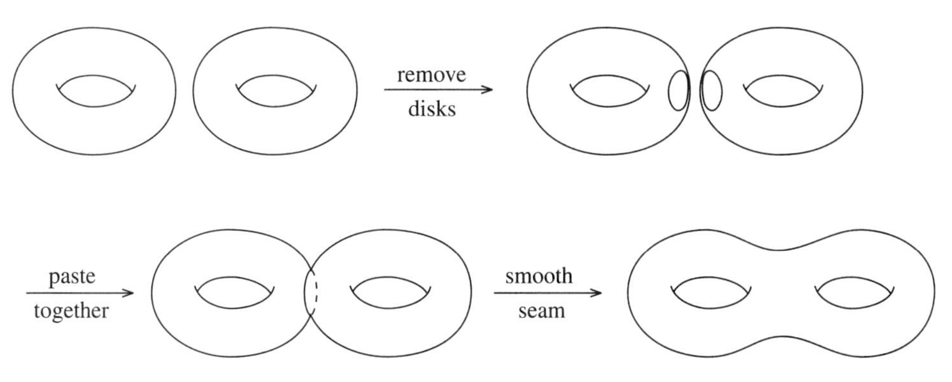

FIGURE 3.14
The connected sum of two tori. ❄

Naturally we only expect the connected sum of two surfaces to be determined up to homeomorphism. But to show the operation is well-defined even to this extent involves some formidable technical details. The ideas are simple enough, however. For example, we need to know that the connected sum is independent of the choice of the disks that are removed. Given two disks in a path-connected surface, we need to find a homeomorphism from the surface to itself that maps the first disk to the second. Such a homeomorphism can be visualized as the result of a sequence of isotopies of the surface. Shrink the disks down to small triangular disks. Then slide the first triangle along a path connecting it to the second triangle (see Exercise 6 of Section 1.3, Exercise 8 of Section 1.4, and Exercise 12 of Section 3.1). Match up the two triangles (see Example 1.52). Finally unshrink the second triangle to restore it to its original shape. Exercises 12 and 13 illustrate some problems that can arise in general. Fortunately, we never encounter this kind of trouble with path-connected surfaces that can be cut apart into a finite number of disks.

When a surface is represented as a polygonal disk with edges to be identified in pairs, it is especially easy to keep track of the results of forming connected sums. Cut along a simple closed curve that meets the boundary of the disk in exactly one of the vertices. Then open up the new boundary component as a new edge of the disk. Finally identify this edge with a similar edge created in the polygonal disk for the other surface. The following example illustrates this operation for the connected sum of two projective planes.

Example 3.15 Show that the connected sum $P^2 \# P^2$ of two projective planes can be represented as a square disk with sides identified according to the word $aabb$.

Solution. Represent the two projective planes as disks with sides identified according to aa and bb. Cut along simple closed curves as illustrated in Figure 3.16. With the orientation of the curves chosen appropriately, this results in triangular disks with sides labeled aac and bbc^{-1}. Paste these together along the c edges to produce the square disk with sides labeled $aabb$. ❄

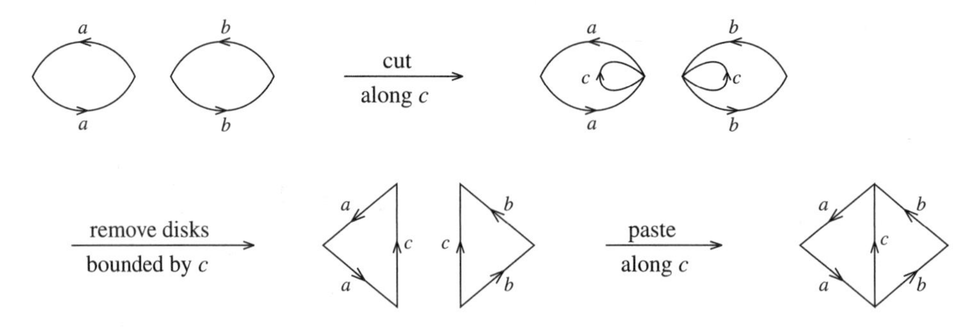

FIGURE 3.16
The connected sum of two projective planes as a square disk with edges identified in pairs.

Exercises 3.2

1. Consider a polygonal disk with an even number of edges. Suppose that each edge has an orientation and that the edges are labeled in pairs. Consider the following three

cases to verify that pasting the disk together along these edges always results in a surface.

(a) Show that any point in the interior of the polygonal disk has a neighborhood homeomorphic to $\mathbb{R}^2$.

(b) Consider a point other than a vertex along of one of the edges. This point is glued to a point in the interior of the other edge in this pair of edges. Find two half-disks that are glued together to form a neighborhood of the point that is homeomorphic to $\mathbb{R}^2$.

(c) Consider a vertex of the disk. Trace a path around the vertex to find little wedges at the corners of the disk that are glued together to form a neighborhood of the point that is homeomorphic to $\mathbb{R}^2$.

2. Show that the 2-sphere S^2 is an identity element for the operation of connected sums.

3. Draw some pictures to convince yourself that the connected sum operation is commutative and associative.

4. **(a)** Show that the connected sum of two tori can be represented by an octagonal disk with edges identified according to the word $aba^{-1}b^{-1}cdc^{-1}d^{-1}$.

(b) Show that with this representation, the eight vertices of the octagon are identified together as a single point of the surface.

(c) Label the interior angles of the octagon and show how they are arranged as pie wedges around this point of the surface.

5. **(a)** Find a representation of the connected sum of g tori as a disk with edges identified in pairs.

(b) Find a representation of the connected sum of n projective planes as a disk with edges identified in pairs.

6. Which of the surfaces in Figure 3.17 are homeomorphic?

7. **(a)** Show that the square disks with edges pasted together according to the words $aabb$ and $aba^{-1}b$ result in homeomorphic surfaces. Suggestion: Cut along one of the diagonals and glue the two triangular disks together along one of the original edges.

(b) Conclude that the connected sum of two projective planes is homeomorphic to a Klein bottle.

8. **(a)** Write $T^2 \# P^2$ and $K^2 \# P^2$ as polygonal disks with pairs of edges to be identified.

(b) Use cut-and-paste techniques to show that the surfaces are homeomorphic. Suggestion: First cut the hexagonal disk for $T^2 \# P^2$ along a diagonal so you can reattach the pieces along the edges that correspond to the projective plane; then cut off a triangular disk and paste it back along another edge.

(c) Bid farewell to the possibility of a cancellation law for the operation of connected sum of surfaces.

9. Exercise 4 of Section 1.1 gives a method for constructing the integers as equivalence classes of ordered pairs of whole numbers. What difficulties arise when you try to follow this program to create objects that serve as inverses for the connected sum operation?

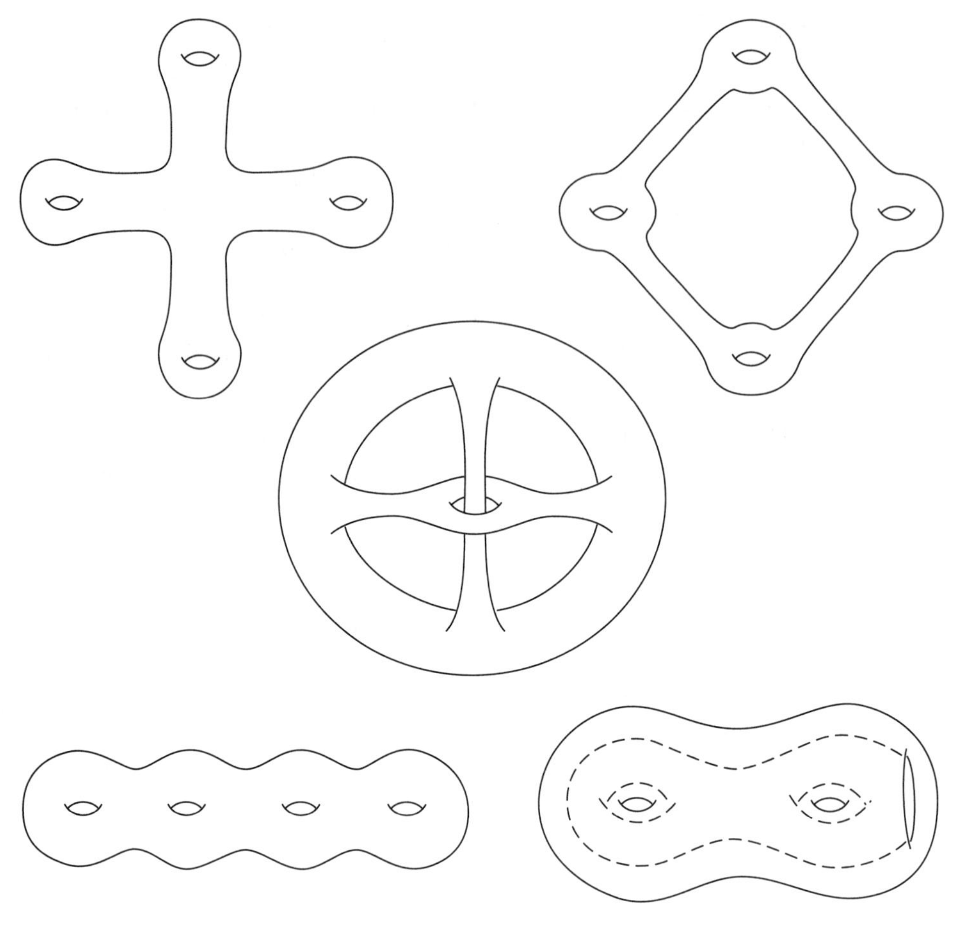

FIGURE 3.17
Which of these surfaces are homeomorphic?

10. Draw a picture to show that gluing an annulus to a surface along a simple closed curve boundary component of each yields a surface homeomorphic to the original surface.

11. Show that cutting a surface along a simple closed curve that bounds a disk in the surface and discarding the disk yields the same result as forming the connected sum of the surface and a disk.

12. Show that the operation of connected sum is not well-defined if we allow surfaces with more than one path component.

13. From the boundary of the unit disk remove the points at angles $\pi, \frac{\pi}{2}, \frac{\pi}{3}, \ldots$ measured counterclockwise from the positive x-axis. Also remove the limit point $(1, 0)$.

(a) Show that a connected sum of two copies of this surface is an annulus with a convergent sequence of points removed from each of its boundary curves.

(b) Show that there are two ways to form the connected sum of two copies of this surface.

3.3 The Euler Characteristic and Orientability

With the basic surfaces (2-sphere, torus, projective plane, and Klein bottle) and the operation of connected sum, we can assemble quite an impressive list of surfaces. It is not entirely clear which of these possibilities are topologically distinct. For example, would you have realized that $T^2\#P^2$ is homeomorphic to $P^2\#P^2\#P^2$ if you had not looked at Exercises 7 and 8 of Section 3.2? We need some properties of surfaces that are preserved under homeomorphisms. These topological invariants will then enable us to distinguish between different surfaces; and once we have assembled a powerful enough set of invariants, they will also enable us to determine when two surfaces are topologically the same.

The primary numerical quantity we want to use to characterize surfaces is derived from a combinatorial structure on a surface. We have previously encountered the idea of breaking a space up into triangles. Here is the official definition as it pertains to surfaces and other 2-dimensional objects.

Definition 3.18 *A **triangulation** of a space is a decomposition of the space into a union of disks, arcs, and points. The disks are called **faces**, the arcs are **edges**, and the points are **vertices** of the triangulation. A face intersects other components of a triangulation only along its boundary; and the boundary of a face consists of three edges and three vertices. An edge intersects other edges and the vertices only at its endpoints; and both endpoints of an edge are vertices.*

In 1925 Tibor Radó presented a proof that any surface can be triangulated. This result justifies our diagrams of surfaces as smooth or polyhedral subsets of Euclidean space. We will continue to rely on this result to define the Euler characteristic of a surface in this section and to prove the Classification Theorems 3.33 and 3.39 in the next section.

We will restrict our consideration to spaces that can be triangulated with a finite number of faces, edges, and vertices. The following definition gives a convenient name to such objects. We shall see in Definition 7.33 of Section 7.4 that this concept can be extended to much more general situations.

Definition 3.19 *A triangulated space is **compact** if and only if it consists of a finite number of faces, edges, and vertices.*

Here is a magic trick to introduce the Euler characteristic. Have a friend draw a few dots on a piece of paper and connect them with arcs. Be sure to explain that since this is a topological trick, the arcs do not have to be straight although they should not cross. When all the dots are connected, have your friend count the number of bounded regions formed by the diagram and add to this the number of dots. Knowing only the total and without seeing the diagram, you can determine the number of arcs, much to the amazement of your audience. The secret is that for any connected network in the plane, the number of vertices minus the number of edges plus the number of faces is always equal to 1. Thus, the number of edges will be one less than the number of regions and dots your friend reports.

It is not hard to see why this magic trick works. Start building the network from a single vertex. Add in edges one at a time. Just make sure to keep the figure connected. Each edge will either extend from a vertex already in place into the interior of one of the regions of the diagram, or else it will run between two vertices already in the diagram. In the first case the new edge is accompanied by a new vertex at the end of the arc. In the other case the edge subdivides a region into two regions. In this way every edge can be paired either with a vertex or a region of the diagram. Since there was one vertex at the beginning, the number of vertices plus the number of regions will always exceed the number of edges by 1.

In 1640 René Descartes observed such a relation among the number of faces, edges, and vertices of a triangulated sphere. In 1752 Leonhard Euler expressed this relation as a formula that generalizes to other polyhedral surfaces.

Definition 3.20 *The* ***Euler characteristic*** *of a compact triangulated space S is the number of vertices minus the number of edges plus the number of faces. The Euler characteristic of S is denoted $\chi(S)$.*

Although the definition of Euler characteristic is in terms of a triangulation, the analysis of the magic trick shows that the faces can actually be polygonal disks with any number of edges, even one or two (although some people prefer to let bigons be bygones). It is always a simple matter to introduce additional vertices in the boundary of a polygonal disk and to subdivide a polygonal disk into triangular disks. A quick glance at the formula shows that neither of these modifications will change the Euler characteristic. It is often useful to introduce additional edges and vertices in some of the faces. Again the analysis of the magic trick shows that this kind of subdivision does not change the Euler characteristic. Thus, for example, if we want to cut along a polygonal curve, we can assume without further comment that the curve is composed of edges and vertices of the triangulation.

Example 3.21 Show that the Euler characteristic of the 2-sphere with any triangulation is 2.

Solution. Cut the 2-sphere along the boundary of one of its faces. With one face removed, the remaining portion of the 2-sphere can be flattened out (with no further change to the numbers of vertices, edges, or faces) to form a connected polygonal region of the plane with no holes. The result of the magic trick shows that the Euler characteristic of this part of the 2-sphere is 1. When we add back the face we removed, the Euler characteristic of the entire 2-sphere is 2. ❄

Example 3.22 Show that the Euler characteristic of the annulus with any triangulation is 0.

Solution. Glue a disk along its boundary to one of the boundary components of the annulus. The resulting surface is a disk and, based on the result of the magic trick, has Euler characteristic 1. The annulus has one less face than this disk. Hence its Euler characteristic is 0. ❄

Example 3.23 Show that the torus and the Klein bottle both have Euler characteristic 0.

Solution. Cut the torus or Klein bottle along a two-sided nonseparating simple closed curve to obtain a cylinder. The two boundary components of this cylinder duplicate the edges and vertices of the curve we cut along. As we travel around any simple closed curve we alternately encounter edges and vertices. Thus, the difference between the number of vertices and the number of edges on a simple closed curve is 0. It follows that cutting along the simple closed curve does not change the Euler characteristic. Therefore the Euler characteristics of the torus and the Klein bottle are the same as the Euler characteristic of the cylinder. As we saw in the previous example, this is 0. ❄

Here is a useful result for computing Euler characteristics of objects in terms of the Euler characteristics of simpler pieces of the object.

Theorem 3.24 *Suppose A and B are triangulated so that $A \cap B$ is also triangulated. Then $\chi(A \cup B) = \chi(A) + \chi(B) - \chi(A \cap B)$.*

Proof. For vertices, edges, and faces, the number in $A \cup B$ is the number in A plus the number in B minus the number in $A \cap B$. Hence the alternating sum of these quantities yields the formula for the Euler characteristic. ❄

With the principles illustrated in the preceding results, your mission in Exercise 4 at the end of this section is to determine the Euler characteristics of the other surfaces we know about. The results are summarized in the following theorem.

Theorem 3.25 *The surface formed by taking the connected sum of g tori and cutting out disks to leave b boundary components has Euler characteristic $2 - 2g - b$. The surface formed by taking the connected sum of n projective planes and cutting out disks to leave b boundary components has Euler characteristic $2 - n - b$.*

Consider a surface formed by taking the connected sum of tori and cutting out disks. Since the Euler characteristic and the number of boundary components are topological invariants of the surface, Theorem 3.25 shows that the number of tori connected together is also an invariant of the surface. This number of handles is known as the **genus** of such a surface.

The Euler characteristic and the number of boundary components do a fairly thorough job of distinguishing among the various examples of surfaces. However, they do not give a complete set of invariants. For example, the torus and the connected sum of two projective planes (the Klein bottle) are both surfaces with Euler characteristic 0 and no boundary components. We need an additional property of surfaces that will allow us to identify surfaces that contain a Möbius band.

Definition 3.26 *An* ***orientation*** *of a polygonal face of a triangulated surface is the choice of one of the two possible orientations of the boundary curve of the face. A surface is* ***orientable*** *if and only if it is possible to choose orientations of all faces of a triangulation of the surface so that whenever two faces share a common edge, the orientations of the faces induce opposite orientations on the edge.*

Example 3.27 Show that the torus is orientable, but that the Klein bottle is not orientable.

Solution. Figure 3.28 shows a torus as a disk with edges identified according to the word $aba^{-1}b^{-1}$. The faces of the triangulation of the torus have been oriented so that any two that meet in an edge will give opposite orientations to the edge (even those that meet along the curves a and b).

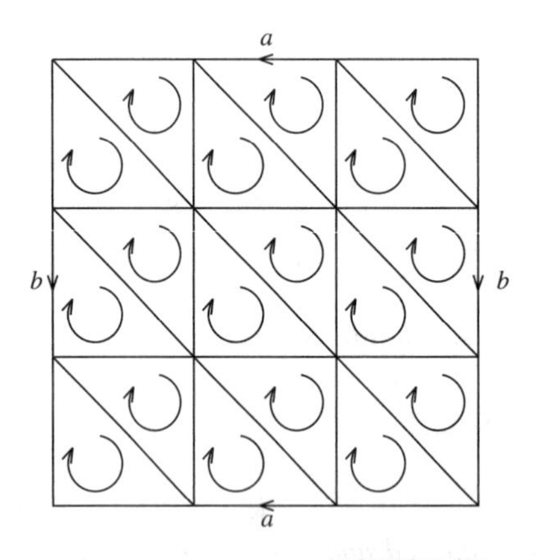

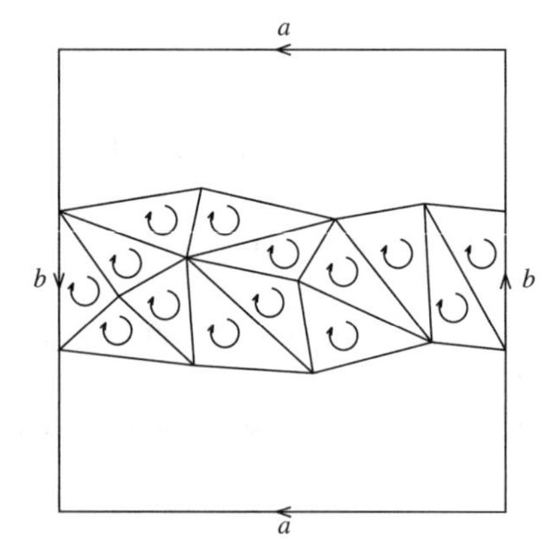

FIGURE 3.28
The torus is orientable; the Klein bottle is not.

The Klein bottle appears as a disk with edges identified according to the word $aba^{-1}b$. Consider a band of triangles across the middle of this disk. If we attempt to satisfy the condition that adjacent faces induce opposite orientations on common edges, an orientation of the leftmost face determines orientations on adjacent faces immediately to its right. These in turn determine orientations on the other faces of this band. We see that the orientations on the two end faces induce the same orientation on the shared edge of the curve b. Thus, it is impossible to choose a consistent orientation of all faces of a triangulation of the Klein bottle. ❄

An intuitive way to check orientability of a surface is to imagine a 2-dimensional bug that lives embedded in the surface. As the bug crawls around in the torus, a flag it carries in its right hand will always remain on the right no matter what path the bug takes. This is consistent with the fact that the torus is orientable. On the other hand, consider a bug embedded in a Klein bottle. Again think of the Klein bottle as a disk with edges identified according to the word $aba^{-1}b$. As the bug crosses the b edge, the hand carrying the flag appears on the left side (at least to us looking down on the unglued disk). Traveling along this kind of path reverses orientation. A surface is nonorientable if and only if it contains such an orientation-reversing path. In any surface, the notions of left and right can be defined locally (within a set homeomorphic to $\mathbb{R}^2$, for example). Only in an orientable surface, however, can the notions of left and right be extended to the entire surface. In a nonorientable surface, traversing certain closed paths will interchange the notions of right and left. Figure 3.29 illustrates the difference between a bug crawling in a torus and its companion crawling in a Klein bottle.

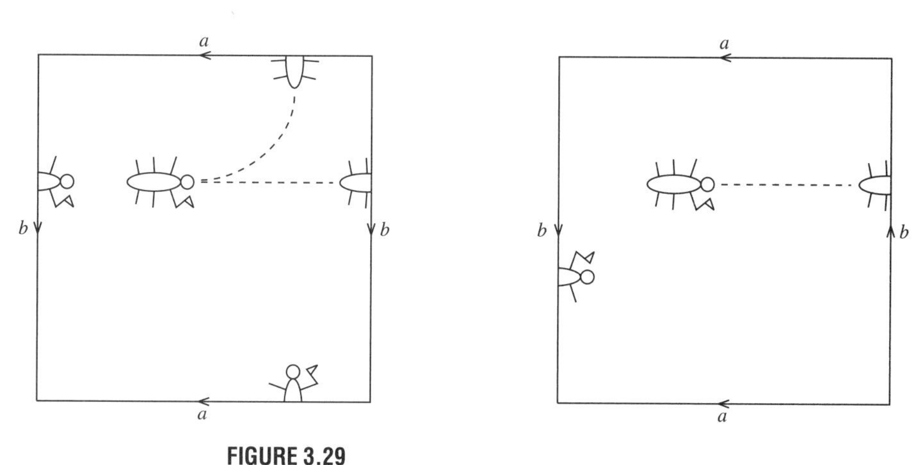

FIGURE 3.29
Bugs crawling around in a torus and a Klein bottle.

The Möbius band is the prototype of a nonorientable surface: Exercise 8 asks you to show that, in fact, a surface is nonorientable if and only if it contains a Möbius band. The Möbius band illustrates another interesting property of nonorientable surfaces when they are embedded in $\mathbb{R}^3$. Suppose you want to paint one side of a paper band red, leaving the other side white. It the band is not twisted, you will have no trouble painting this cylinder. However, if the band is twisted to form a Möbius band, when you have painted once around the band you will be on the other side; and as you continued to paint, the whole band will end up red. A Möbius band in $\mathbb{R}^3$ really has only one side. On the other hand, a cylinder in $\mathbb{R}^3$ is a two-sided surface.

It turns out that a surface in $\mathbb{R}^3$ is nonorientable if and only if it is one-sided; hence a surface in $\mathbb{R}^3$ is orientable if and only if it is two-sided. One way to see this is to consider a 3-dimensional bug walking on one side of the surface in $\mathbb{R}^3$ carrying a flag in its right hand. Think of the shadow of the bug on the surface as a 2-dimensional bug in the surface. As the 3-dimensional bug walks around on the surface along a closed path, the flag stays in its right hand. But if the bug finds itself on the other side of the surface when it completes one lap around the closed path, the shadow in the surface will have the shadow of the flag in its opposite hand.

The idea of one-sidedness and the argument above still make sense even if we allow the surface to intersect itself when it is placed in $\mathbb{R}^3$. Exercise 7 asks you check that the Klein bottle as depicted in Figure 3.7 and the projective plane in Figure 3.8 are really one-sided.

Be sure to note that our discussion of this correspondence between orientability and two-sidedness has been restricted to surfaces in $\mathbb{R}^3$. In Chapter 4, we will construct 3-dimensional spaces where other possibilities occur. There we will encounter both two-sided Klein bottles and one-sided tori.

Exercise 6 asks you to show that any surface formed by taking the connected sum of tori is orientable, whereas any surface formed by taking the connected sum of projective planes is nonorientable. Thus, the Euler characteristic, the number of boundary components, and the property of orientability allow us to distinguish among all of our basic examples. The Classification Theorem 3.39 shows that this set of invariants is indeed powerful enough to distinguish among all compact surfaces.

Exercises 3.3

1. Leonhard Euler was a Swiss mathematician, not to be confused with the Greek geometer Euclid. Practice saying his last name with the correct German pronunciation.

2. Draw pictures of the various ways that two triangular disks can intersect. Which of these is permitted in a triangulation?

3. **(a)** Show that the Möbius band has Euler characteristic 0.
   (b) Show that the projective plane has Euler characteristic 1.

4. **(a)** For surfaces S and T show that $\chi(S\#T) = \chi(S) + \chi(T) - 2$.
   (b) Show that the connected sum of g tori has Euler characteristic $2 - 2g$.
   (c) Show that the connected sum of n projective planes has Euler characteristic $2 - n$.
   (d) Show that removing a disk from a surface to create a boundary component changes the Euler characteristic by -1.

5. Suppose the surface S is created by gluing together two surfaces S_1 and S_2 along arcs in their boundaries. Give a formula for $\chi(S)$ in terms of $\chi(S_1)$ and $\chi(S_2)$.

6. **(a)** Show that the 2-sphere is orientable.
   (b) Show that the projective plane is not orientable.
   (c) Show that the connected sum of two orientable surfaces is orientable.
   (d) Show that the connected sum of any surface with a nonorientable surface is nonorientable.

7. **(a)** Find an orientation-reversing path in the Klein bottle in Figure 3.7. Show that this surface is one-sided.

   (b) Find an orientation-reversing path in the projective plane in Figure 3.8. Show that this surface is one-sided.

8. Show that a surface is nonorientable if and only if it contains a Möbius band.

9. **(a)** Give a triangulation of the court-jester hat obtained from identifying the three sides of a triangular disk according to the word $aa^{-1}a$.

   (b) Use your triangulation to compute the Euler characteristic of the court-jester hat.

10. Figure 3.30 is a diagram of Bing's house with two rooms. It is a 2-dimensional object consisting of an upper room, a lower room, a passageway through the upper room from the outside to the lower room, a passageway through the lower room from the outside to the upper room, and panels that connect each passageway to an adjacent wall. Each passageway is a cylinder that connects a hole in the ceiling to a hole in the floor of the room through which it passes. Determine the Euler characteristic of Bing's house with two rooms.

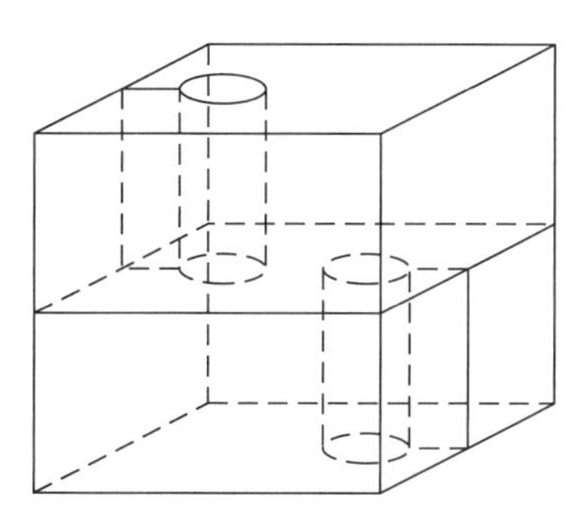

FIGURE 3.30
Bing's house with two rooms.

11. What is nonorientable and lives in the sea? See the March 1984 issue of *Mathematics Magazine* for a picture and the answer.

3.4 Classification of Surfaces

One of the fundamental problems in any discipline is the classification of the objects under study. In the field of chemistry, the periodic table lists all the elements with their atomic numbers and other features that distinguish the various entries. Physicists are actively working on a scheme for classifying the elementary particles. Biologists face the challenge of classifying living organisms subject to mutations and genetic drift under the pressures of natural selection. The tasks become even more difficult in the social sciences and humanities. Nevertheless, linguists classify parts of speech, musicians classify compositions by key and mode, and historians identify various movements in the development of human culture.

The classification theorems of mathematics are among the ultimate triumphs of human intellectual achievement. A classification theorem provides a complete list of all objects in a given category as well as a scheme for matching an unknown object from the category with exactly one of the canonical examples. The classification of the five Platonic solids dates back to the ancient Greek civilization. Finite-dimensional vector spaces can be classified by a single nonnegative number, the dimension. If you take a course in abstract algebra, you may encounter the classification theorem for finitely generated Abelian groups. The classification of finite simple groups is a monumental success story of twentieth-century mathematics. See Daniel Gorenstein's article "The Enormous Theorem" [40] for an overview of this result.

The following definition introduces the conditions on the surfaces we will focus on in this section. We will continue to assume that surfaces are triangulated. Recall from Definition 3.19 that a compact surface is one whose triangulation consists of a finite number of faces, edges, and vertices.

Definition 3.31 *A surface is* ***closed*** *if and only if it is compact and has no boundary points.*

This definition provides the key conditions for the following lemma as the starting step in the classification theorem for surfaces.

Lemma 3.32 *A closed, path-connected surface can be represented as a polyhedral disk with edges identified in pairs.*

Proof. Cut the surface apart along the edges of a triangulation, keeping track of the pairs of edges that need to be identified to reconstitute the original surface. We will paste the triangles back together in a sequence that will ensure we have a disk at each stage.

We can start with any triangle to form the initial disk. Suppose we have a disk constructed from pasting together some of the triangles along matching pairs of edges. If there are triangles left over, go back to the original surface and run a path from one of the left-over triangles to a triangle that has been pasted into the disk. Adjust this path if necessary to push it off any vertices of the triangulation. Find a place where this path leaves the left-over triangles and enters the triangles of the disk. This point will mark the edge of a triangle that can be glued to a matching edge of the disk. Topologically, this gluing is just like attaching two square regions of the same size along an edge of each to form a rectangle that can then be squeezed to the size of the original square. Thus, the result is a disk that incorporates an additional triangle.

When all the triangles have been pasted on, the boundary of the resulting disk will be a sequence of edges of triangles that can be matched up in pairs to reconstitute the original surface. ❄

In Section 3.2 we developed an extensive list of examples of compact surfaces: the 2-sphere, the torus, the projective plane, and various connected sums of these basic surfaces (including the Klein bottle and surfaces of higher genus). From Section 3.2 we have invariants to distinguish among various topological types of surfaces: the Euler characteristic and orientability. The following theorem provides the final ingredient needed for a classification of closed surfaces.

Classification Theorem for Closed Surfaces 3.33 *A closed, path-connected surface is homeomorphic to a 2-sphere, a connected sum of tori, or a connected sum of projective planes. The Euler characteristic and orientability of the surface distinguish among these possibilities.*

Proof. By Example 3.27 and Exercises 4 and 6 of Section 3.3, the connected sum of g tori is orientable and has Euler characteristic $2 - 2g$. Also, the connected sum of p projective planes is nonorientable and has Euler characteristic $2 - p$. Thus, the Euler characteristic and orientability of a closed, path-connected surface suffice to distinguish among these various possibilities.

By Lemma 3.32 we can represent the surface as a polygonal disk with edges identified in pairs. The standard proof of the classification theorem involves a series of steps to rearrange the edges until it is possible to read off the identifications as producing connected sums of tori and projective planes. A proof by Jerome Dancis simplifies this by removing connected sum components of tori and projective planes as soon as they are identified. This proof is an adaptation of that approach.

We proceed by induction on the number of pairs of edges to be identified in the boundary of the polygonal disk. If there is only one pair of edges, the edges in the boundary are identified according to the word aa^{-1} or the word aa. In the first case, the surface is a sphere; in the second case, it is a projective plane. If there are more than one pair of edges, we will consider five possibilities in the following paragraphs.

The simplest case is a pair of adjacent edges to be identified with opposing orientations. As illustrated in Figure 3.34, this pair of edges, corresponding to the sequence aa^{-1}, can be folded in and glued together to yield a polyhedral disk for the same surface, but with one less pair of edges.

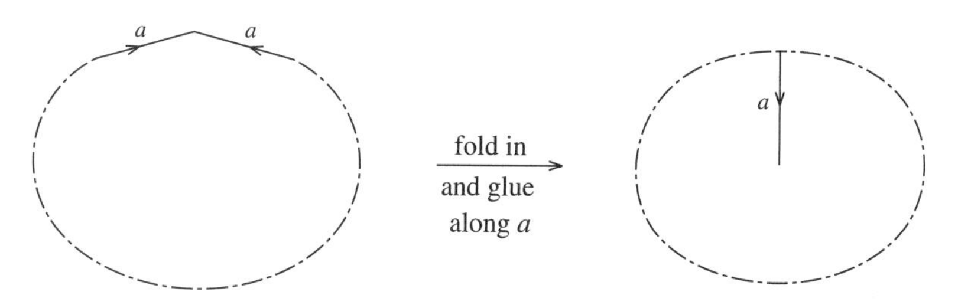

FIGURE 3.34
Identifying adjacent edges with opposing orientations.

An adjacent pair of edges to be identified with the same orientation (such as aa in Figure 3.35) would require some twisting of the disk to perform this glue job. Instead, the following argument allows us to recognize the surface as the connected sum of a projective plane and surface formed from the disk with this pair of edges removed. Notice that since the vertex at the head of a is the same as the vertex at the tail of a, the curve b in Figure 3.35 is actually a circle in the surface. Cut along this curve and glue in disks along the copies of b in each of the two resulting components. One of the pieces will be a disk with one less pair of edges to be identified. The other piece will be a disk with an aa pair to be identified; that is, it will be a projective plane. Notice that the operations we have performed are the reverse of the steps for forming the connected sum of the surface and a projective plane. Thus, we can reconstruct the original surface by forming this connected sum.

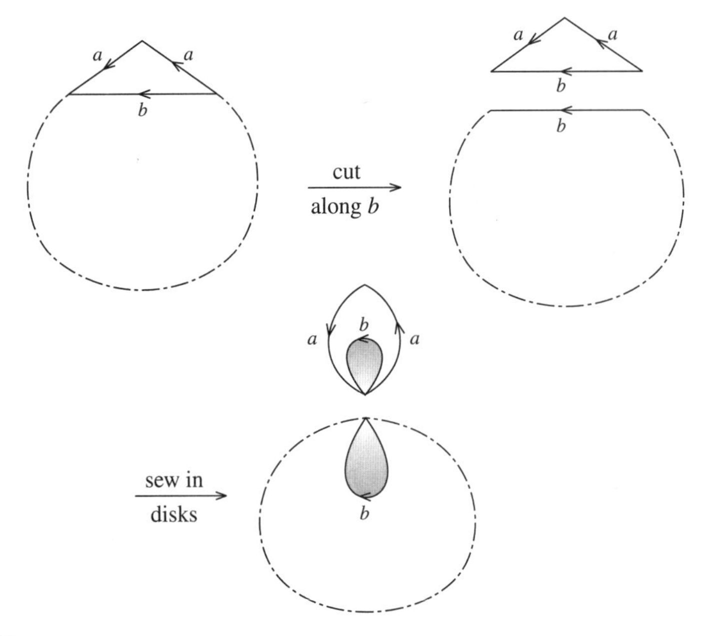

FIGURE 3.35
An adjacent pair of edges oriented in the same direction gives a projective plane as a connected sum component.

Next consider a nonadjacent pair of edges with the same orientation. Figure 3.36 illustrates this case with edges labeled a. Cut the disk along an arc b joining the heads of the two edges. Glue the two pieces together along a. This disk has the same number of pairs of edges, but it has an adjacent pair of edges bb with the same orientation. Thus, we can proceed as in the previous case to cut off a projective plane as a connected sum component leaving a disk with one less pair of edges.

The final two cases arise from nonadjacent pairs oriented in opposing directions. The two edges (labeled a in Figures 3.37 and 3.38) separate the boundary of the disk into two arcs, one joining the heads of the two a edges and one joining their tails.

For the first of these two cases, suppose that none of the edges in either of these arcs is identified with any edge in the other arc. Let c_1 and c_2 denote these two arcs. Glue the disk

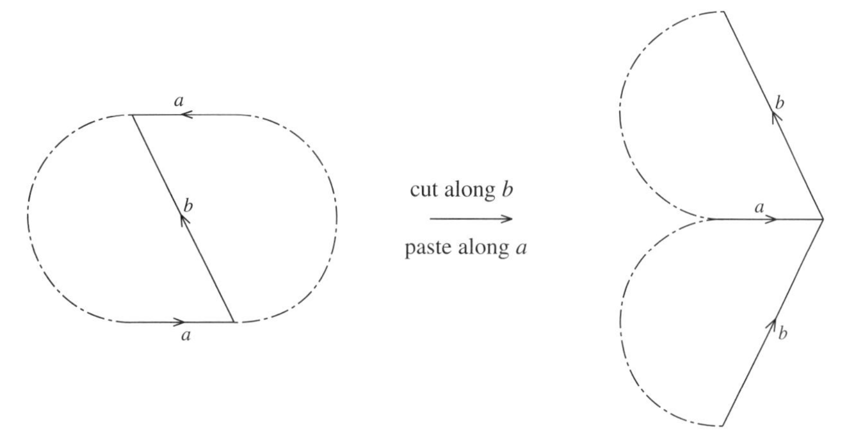

FIGURE 3.36
A nonadjacent pair of edges oriented in the same direction converted into an adjacent pair of edges.

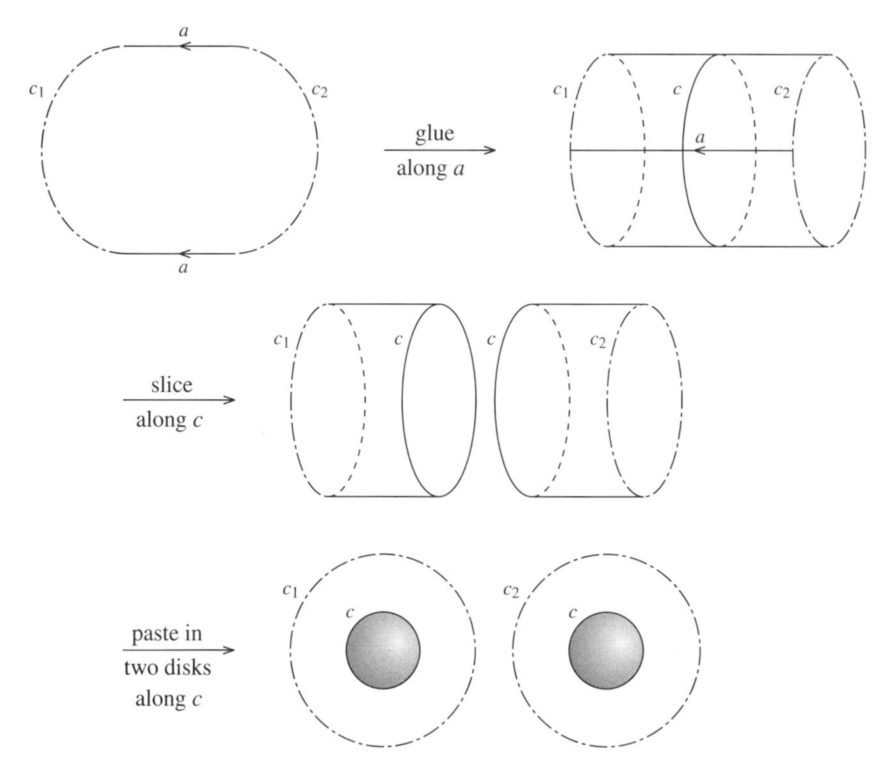

FIGURE 3.37
A nonadjacent pair of edges oriented in opposing directions.

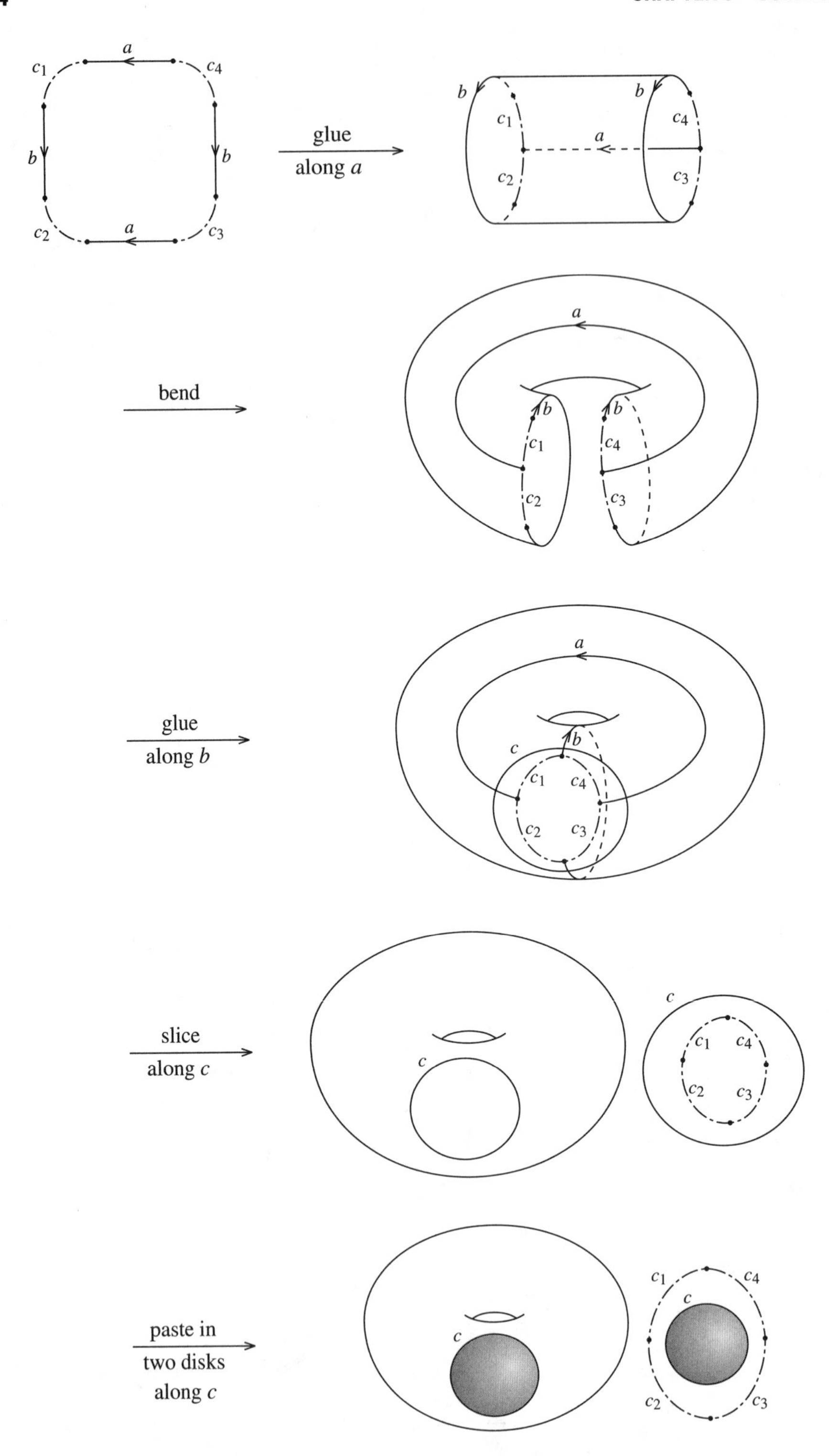

FIGURE 3.38
Two pairs of edges that alternate along the boundary of the disk.

together along the edges labeled a to form a cylinder. Notice how the two ends of c_1 and the two ends of c_2 have been glued together so that they form the boundary circles of the cylinder. Slice the cylinder along a curve c that separates c_1 and c_2. Finally, paste in two disks along the copies of c in the two pieces of the cylinder. The result is two disks whose boundaries c_1 and c_2 consist of edges to be identified in pairs to form two surfaces. The operation of slicing and pasting in the two disks is the inverse of the operation of forming a connected sum. Thus, the two disks represent surfaces whose connected sum is the original surface. Notice that each of the two disks has fewer pairs of edges than the original disk.

Last, but not least, is the case of nonadjacent pairs of edges oriented in opposing directions that are interlaced with another pair of edges as we go around the boundary of the disk. For the sake of simplicity, we can assume that all pairs with the same orientation have been removed by the procedures described in the second and third cases. Thus, the two pairs of edges are as indicated by a and b in Figure 3.38 with four arcs c_1, c_2, c_3, and c_4 connecting them to form the boundary of the disk. Glue the disk together along the edges labeled a and b to form a torus with a hole. Notice how the four arcs c_1, c_2, c_3, and c_4 join together to form the boundary curve of the hole. Slice along a curve c that separates this boundary curve from the rest of the torus. Finally, paste in two disks along the copies of c in the two pieces. The result is a torus and a disk whose boundary consist of edges to be identified in pairs to form a surface. The operation of slicing and pasting in the two disks is the inverse of the operation of forming a connected sum. Thus, the torus and disk represent surfaces whose connected sum is the original surface. Notice that the disk has fewer pairs of edges than the original disk.

Iteratively applying processes described in the previous five cases will decompose the original surface as a connected sum of tori and projective planes. If any projective planes appear, we can use Exercises 7 and 8 of Section 8 to replace $T^2\#P^2$ by $P^2\#P^2\#P^2$. Repeat this as often as necessary to write the surface as a connected sum of projective planes. ❄

The Classification Theorem for Closed Surfaces leads to a similar classification theorem for surfaces with boundary. Basically, the surfaces with boundary are obtained by removing the interiors of disjoint disks from closed surfaces.

Classification Theorem for Surfaces with Boundary 3.39 *A compact, path-connected surface with boundary is homeomorphic to a 2-sphere, a connected sum of tori, or a connected sum of projective planes, from which the interiors of a finite number of disjoint disks have been removed. The Euler characteristic, orientability, and number of boundary curves of the surface distinguish among these possibilities.*

By Example 3.27 and Exercises 4 and 6 of Section 3.3, the connected sum of g tori with the interiors of b disjoint disks removed is orientable and has Euler characteristic $2 - 2g - b$. Also, the connected sum of p projective planes with the interiors of b disjoint disks removed is nonorientable and has Euler characteristic $2 - p - b$. Thus, the Euler characteristic, orientability, and number of boundary components of a compact, path-connected surface suffice to determine the topological type of the surface.

Exercise 14 gives you some suggestions for filling in the technical details for extending the classification theorem to surfaces with boundary.

The following examples illustrate the process of identifying a surface from its Euler characteristic, orientability, and number of boundary components. In calculating the Euler characteristic, we will use the more general form in which the faces can be polygonal disks with any number of edges.

Example 3.40 Suppose that two pairs of edges of a 7-sided polygonal disk are glued together according to the word $acabdb^{-1}e$ (so edges c, d, and e are not glued). Determine the topological type of the resulting surface with boundary.

Solution. First, let's see how gluing the edges matches up the vertices of the polygonal disk. Let P be the vertex where the tail of the c edge joins the head of the first a edge. Follow a curve around this vertex from the tail of c to the head of a. Then continue around the vertex where the head of a joins the tail of b. This vertex will be identified with P. Continue once more around the vertex where the tail of b meets the tail of e. We have another vertex that will be identified with P. Next let Q be the vertex between the head of c and the tail of a. Follow a curve around this vertex going from the head of c to the tail of a, ending at the head of e. Finally, let R be the vertex between the tail of d and the head of b. A curve around R goes from the tail of d to the head of b and ends at the head of d. These vertices are labeled in Figure 3.41.

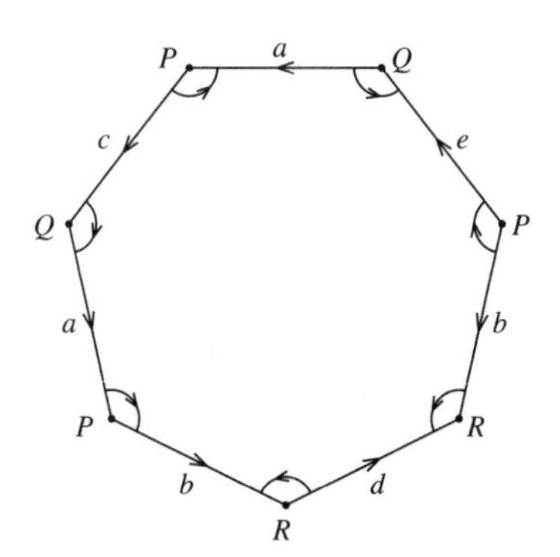

FIGURE 3.41
What surface is formed when this 7-sided disk is glued together?

We now see that the surface has two boundary circles: one formed by the edges c and e glued at P and Q, and one formed by the edge d glued into a loop at R.

Since the glued polygonal disk has 3 vertices, 5 edges, and 1 face, the Euler characteristic of the resulting surface is $3 - 5 + 1 = -1$. A path crossing the edge a reverses orientation, so the surface is nonorientable. Hence, the surface is the connected sum of a positive number p of projective planes with the interiors of two disjoint disks removed. By Exercise 4 of Section 3.3, the Euler characteristic is $2 - p - 2 = -p$. From these two expressions for the Euler characteristic, we conclude that $p = 1$. That is, the surface is a projective plane with two disks removed. ❄

Example 3.42 Consider the surface consisting of a disk with two holes that are connected by a tube as shown in Figure 3.43. Identify this surface.

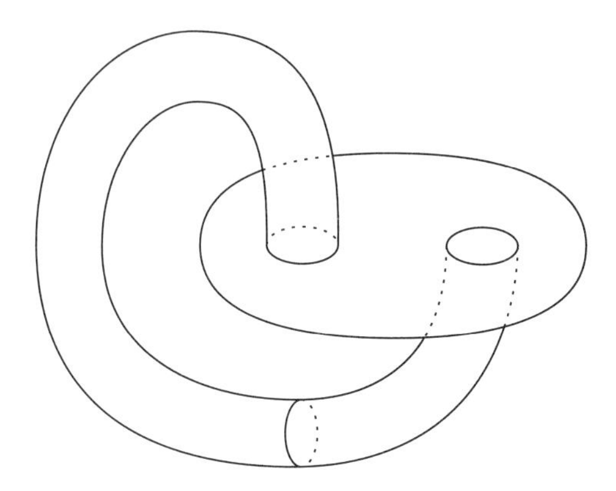

FIGURE 3.43
Identify this surface.

Solution. Notice that the surface has a single boundary circle: the edge of the disk. Since the tube connects the upper side of the disk to the lower side, this is a one-sided surface in $\mathbb{R}^3$. Hence, it is nonorientable. To calculate its Euler characteristic, cut the tube along the circle indicated in Figure 3.43. This yields a disk with two holes, a surface with Euler characteristic -1. Since cutting along a circle does not change the Euler characteristic, the original surface also has Euler characteristic -1. By Exercises 4 of Section 3.3, the number p of projective planes satisfies $2 - p - 1 = -1$. So $p = 2$, and the surface is the connected sum of two projective planes with one disk removed. ❄

Exercises 3.4

1. Consider four triangular disks glued together to form a surface. The sides of the disks are labeled abd, bce, caf, and def to show how the edges are to be glued together. Identify the resulting surface.

2. Consider polygonal disks with edges identified according to the following words. Follow the steps in the proof of the classification theorem to identify the resulting surfaces.
   (a) $abcddc^{-1}a^{-1}b^{-1}$
   (b) $abcabc$
   (c) $abccba$
   (d) $abca^{-1}b^{-1}c^{-1}$
   (e) $abcda^{-1}b^{-1}c^{-1}d^{-1}$
   (f) $abca^{-1}db^{-1}c^{-1}d^{-1}$

3. Consider a polygonal disk with pairs of edges identified to form a closed surface. Determine a way to know instantly whether the surface is orientable or nonorientable from reading off the word that tells how the edges are to be identified.

4. Use the techniques of Example 3.40 to identify the surfaces in Exercise 2 formed by gluing the edges of a polygonal disk according to the following words.

(a) $abcddc^{-1}a^{-1}b^{-1}$

(b) $abcabc$

(c) $abccba$

(d) $abca^{-1}b^{-1}c^{-1}$

(e) $abcda^{-1}b^{-1}c^{-1}d^{-1}$

(f) $abca^{-1}db^{-1}c^{-1}d^{-1}$

5. Consider ten triangular disks glued together to form a surface. The vertices of the disks are labeled ABC, ABE, ACF, ADE, ADF, BCD, BDF, BEF, CDE, and CEF. As vertices are matched up, the corresponding pairs of edges between any two vertices are glued together. Identify the resulting surface.

6. Identify the surface obtained when some pairs of edges of a 15-sided polygonal disk are glued together according to the word $afbc^{-1}id^{-1}ejae^{-1}c^{-1}dgb^{-1}h$.

7. Identify the surface obtained when the pairs of edges of a triangular annulus are glued together as shown in Figure 3.44.

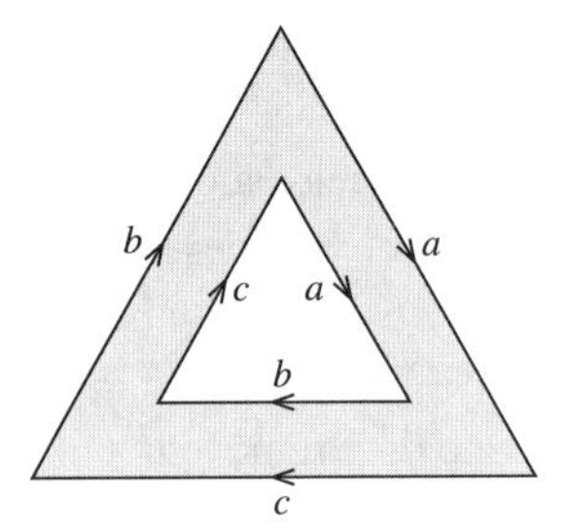

FIGURE 3.44
A surface formed by gluing pairs of edges of a triangular annulus.

8. Consider a polygonal disk with pairs of edges identified to form a closed surface. Show that four consecutive edges identified according to the word $aba^{-1}b$ lead to a Klein bottle as a connected sum component of the surface.

9. Where does the Klein bottle fit into the Classification Theorem for Closed Surfaces? Why is it not necessary in the Classification Theorem to mention the connected sum of Klein bottles?

10. In Lewis Carroll's novel *Sylvie and Bruno Concluded*, Mein Herr teaches Lady Muriel how to construct the "Purse of Fortunatus" by sewing together three handkerchiefs as indicated in Figure 3.45.

(a) What surface is this?

(b) He justifies the name of the purse as follows: "Whatever is inside that purse, is outside it; and whatever is outside it, is inside it. So you have all the wealth of the world in that leetle purse!" Explain.

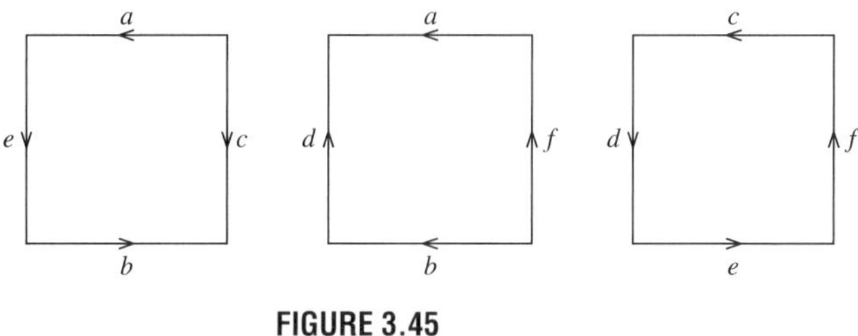

FIGURE 3.45
The Purse of Fortunatus.

(c) Lady Muriel sews together two of the handkerchiefs and examines how she is to sew the third handkerchief along the remaining four edges of the first two. She states, "But it will take time. I'll sew it up after tea." Why was that a wise decision?

11. Start with a sheet of paper and a collection of paper strips. Attach any number of strips to the sheet in the three ways illustrated in Figure 3.46.

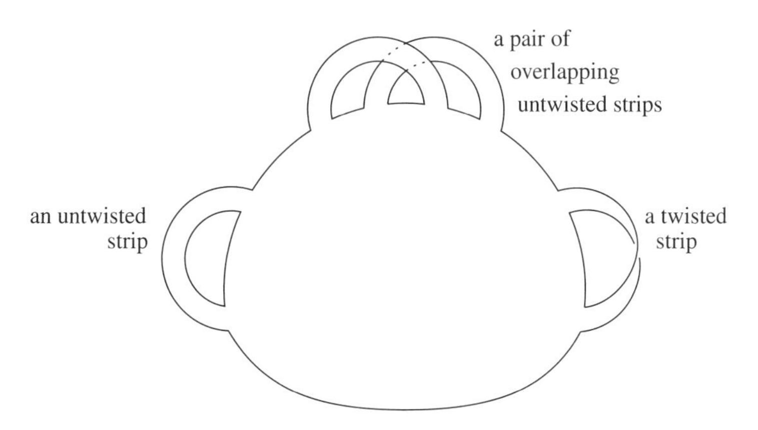

FIGURE 3.46
Three ways to attach strips of paper.

(a) Investigate how attaching strips in each of these ways affects the Euler characteristic, orientability, and number of boundary curves of the surface.

(b) Use the Classification Theorem for Surfaces with Boundary to show that any compact surface with nonempty boundary can be obtained by attaching strips to a sheet of paper in these three ways. State explicitly how to attach strips to get the given surface.

12. Here is another way to handle the case of a nonadjacent pair of edges oriented in the same direction.

(a) Draw an arc from the head of one edge of the pair to the tail of the matching edge and another arc from the tail of the first to the head of the mate. Show that these two arcs form a circle in the surface.

(b) Cut along these arcs to separate the disk into three pieces.

(c) Show that the piece with the pair of edges is a Möbius band. Thus, pasting a disk along the cut edge will yield a projective plane.

(d) Paste the cut edges of the two other pieces along the boundary of a disk. Observe that the result is a disk with fewer pairs of edges than the original disk.

13. Show that the 2-sphere is the only additive identity for the connected sum operation.

14. Fill in the details to prove the Classification Theorem for Surfaces with Boundary: Start with a compact, path-connected surface. Cap off the boundary curves with disks. Determine the Euler characteristic and orientability of the resulting closed surface in terms of the Euler characteristic and orientability of the original surface. Apply the Classification Theorem for Closed Surfaces to identify the capped-off surface. Remove the disks to recover the original surface.

3.5 Surfaces Bounded by Knots

It is quite possible for a surface embedded in $\mathbb{R}^3$ to have a boundary that is a nontrivial knot or link. For example, the diagram on the left in Figure 3.47 shows a surface whose boundary is a trefoil knot. In looking at the picture, you should think of the surface as having a twist where the knot has a crossing. A 2-dimensional bug traveling around in the surface will be flipped over three times before returning to its starting place with its orientation reversed. Hence this surface is non-orientable. Indeed, we have seen in Exercise 10 of Section 2.3 that this band with three twists is homeomorphic to a Möbius band, although it is not isotopic to the standard Möbius band in $\mathbb{R}^3$.

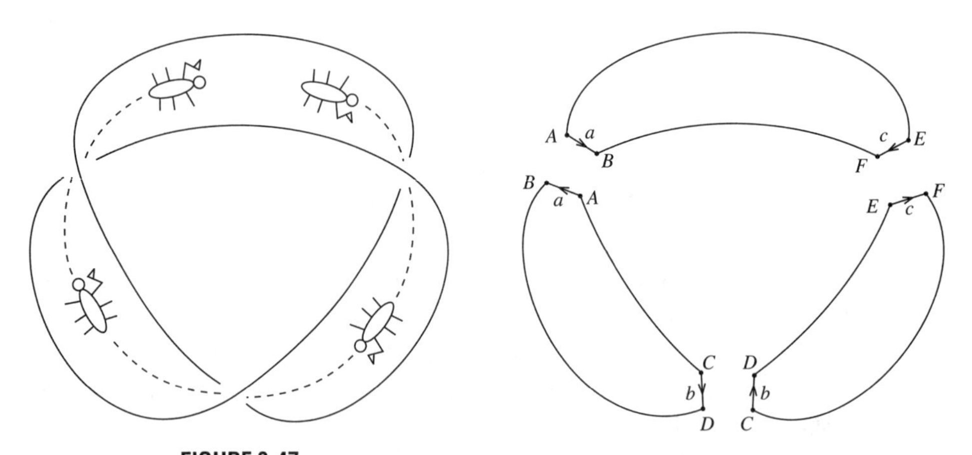

FIGURE 3.47
Cutting a Möbius band bounded by a trefoil knot into three pieces.

We can also use the Classification Theorem for Surfaces with Boundary (Theorem 3.39) to identify this surface. It is nonorientable and has one boundary curve. We can easily calculate its Euler characteristic by cutting it apart at each crossing as illustrated in the diagram on the right of Figure 3.47. With this decomposition into 3 disks, 9 edges, and 6 vertices, we can determine the Euler characteristic to be $6-9+3=0$. By the Classification Theorem for Surfaces with Boundary, this surface is the connected sum of some number p of projective planes with a disk cut out to form the boundary curve. From this description, we know that its Euler characteristic is $2 - p - 1$. The equation $2 - p - 1 = 0$ reveals that $p = 1$. That is, the surface is a projective plane with a disk removed. We know from Exercise 3 in Section 3.1 that this is a Möbius band.

Figure 3.48 shows another surface whose boundary is a trefoil knot. Think of it as a sphere with a trefoil hole. This surface is orientable because any closed path in the surface must pass through an even number of crossings. Hence, a bug traveling a closed path in the surface is flipped an even number of times and returns with its original orientation.

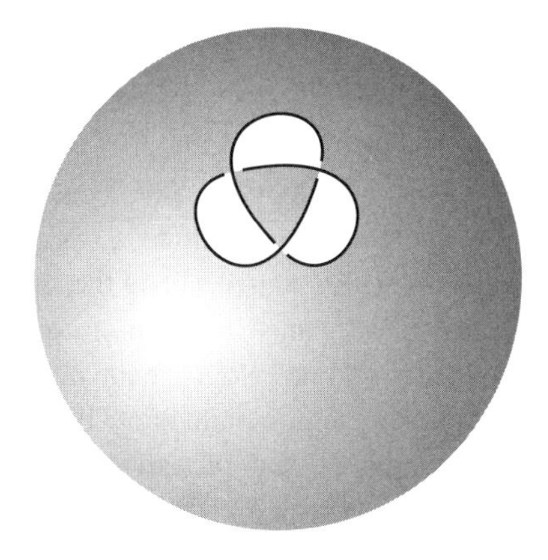

FIGURE 3.48
An orientable surface bounded by a trefoil knot.

To identify this surface, we can again use the technique of cutting it apart at the crossings. This is illustrated in Figure 3.49 where the large back piece has been flattened out.

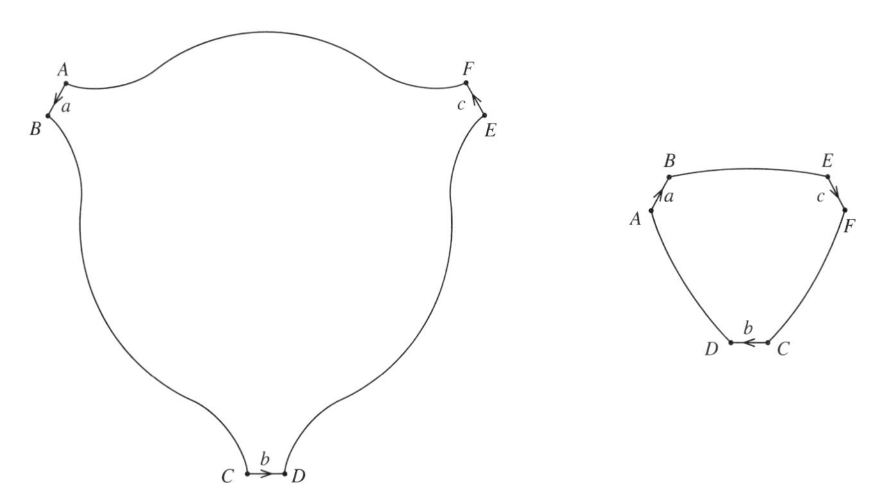

FIGURE 3.49
The surface bounded by a trefoil knot cut and flattened out.

Here we have 2 faces, 9 edges, and 6 vertices. So the Euler characteristic of the surface is $2 - 9 + 6 = -1$. Since the surface is orientable and has one boundary component, $2 - 2g - 1 = -1$ where g is the genus of the surface. This equation yields that $g = 1$. Thus, the surface is a sphere with one handle and one disk removed: a torus with a hole. This is not obvious! We are seeing the power of the Classification Theorem.

In 1934, Herbert Seifert introduced the following idea of using the genus of an orientable surface to define a new knot invariant.

Definition 3.50 *A surface embedded in* $\mathbb{R}^3$ ***spans*** *a knot if and only if the knot is the boundary of the surface. The* **genus** *of a knot* K *is the smallest genus of the orientable surfaces that span* K.

The trivial knot is spanned by a disk (see Exercise 5 of Section 2.1). This disk is a sphere with a disk removed. So the trivial knot has genus 0. In fact, Exercise 5 of Section 2.1 shows that the trivial knot is the only knot with genus 0. Since the trefoil knot is spanned by a torus with a disk removed, it must have genus less than or equal to 1. But we know the trefoil knot is not equivalent to the trivial knot. So its genus must be exactly 1.

What about the next simplest knot, the figure-eight knot? Figure 3.51 shows two obvious surfaces spanning the figure-eight knot. You can picture the surface on the right of Figure 3.51 as bending around in back to form a sphere with a figure-eight hole, similar to Figure 3.48. These surfaces are sometimes called **checkerboard surfaces** because we can get pictures of them by coloring the regions into which the knot diagram divides the plane alternately black and white, as on a checkerboard. Notice that in this case, both of the checkerboard surfaces are nonorientable since in each surface you can find a closed path in the surface that goes through an odd number of crossings (the dotted curves in Figure 3.51).

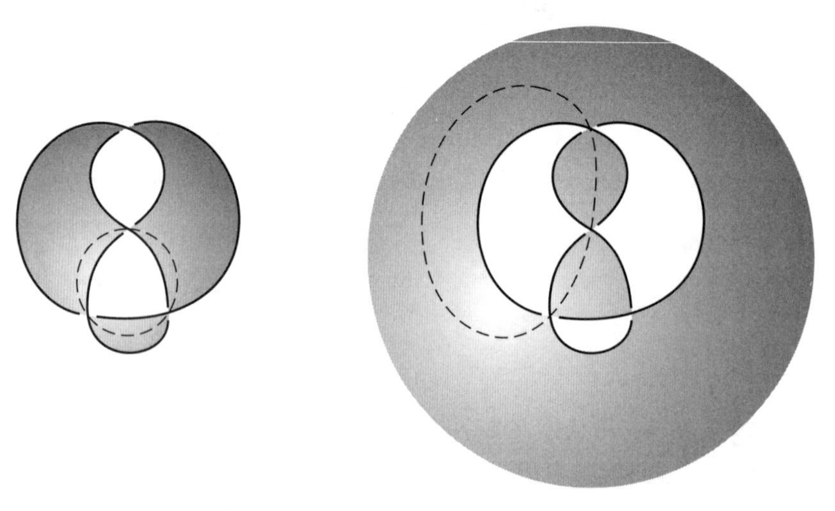

FIGURE 3.51
Two surfaces bounded by a figure-eight knot.

Can we find an orientable surface that spans the figure eight knot? Indeed, for Seifert's definition to make sense for an arbitrary knot K, we must always be able to find some orientable surface that spans K. Seifert found a clever way to do this, starting from any knot diagram for K. The resulting orientable surface is called the **Seifert surface** associated with that knot diagram. The following example illustrates Seifert's construction for the figure-eight knot.

Example 3.52 Construct the Seifert surface for the figure-eight knot based on the diagram in Figure 3.53. Calculate the genus of the surface. What can we conclude about the genus of the figure-eight knot?

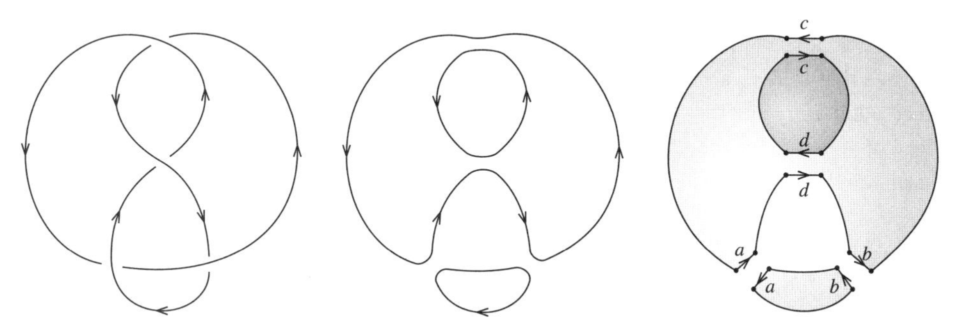

FIGURE 3.53
Constructing a Seifert surface for a figure-eight knot.

Solution. First, orient the knot. This has been done in the diagram for the figure-eight knot in Figure 3.53. Now travel around the knot following the orientation, but at each crossing, jump over to the other strand of the knot, still following the orientation. We obtain a collection of disjoint circles, as in the center diagram of Figure 3.53. These are called **Seifert circles**.

Now span each Seifert circle by a disk, making sure that the disks are disjoint from each other. In our example, the right diagram of Figure 3.53 depicts a small disk in the middle as being in front of the large disk. Finally, at each knot crossing, glue two disks together along arcs in their boundaries, with a twist, to obtain a surface whose boundary is the figure-eight knot. The necessary gluings are shown on the right of Figure 3.53. If you actually try to make Seifert surfaces from paper, as you are asked to do in Exercise 2 at the end of this section, you may find it helpful to use paper strips to attach disks to each other at the knot crossings. You must pay careful attention to whether the necessary twist is clockwise or counter-clockwise.

You should check that this surface is indeed orientable. Exercise 5 asks you to fill in the details of an argument that a surface constructed by Seifert's method is always orientable.

The decomposition of the Seifert surface in Figure 3.53 has 3 faces, 12 edges, and 8 vertices. Thus, its Euler characteristic is $3 - 12 + 8 = -1$. Since the surface is orientable with one boundary component, the genus g of the surface satisfies $2 - 2g - 1 = -1$.

So $g = 1$, and we discover that this Seifert surface is another non-obvious picture of a torus with a hole. By the same reasoning as for the trefoil knot, the figure-eight knot has genus 1. ❄

Seifert's construction is clever, but it does not make calculating the genus of a knot easy. The problem is that if you construct the Seifert surface for one diagram of a knot K and find that it has genus n, all you know is that K has genus less than or equal to n. Perhaps there is a different diagram for K that will yield a Seifert surface with genus less than n. There is even a more disheartening possibility: perhaps K spans some orientable surface of smaller genus that cannot be obtained by Seifert's construction from any knot diagram for K. In 1987, Israeli knot theorist Yoav Moriah actually found a family of knots for which this nightmarish situation occurs.

There is, however, some good news. In 1986, David Gabai used the Jones polynomial to prove that if K is an alternating knot, then the Seifert construction applied to any reduced alternating diagram of K yields an orientable surface of minimal genus spanning K. Thus, as was true for the crossing number, it is now possible to determine the genus of an alternating knot efficiently.

Exercises 3.5

1. For each of the following knot diagrams construct the Seifert surface, use the classification theorem to identify the surface, and compute the genus of the knot. (Notice that all of these knots are drawn with reduced alternating diagrams.)
   (a) The two knots of five crossings in Figure 2.41.
   (b) The three knots of six crossings in Figure 2.42.
2. Choose one of the knot diagrams from Exercise 1 and construct its Seifert surface by taping together paper disks and twisted strips of paper. Be sure you get your twists in the right directions. Then check your construction by cutting off a thin strip along the boundary of your surface and verifying that it really is the knot you started with.
3. Describe an infinite family of distinct knots that all have genus 1. To get started, consider examples in the text and Exercise 1. Be sure you give an argument to prove that your knots have genus 1.
4. Suppose we perform Seifert's construction on a knot diagram with c crossings and s Seifert circles, and then cut the resulting surface apart at the knot crossings as we did in the examples in this section.
   (a) Express the number of vertices, edges, and faces in the resulting decomposition in terms of c and s.
   (b) Express the Euler characteristic and the genus of the Seifert surface in terms of c and s. Check your answer by seeing if it is consistent with your results in Exercise 1.
   (c) From the result in part (b), what can you say about relations which must hold between c and s?

5. One argument that Seifert's construction always results in an orientable surface is based on the observation in Section 3.3 that a surface embedded in $\mathbb{R}^3$ is orientable if and only if it is two-sided. To show a Seifert surface is two-sided, we will color one of its sides black and the other white. To do this, notice that each Seifert circle inherits either a clockwise or a counter-clockwise orientation from the orientation of the original knot. If a Seifert circle is oriented clockwise, color the top of the disk that spans it black and the bottom white; if a Seifert circle is oriented counter-clockwise, color the top of the disk that spans it white and the bottom black. We must now check that when we glue disks together with twists at the knot crossings, the colors match up appropriately. There are two cases.

(a) Consider two Seifert circles whose spanning disks are to be glued together, and suppose these circles lie next to each other. What must be true of the orientations of these circles? Why? When we glue them with a twist, do the colors of the sides match up appropriately?

(b) Consider two Seifert circles whose spanning disks are to be glued together, and suppose one of these circles lies inside the other. What must be true of the orientations of these circles? Why? When we glue them with a twist, do the colors of the sides match up appropriately?

6. Investigate how the Seifert construction works for a link diagram. Does it always give an orientable surface whose boundary is the link? For a knot, it is clear that reversing the orientation of the knot will not affect the Seifert surface, but for a link it is possible to reverse the orientation of only some of the link components. Investigate how this affects the Seifert surface. Use the links in Figure 2.8 as examples in your investigation.

7. One could define the **nonorientable genus** of a knot to be zero for the trivial knot, and for any other knot K to be the smallest number p such that the surface formed by taking the connected sum of p projective plans and removing one disk will span K.

(a) Give a simple argument that every knot must bound some nonorientable surface, so that the nonorientable genus is well-defined.

(b) Investigate the nonorientable genus of some simple knots.

(c) It is a remarkable fact that for any knot diagram, at least one of the two checkerboard surfaces of the diagram must be nonorientable. Check this for some simple cases. Can you give an argument for why this is true in general?

References and Suggested Readings for Chapter 3

30. V. G. Boltyanskiĭ and V. A. Efremovich, *Intuitive Combinatorial Topology*, Springer-Verlag, New York, 2001.

This book covers the topology of curves and surfaces and introduces the basic concepts of homotopy and homology.

31. Henry Brahana, "Systems of circuits on two-dimensional manifolds," *Annals of Mathematics*, 23 (1922), 144–168.

The first combinatorial proof of the classification theorem for surfaces is in this publication of the author's doctoral dissertation completed in 1920 under Oswald Veblen.

32. J. Scott Carter, *How Surfaces Intersect in Space: An Introduction to Topology* (second edition), World Scientific Publishing Company, Singapore, 1995.

Lots of diagrams describe the geometric consequences of trying to embed surfaces in Euclidean space.

33. David Farmer and Theodore Stanford, *Knots and Surfaces: A Guide to Discovering Mathematics*, American Mathematical Society, Providence, RI, 1996.

Many do-it-yourself investigations of graphs, surfaces, and knots provide the basis for exploring these topics and discovering their basic properties.

34. John Fauvel, Raymond Flood, and Robin Wilson (editors), *Möbius and His Band: Mathematics and Astronomy in Nineteenth-Century Germany*, Oxford University Press, Oxford, England, 1993.

The context, life, work, and influence of August Möbius are presented in fascinating detail.

35. Peter Firby and Cyril Gardiner, *Surface Topology* (third edition), Horwood Publishing, Chichester, England, 2001.

An introduction to surfaces with some instructions for paper models of surfaces.

36. George Francis and Jeffrey Weeks, "Conway's ZIP proof," *American Mathematical Monthly*, 106 (1999), 393–399.

John Conway's Zero Irrelevancy Proof of the classification of compact surfaces.

37. Maurice Fréchet and Ky Fan, *Initiation to Combinatorial Topology*, Prindle, Weber, and Schmidt, Boston, 2003.

An introduction to topology leading to the classification of surfaces.

38. Martin Gardner, "Doughnuts: linked and knotted" in *Knotted Doughnuts and Other Mathematical Entertainments*. W. H. Freeman, New York, 1986.

Some recreational observations and puzzles about tori embedded in $\mathbb{R}^3$ *including Conway's proof that two knots cannot cancel each other.*

39. N. D. Gilbert and T. Porter, *Knots and Surfaces*, Oxford University Press, Oxford, England, 1994.

Knot theory interacts with the theory of surfaces, groups, and graphs.

40. Daniel Gorenstein, "The enormous theorem," *Scientific American*, 253 (1985), 104–115.

An overview of the work leading to the classification of finite simple groups.

41. H. B. Griffiths, *Surfaces* (second edition), Cambridge University Press, Cambridge, England, 1981.

An intuitive exploration of surfaces.

42. L. Christine Kinsey, *Topology of Surfaces*, Springer-Verlag, New York, 1993.

A rigorous presentation of the principles of topology are applied to develop the geometric and algebraic topology of surfaces.

4

Three-dimensional Manifolds

Surfaces are modeled after the Euclidean plane: each point has a neighborhood that is a plane subject to a little bending and other topological distortions. However, surfaces exhibit many kinds of global structure quite unlike the plane they were modeled upon. Likewise, the world we live in appears to be a giant three-dimensional Euclidean space. But perhaps that is due to the local view we have of the universe. In this chapter we will open our minds to the variety of global structures possible with three dimensions. These are the three-dimensional manifolds.

4.1 Definitions and Examples

The simplest kind of three-dimensional manifolds are the solid objects in three-dimensional Euclidean space. We usually want to consider the boundary at the interface between an embedded object and its complement as part of the manifold. Just as a surface (a two-dimensional manifold) can have one-dimensional boundary components, a three-dimensional manifold can have boundary components that are surfaces. The following definition is entirely analogous to Definition 3.2 of a surface. As in that definition, the condition that the space can be embedded in a Euclidean space will avoid some pathological examples. Recall that a neighborhood of a point in a topological space is a subset that contains all points within some positive distance of the point.

Definition 4.1 *A* ***three-dimensional manifold*** *(or* ***3-manifold****) is a space that is homeomorphic to a nonempty subset of a finite-dimensional Euclidean space and in which every point has a neighborhood homeomorphic to* $\mathbb{R}^3$*. We sometimes also wish to admit* **boundary** *points, which have neighborhoods homeomorphic to* $\{(x, y, z) \in \mathbb{R}^3 \mid z \geq 0\}$.

Example 4.2 Show that the ball $B^3 = \{(x, y, z) \in \mathbb{R}^3 \mid x^2+y^2+z^2 \leq 1\}$ is a 3-manifold.

Solution. Radial stretching of the radii of the interior of the ball onto the corresponding rays from the origin of $\mathbb{R}^3$ is a homeomorphism of the interior onto $\mathbb{R}^3$. Thus, for any point (x, y, z) with $x^2 + y^2 + z^2 < 1$, the interior of the ball is a neighborhood homeomorphic to $\mathbb{R}^3$.

For a point on the boundary of the ball, we can rotate the ball so the point is at the south pole. Slide vertical chords upward so that the southern hemisphere is flattened onto a unit disk in the xy-plane. The resulting half of an ellipsoidal solid can then be compressed toward the xy-plane by dividing all z-coordinates by a factor of two. The origin is now the image of the original point. It has a neighborhood $\{(x, y, z) \in \mathbb{R}^3 \mid x^2 + y^2 + z^2 < 1, z \geq 0\}$ in the image of the ball that (again by radial stretching) is homeomorphic to $\{(x, y, z) \in \mathbb{R}^3 \mid z \geq 0\}$. Thus, points on the boundary of the ball form the boundary of this space as a 3-manifold. ❄

One way to generate 3-manifolds is to take the Cartesian product of a surface with a circle or a closed interval.

Example 4.3 Describe the 3-manifold $D^2 \times S^1$ formed by taking the Cartesian product of a disk and a circle. Determine the boundary of this 3-manifold.

Solution. This space looks like a doughnut. We can picture it as the region bounded by a torus in $\mathbb{R}^3$, such as the surface illustrated in Figure 1.12. Let the z-axis be the axis of symmetry through the hole in the doughnut. Then each polar angle (corresponding to a point in the S^1 factor) in the xy-plane gives a disk as a slice of the set in that direction.

The boundary of this solid torus is a two-dimensional torus formed by the Cartesian product of the circle that bounds D^2 and the factor S^1. ❄

Example 4.4 Describe the 3-manifold $S^2 \times [0, 1]$ formed by taking the Cartesian product of a 2-sphere and a closed, bounded interval. Determine the boundary of this 3-manifold.

Solution. This space looks like a spherical shell. We can picture it as the region between two concentric 2-spheres in $\mathbb{R}^3$. It has two boundary components. The inner sphere of this shell can correspond to $S^2 \times \{0\}$, and the outer sphere to $S^2 \times \{1\}$. Spheres with intermediate radii correspond to the S^2 slices at intermediate points of the interval $[0, 1]$. ❄

The idea in the previous example works with surfaces other than just the 2-sphere. Let S be any closed orientable surface embedded as a polyhedral subset of $\mathbb{R}^3$. Thicken up the surface to include the region between two roughly parallel copies of the surface. This region will be homeomorphic to the Cartesian product $S \times [0, 1]$. One boundary component will correspond to $S \times \{0\}$; the other to $S \times \{1\}$. Parallel copies of the surface will form the slices of the product at intermediate levels between 0 and 1.

The examples so far have all been 3-manifolds with boundary. As is the case with surfaces, 3-manifolds without boundary are more basic spaces. Since closed 3-manifolds cannot be embedded in $\mathbb{R}^3$, they are more difficult to visualize. However, we can still get a good idea of what some simple examples look like. We begin by considering three ways to view the three-dimensional analog of the 2-sphere.

Recall from Definition 1.32 that the standard 3-sphere is the set $S^3 = \{(w, x, y, z) \in \mathbb{R}^4 \mid w^2 + x^2 + y^2 + z^2 = 1\}$. Any point of S^3 will have at least one nonzero coordinate. If the first coordinate is positive, the projection that maps (w, x, y, z) to $(0, x, y, z)$ will be a homeomorphism from the open hemisphere $\{(w, x, y, z) \in S^3 \mid w > 0\}$ to the interior of the unit ball in the xyz-hyperplane. Radial stretching gives a homeomorphism between the open ball and $\mathbb{R}^3$. A similar argument shows that no matter which coordinate is nonzero, the point has a neighborhood homeomorphic to $\mathbb{R}^3$. Thus, S^3 is a 3-manifold without boundary.

The projection described in the previous paragraph extends to a homeomorphism from each of the two closed hemispheres $\{(w, x, y, z) \in S^3 \mid w \geq 0\}$ and $\{(w, x, y, z) \in S^3 \mid w \leq 0\}$ (including the 2-sphere boundary $\{(w, x, y, z) \in S^3 \mid w = 0\}$ they share) to the unit ball $\{(w, x, y, z) \in \mathbb{R}^4 \mid w = 0 \text{ and } x^2 + y^2 + z^2 \leq 1\}$ in the xyz-hyperplane. Thus, we can envision the 3-sphere as a pair of three-dimensional balls glued together along their spherical boundaries.

A third way to picture the 3-sphere is to imitate the process of forming a 2-sphere by adjoining a point at infinity to a plane. Stereographic projection (see Exercise 5) gives a homeomorphism from $S^3 - \{(0, 0, 0, 1)\}$ onto $\mathbb{R}^3$ (thought of as the wxy-hyperplane of $\mathbb{R}^4$). Thus, we can extend this function to map $(0, 0, 0, 1)$ to a point we adjoin to $\mathbb{R}^3$. Think of this point at infinity as joining together the two ends of every straight line in $\mathbb{R}^3$ to form a circle. Thus, the points far from the origin of $\mathbb{R}^3$ will be close to the point at infinity just as the points far from $(0, 0, 0, -1)$ (which maps to the origin) are close to $(0, 0, 0, 1)$ in S^3.

Gluing two 3-balls together along their boundaries to form a 3-sphere suggests an easy way to eliminate boundary components of a 3-manifold. We can even glue together two homeomorphic boundary components of a single manifold. We saw in Example 4.4 that $S^2 \times [0, 1]$ has two boundary components. We can glue together these two surfaces so each point on the 2-sphere at the $\{0\}$ level matches the corresponding point on the 2-sphere at the $\{1\}$ level. For every point of S^2 this joins the two ends of the $[0, 1]$ factor into a circle. Thus, we recognize that we have formed the Cartesian product $S^2 \times S^1$.

We can perform the construction described in the previous paragraph with any closed surface S. The Cartesian product $S \times [0, 1]$ has $S \times \{0\}$ and $S \times \{1\}$ as boundary components. Gluing together these two surfaces by matching corresponding points will change the $[0, 1]$ factor into a circle. The resulting 3-manifold will thus be the Cartesian product $S \times S^1$. This is especially interesting when the surface we start with is a torus T^2. Since T^2 is the Cartesian product of two circles (see Example 1.13), the resulting 3-manifold is the triple Cartesian product $S^1 \times S^1 \times S^1$ of circles. This space is known as the 3-torus, and is denoted T^3.

Exercises 4.1

1. **(a)** Draw pictures to convince yourself that a closed interval cannot be a boundary component of a 2-manifold.

(b) Draw pictures to convince yourself that a surface with nonempty boundary cannot be a boundary component of a 3-manifold.

2. **(a)** Extend Definitions 3.2 and 4.1 to give a definition of a four-dimensional manifold.

(b) Suppose M is a 3-manifold without boundary. Show that the Cartesian product $M \times [0, 1]$ is a 4-manifold. Determine the boundary of $M \times [0, 1]$.

(c) Suppose M is a 3-manifold with boundary. Show that the Cartesian product $M \times S^1$ of M and a circle is a 4-manifold. Determine the boundary of $M \times S^1$.

(d) Suppose S is a surface without boundary and T is a surface with boundary. Show that the Cartesian product $S \times T$ is a 4-manifold. Determine the boundary of $S \times T$.

(e) What happens when you form the Cartesian product of two surfaces with boundary?

3. **(a)** Generalize Definitions 3.2 and 4.1 to define an n-manifold for any nonnegative integer n.

(b) Suppose M is an m-manifold without boundary and N is an n-manifold with boundary. Show that the Cartesian product $M \times N$ is a manifold. What is the dimension of $M \times N$? Determine the boundary of $M \times N$.

(c) What happens when you form the Cartesian product of two manifolds with boundary?

4. When we look out into the clear night sky, we receive light that has traveled through space for a certain amount of time. Assume the amount of time the light has traveled is proportional to the distance from the source of the light. Then we are at the center of nested spherical shells, and the light we see from each shell was emitted at the same time. Light we see from shortly after the big bang is from some of the most distant spheres. But the objects we see at these distances appear relatively close together (since they were close together when they emitted the light). If you extrapolate this model back to the big bang, what 3-manifold do we see as we look out into space?

5. Example 1.35 of Section 1.4 introduced stereographic projection from $S^2 - \{(0, 0, 1)\}$ to the xy-plane in $\mathbb{R}^3$. We want to extend this idea to S^3 in $\mathbb{R}^4$. Begin by considering the plane in $\mathbb{R}^4$ passing through $(0, 0, 0, 0)$, $(0, 0, 0, 1)$, and a point $(w, x, y, z) \in S^3$ with $z < 1$.

(a) Use similar triangles in this plane to show that the line through $(0, 0, 0, 1)$ and (w, x, y, z) will intersect the wxy-hyperplane at a distance

$$\frac{\sqrt{w^2 + x^2 + y^2}}{1 - z}$$

from the origin.

(b) Conclude that stereographic projection from the point $(0, 0, 0, 1)$ of S^3 onto the wxy-hyperplane maps (w, x, y, z) to

$$\left(\frac{w}{1 - z}, \frac{x}{1 - z}, \frac{y}{1 - z}, 0\right).$$

(c) Find a formula for the inverse of this function. Conclude that stereographic projection is a homeomorphism from $S^3 - \{(0, 0, 0, 1)\}$ to the wxy-hyperplane in $\mathbb{R}^4$.

(d) Generalize the idea of stereographic projection to S^n in $\mathbb{R}^{n+1}$ for any $n \geq 1$.

6. Consider the cubic region $\{(x, y, z) \in \mathbb{R}^3 \mid -1 \leq x \leq 1,\ -1 \leq y \leq 1,\ -1 \leq z \leq 1\}$. Glue the top to the bottom by matching points with the same xy-coordinates. Glue the left face to the right face by matching points with the same xz-coordinates. Glue the front to the back by matching points with the same yz-coordinates. Show that the resulting space is a 3-manifold.

7. Glue together two copies of a solid torus $D^2 \times S^1$ by matching corresponding points on the boundaries of the two solids. Describe the resulting 3-manifold as a Cartesian product.

8. Here are four descriptions of 3-manifolds. Show that these spaces are homeomorphic. This 3-manifold is known as projective three-space and is denoted P^3.

(a) Consider the configuration space of the lines through the origin of $\mathbb{R}^4$.

(b) In S^3 identify pairs of antipodal points (w, x, y, z) and $(-w, -x, -y, -z)$ as a single point.

(c) Glue the boundary of the 3-ball $B^3 = \{(x, y, z) \in \mathbb{R}^3 \mid x^2 + y^2 + z^2 \leq 1\}$ to itself by matching a point (x, y, z) with its antipodal point $(-x, -y, -z)$.

(d) Consider the configuration space of rotations of the 2-sphere $S^2 = \{(x, y, z) \in \mathbb{R}^3 \mid x^2 + y^2 + z^2 = 1\}$ in $\mathbb{R}^3$. Suggestion: Any rotation of S^2 can be described by its axis and the angle (between $-\pi$ and π) of the rotation. Identify a rotation with a point on the axis inside B^3 with the distance from the origin corresponding to the angle of the rotation.

9. Show that the three descriptions of the 3-sphere given in this section can be generalized to give three descriptions of the n-sphere for any positive integer n.

10. **(a)** The Alexander trick: Consider the $(n - 1)$-sphere S^{n-1} as the boundary of the n-cell B^n. Show that any homeomorphism $h : S^{n-1} \to S^{n-1}$ extends to a homeomorphism $H : B^n \to B^n$. Suggestion: Each point in the interior of B^n lies on an $(n-1)$ sphere concentric with the boundary sphere. Use h to guide you in defining the image of such a point.

(b) Show that gluing together two n-cells by any homeomorphism of their boundaries will yield a topological n-sphere.

4.2 Euler Characteristic

The Euler characteristic is a useful invariant for distinguishing among the various topological types of surfaces. The surprising result of Theorem 4.11 makes the Euler characteristic of little use as an invariant for 3-manifolds. But the theorem itself is nevertheless interesting and useful in other ways. We will define the Euler characteristic of a 3-manifold in terms of a polyhedral structure on the space. The following definition extends Definition

3.18 to 3-manifolds. It also relaxes the conditions on the geometry of the cells involved, thereby simplifying the computation of the Euler characteristic.

Definition 4.5 *A* ***cellular decomposition*** *of a 3-manifold is a subdivision of the space into a union of 3-cells. The boundary of each of these 3-cells consists of polygonal* ***faces****. The perimeter of each face consists of* ***edges*** *whose endpoints are the* ***vertices****. The intersection of two distinct cells (3-cells, faces, edges, or vertices) is required to consist of faces, edges, and vertices of each. A* ***triangulation*** *is a cellular decomposition for which each 3-cell has four triangular faces arranged as the sides of a tetrahedron, with six edges and four vertices.*

Notice that in a cellular decomposition of a 3-manifold, a face on the boundary of the 3-manifold will belong to exactly one of the 3-cells, whereas an interior face will be a face common to exactly two of the 3-cells.

Also notice how 3-cells cluster around any vertex of a cellular decomposition of a 3-manifold. Near any vertex we can span little polygonal 2-cells across the 3-cells containing the vertex. These polygonal 2-cells will intersect the edges and faces of the 3-cell that emanate from the vertex. As illustrated in Figure 4.6, they can be chosen to fit together to form a surface. We can find the surface on a smaller scale closer to the vertex if necessary to get this to take place in a Euclidean neighborhood of the vertex. In any case, for this neighborhood to function like Euclidean space, this surface must be a 2-sphere (or a 2-cell if the vertex is on the boundary of the 3-manifold).

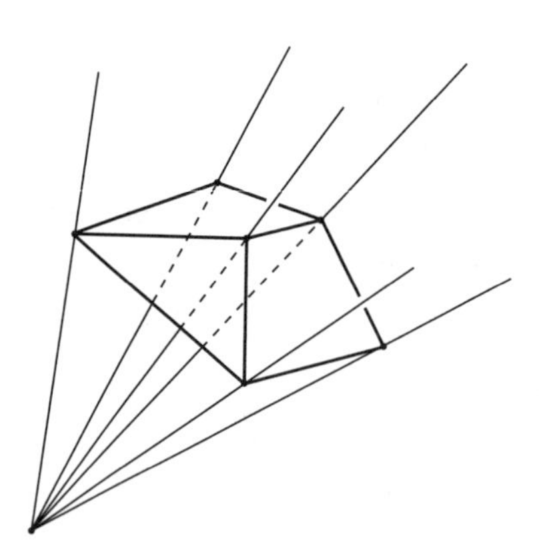

FIGURE 4.6
Little polyhedral 2-cells form a surface around the vertex of a cellular decomposition of a 3-manifold.

In 1952 Edwin Moise proved that all 3-manifolds can be triangulated. Based on this result, we can proceed without loss of generality in assuming that 3-manifolds come with triangulations. We will, however, restrict our consideration to 3-manifolds that are triangulated with a finite number of 3-cells.

Definition 4.7 *A* 3*-manifold that is triangulated with a finite number of* 3*-cells is* ***compact****. A compact* 3*-manifold with no boundary points is* ***closed****.*

In Section 3.3 we defined the Euler characteristic of a surface and gave geometric arguments that this was a topological invariant for surfaces. Although the definition of Euler characteristic extends easily to triangulated spaces of higher dimensions, the fact that it is a topological invariant becomes more difficult to prove. This issue is best handled in general with the tools of algebraic topology. For now however, we can rely on Theorem 4.11 to alleviate any worry about different cellular decompositions of a 3-manifold giving different Euler characteristics. Exercises 3 and 4 relate the Euler characteristic of a compact 3-manifold to the Euler characteristic of its boundary, thus extending the topological invariance of the Euler characteristic to compact 3-manifolds.

Definition 4.8 *The* ***Euler characteristic*** *of a cellular decomposition of a* 3*-manifold is the number of vertices, minus the number of edges, plus the number of faces, minus the number of* 3*-cells in the decomposition. The Euler characteristic of a* 3*-manifold* M *is denoted* $\chi(M)$.

Example 4.9 Show that the Euler characteristic of S^3 is zero.

Solution. As we saw in Section 4.1, we can obtain S^3 by gluing two 3-cells along their boundaries. Triangulate each of the 3-cells as a tetrahedral solid with four vertices, six edges, four faces, and one 3-cell. Since the boundaries of the 3-cells are glued together, we have a cellular decomposition of S^3 with four vertices, six edges, four faces, and two 3-cells. Thus, the Euler characteristic is $4 - 6 + 4 - 2 = 0$. ❄

Example 4.10 Show that the Euler characteristic of $S^2 \times S^1$ is zero.

Solution. Triangulate S^2 as a tetrahedron, with four faces, six edges, and four vertices. Triangulate S^1 as a triangle, with three edges and three vertices. Triangulate $S^2 \times S^1$ by taking the Cartesian product of each component of the triangulation of S^2 with each component of the triangulation of S^1. This will result in 2-cells that are rectangles and 3-cells that are triangular prisms. The three vertices of S^1 give $3 \cdot 4$ vertices, $3 \cdot 6$ edges, and $3 \cdot 4$ faces of $S^2 \times S^1$. The three edges of S^1 give $3 \cdot 4$ edges, $3 \cdot 6$ faces, and $3 \cdot 4$ prisms. All together we have 12 vertices, 30 edges, 30 faces, and 12 prisms in a cellular decomposition of $S^2 \times S^1$. Thus, the Euler characteristic of $S^2 \times S^1$ is $12 - 30 + 30 - 12 = 0$. ❄

These two examples are a little disappointing if we harbor hope of using the Euler characteristic to distinguish among closed 3-manifolds. The bad news is that these are not just isolated examples. The good news is that we will never have trouble computing the Euler characteristic of a closed 3-manifold.

Theorem 4.11 *The Euler characteristic of a closed 3-manifold is zero.*

Proof. For a cellular decomposition of a closed 3-manifold, let v denote the number of vertices, e the number of edges, f the number of faces, and t the number of 3-cells. Now cut the 3-manifold apart along the faces of the decomposition. There will be t 3-cells and $2f$ faces (since each face in the manifold is a common face of two 3-cells). Let e' be the number of edges and let v' be the number of vertices in the cut-apart 3-cells.

In the original manifold, find little 2-spheres around each of the v vertices. They will be formed from polygonal disks that span across the 3-cells near each of the vertices. These 2-spheres should be small enough that no two of them intersect. We want to compute the Euler characteristic of the union of these little spheres. Since each of the e edges of the cellular decomposition contains two vertices of these spheres (one near each of its endpoints), the spheres have $2e$ vertices. Now trace around the edges of each of the polygonal disks that make up these little 2-spheres. You will encounter one edge of the cut-apart cellular decomposition (passing through a vertex of the polygonal disk) for each edge of the polygonal disk. Since each edge of a polygonal disk is shared with a neighboring polygonal disk and since each edge of the cut-apart cellular decomposition will have polygonal disks near both ends, it follows that the spheres have e' edges. Also, near each of the v' vertices of the cut-apart cells, one of the polygonal disks will cut off a little corner of the 3-cell that contains the vertex. So the spheres have v' faces. Thus, $2v$, the Euler characteristic of v of these little 2-spheres, is equal to $2e - e' + v'$.

The boundaries of the 3-cells of the decomposition are also 2-spheres. Thus, $2t$, the Euler characteristic of t of these 2-spheres, is equal to $v' - e' + 2f$.

We can subtract to obtain

$$\begin{aligned} 2v - 2t &= (2e - e' + v') - (v' - e' + 2f) \\ &= 2e - 2f. \end{aligned}$$

Thus, the Euler characteristic of the 3-manifold is $v - e + f - t = 0$. ❄

Exercises 4.2

1. Use the principle of general position to show that any 3-manifold can be embedded as a subset of $\mathbb{R}^7$.

2. Show that any polyhedral 3-cell can be triangulated. Suggestion: First triangulate the faces of the 3-cells, and then introduce a new vertex in the interior of the 3-cell.

3. Show that the Euler characteristic of the boundary of a compact 3-manifold is equal to two times the Euler characteristic of the 3-manifold. Suggestion: Glue two copies of the 3-manifold along the common boundary to form a closed 3-manifold.

4. Show that the Euler characteristic of a compact 3-manifold is a topological invariant. Feel free to use the facts that the boundary of a manifold is a topologically invariant set and that the Euler characteristic is a topological invariant of a surface.

5. Show that a projective plane cannot be the boundary of a compact 3-manifold.

6. Construct a 3-manifold whose boundary is a Klein bottle.

7. Consider a triangulation of a projective plane. Adjoin a new vertex disjoint from the projective plane; adjoin new edges from the new vertex to the vertices of the projective plane; adjoin new faces spanned by the new vertex and the edges of the projective plane; and adjoin new 3-cells spanned by the new vertex and the faces of the projective plane. Compute the Euler characteristic of the resulting 3-dimensional space. Why is this space not a 3-manifold?

8. **(a)** Extend the definition of Euler characteristic to cellular decompositions of spaces of arbitrary dimensions. Cheerfully assume that the Euler characteristic is a topological invariant.

(b) Determine the Euler characteristic of the 4-sphere (see Definition 1.32).

(c) Determine the Euler characteristic of the n-sphere.

(d) Use a simple triangulation of S^2 to obtain a cellular decomposition of the 4-manifold $S^2 \times S^2$. Determine the Euler characteristic of $S^2 \times S^2$.

(e) Explore the Euler characteristics of other Cartesian products.

(f) Determine a relation between the Euler characteristics of any two spaces and the Euler characteristic of their Cartesian product.

9. Use the difference in the separating ability of embedded 2-spheres to distinguish between S^3 and $S^2 \times S^1$.

4.3 Gluing Polyhedral Solids

In Chapter 3 we saw that all closed, path-connected surfaces can be obtained by starting with a polygonal disk with an even number of edges and gluing its edges together in pairs. It is also true that any closed, path-connected 3-manifold can be obtained by starting with a 3-dimensional polyhedral ball with an even number of faces and gluing its faces together in pairs. Indeed, this follows from Moise's triangulation theorem, which allows us to cut a 3-manifold apart into tetrahedral solids. We can reassemble these tetrahedra, adding one at a time by gluing just one of its faces to the corresponding face in the collection already assembled. When all of the tetrahedra have been joined, the result is a polyhedral 3-ball with many pairs of faces yet unglued. The original 3-manifold is obtained by gluing those remaining faces in pairs.

On the other hand, suppose we start with any polyhedral solid that has an even number of triangular faces, an even number of quadrilateral faces, and so on, and glue its faces together in pairs, being sure to glue triangles to triangles, quadrilaterals to quadrilaterals, and so on. We will get a closed space that looks like it might be a closed 3-manifold. Indeed, such a space is called a 3-dimensional **pseudo-manifold**. Sometimes it really is a 3-manifold.

Example 4.12 Suppose we glue the faces of the solid cube $\{(x, y, z) \mid x, y, z \in [-1, 1]\}$ as follows:

$$(1, y, z) \longleftrightarrow (-1, y, z)$$

$$(x, 1, z) \longleftrightarrow (-x, -1, z)$$

$$(x, y, 1) \longleftrightarrow (x, y, -1).$$

Show that the result is a 3-manifold.

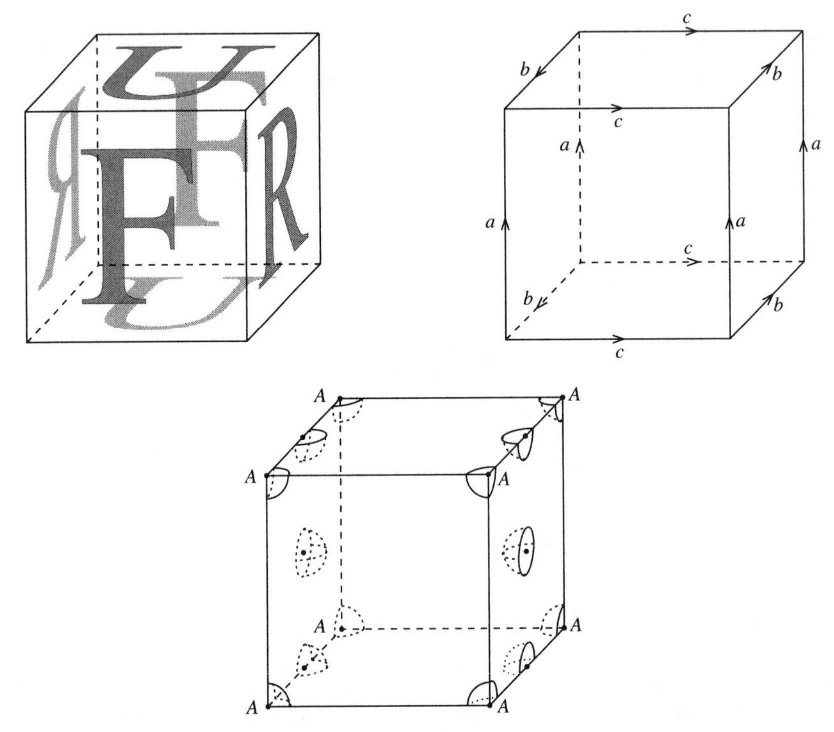

FIGURE 4.13
Gluing the faces, edges, and vertices of a cubical solid to form a 3-manifold.

Solution. Notice that the face gluings illustrated in the diagram in the upper left of Figure 4.13 force edge gluings as shown in the diagram in the upper right. This in turn forces all eight vertices of the cube to be glued to a single vertex labeled A in the diagram at the bottom. We can check that every point in the resulting space has a neighborhood homeomorphic to a 3-dimensional ball with the interior of the ball providing a neighborhood homeomorphic to $\mathbb{R}^3$. For a point in the interior of the cube, a small ball around the point that misses the faces of the cube is such a neighborhood. A point in the interior of a face of the cube has a neighborhood made up of two half-balls glued together into a whole ball. A point on the edge of a cube has a neighborhood made up of a number of pieces (in this example, four pieces) that look like slices of an orange, glued together around a common core—the edge—to form a ball. The single vertex A has a neighborhood that is made up of eight pieces glued together into a ball. This is a little harder to see. One way to see it is to consider the eight little triangles that bound the neighborhood of A, label their edges to show how they are glued, and verify that the triangles are glued into a 2-sphere, the boundary of a 3-ball. You are asked to check this in Exercise 1.

You may have recognized this 3-manifold. Each horizontal cross-section is a square disk glued into a Klein bottle. This gives a stack of Klein bottles. When we glue the top to the bottom, we have a circle of Klein bottles. Thus, the 3-manifold is $K^2 \times S^1$. Notice that with 1 vertex, 3 edges, 3 faces, and 1 solid cube, the Euler characteristic of this 3-manifold is $1 - 3 + 3 - 1 = 0$, as we proved must be true for any 3-manifold. ❄

However, a pseudo-manifold may fail to be a manifold. To see what can go wrong, let's look at a slightly more complicated gluing.

Example 4.14 Suppose we glue the faces of the solid cube $\{(x, y, z) \mid x, y, z \in [-1, 1]\}$ as follows:

$$(1, y, z) \longleftrightarrow (-1, -y, z)$$
$$(x, 1, z) \longleftrightarrow (x, -1, -z)$$
$$(x, y, 1) \longleftrightarrow (-x, y, -1).$$

In other words, instead of gluing faces straight across, we first flip each face as illustrated in the diagram on the left of Figure 4.15. Show that the result is not a 3-manifold.

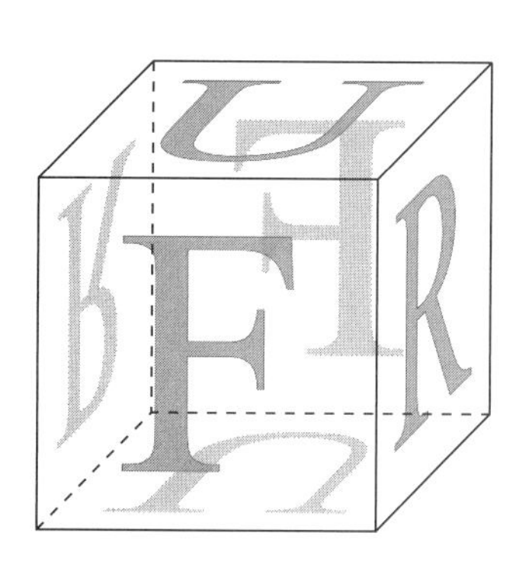

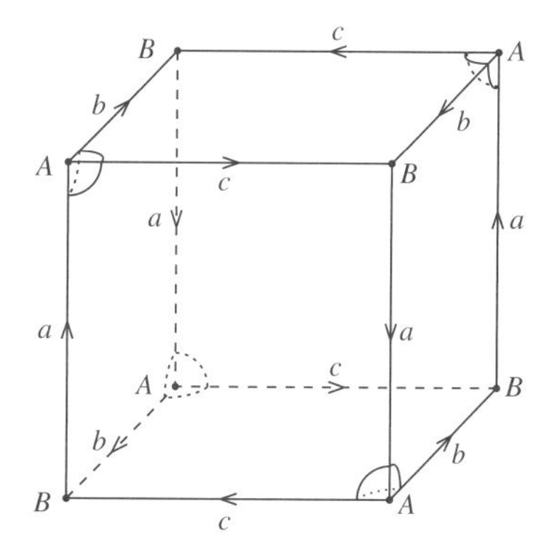

FIGURE 4.15
A pseudo-manifold that is not a manifold.

Solution. The face gluings illustrated in the diagram on the left of Figure 4.15 force edge gluings as shown in the diagram on the right. This in turn forces the vertices of the cube to be glued into two vertices, labeled A and B in the diagram.

One way to see that this is not a 3-manifold is to compute its Euler characteristic: $2 - 3 + 3 - 1 = 1 \neq 0$. What went wrong? The arguments that interior points, face points, and edge points of the cube have neighborhoods homeomorphic to 3-balls (after the gluings are done) still work. In fact, these arguments work for any gluing of the faces of a polyhedral solid together in pairs. The problem is with the vertices. Consider point A. It has a neighborhood that is bounded by four triangles with their edges glued together in pairs. Exercise 2 asks you to check that the resulting surface is not a sphere, but rather a projective plane. Point A (and also point B) has a neighborhood whose boundary is a projective plane, and such a neighborhood is most certainly not a 3-ball! This pseudo-manifold fails to be a 3-manifold at two points. ❄

Fortunately, there is an easy way to tell if a 3-dimensional pseudo-manifold is a 3-manifold. As you might guess from the above examples, it depends on the Euler characteristic.

Theorem 4.16 *A 3-dimensional pseudo-manifold is a 3-manifold if and only if its Euler characteristic is zero.*

Proof. Theorem 4.11 tells us that if the pseudo-manifold is a 3-manifold, the Euler characteristic is zero. Conversely, consider a 3-dimensional pseudo-manifold M whose Euler characteristic is zero. We must show that the vertices have neighborhoods bounded by 2-spheres, and hence the neighborhoods are 3-balls. Using the notation and ideas from the proof of Theorem 4.11, we have that the total Euler characteristic of all of the surfaces bounding neighborhoods of the v vertex points is $2e - e' + v'$. Since the Euler characteristic of a surface is always less than or equal to 2, and is equal to 2 only if the surface is a sphere, we have

$$2e - e' + v' \leq 2v, \text{ and } M \text{ is a 3-manifold if equality holds.}$$

Now, as in the proof of Theorem 4.11, subtract the equality $v' - e' + 2f = 2t$ to get

$$2e - 2f \leq 2v - 2t, \text{ and } M \text{ is a 3-manifold if equality holds.}$$

Bring all the terms to the right side and divide by 2. Thus $\chi(M) \geq 0$, and M is a 3-manifold if $\chi(M) = 0$. ❄

Example 4.17 Glue the faces of a cube as shown in Figure 4.18. Is this a 3-manifold? What about the gluing illustrated in Figure 4.19?

Solution. To calculate the Euler characteristic, we need to see what edge and vertex gluings are forced by the face gluings. These are shown for the two pseudo-manifolds in the figures. In Figure 4.18, the Euler characteristic is $1 - 3 + 3 - 1 = 0$, so this is a 3-manifold. It is called the half-twist cube. In Figure 4.19, the Euler characteristic is $1-1+3-1 = 2 > 0$, so it is not a 3-manifold. If you feel adventurous, you might like to try your hand at Exercise 4 and check that the boundary of a neighborhood of the vertex A is the connected sum of four projective planes! ❄

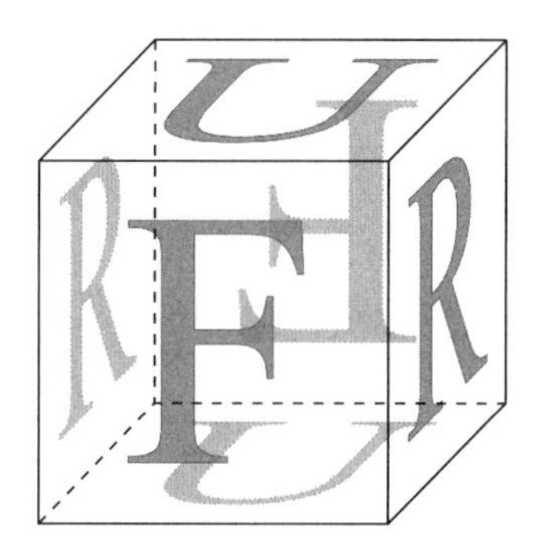
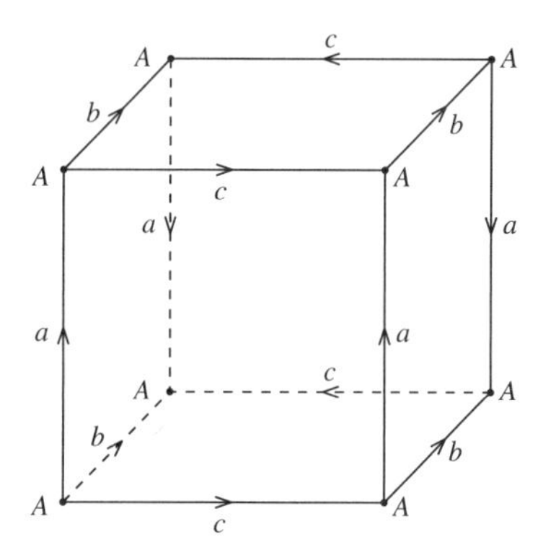

FIGURE 4.18
Is the half-twist cube a manifold?

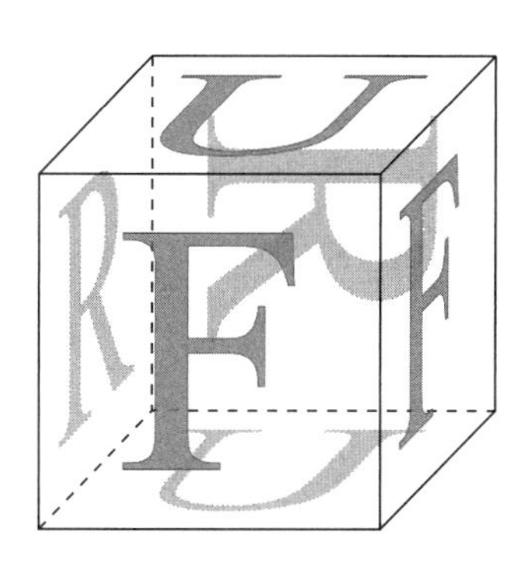
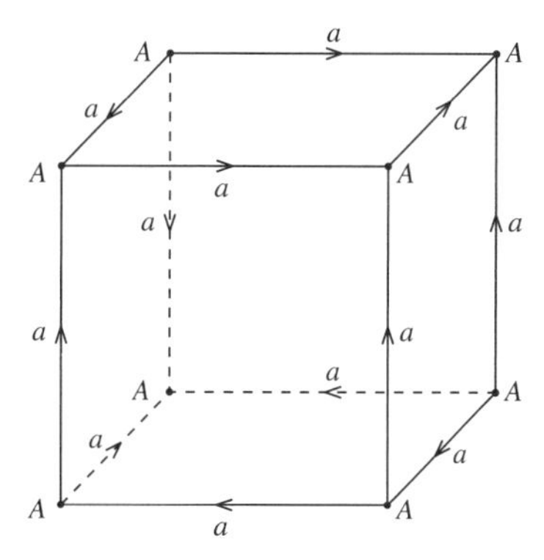

FIGURE 4.19
Is this a manifold?

In some 3-manifolds it is possible for a bug (3-dimensional this time) to fly around a closed path and find that when it returns to its original position, its orientation has been reversed.

Definition 4.20 *A 3-manifold is* ***nonorientable*** *if it contains an orientation-reversing path. It is* ***orientable*** *if it contains no such path.*

As is true for surfaces, orientability can also be defined in terms of consistent orientations of tetrahedra in a triangulation of the manifold.

Example 4.21 Show that $K^2 \times S^1$ is nonorientable.

Solution. A bug flying out the front of the cube and returning through the back will not reverse orientation. Neither will a bug flying out the top of the cube and returning through the bottom. However, consider a bug flying off the right side and returning through the left

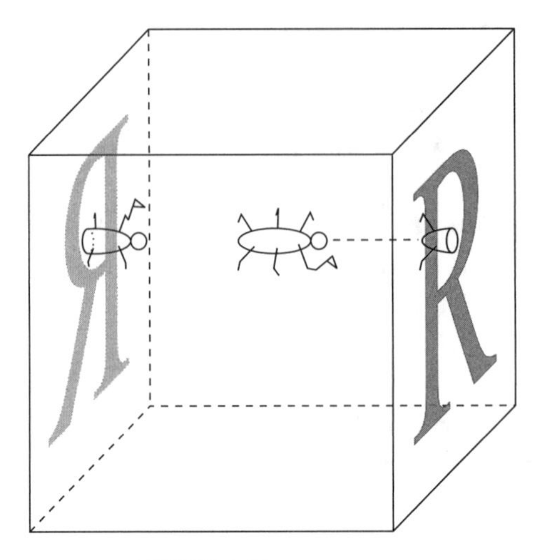

FIGURE 4.22
$K^2 \times S^1$ is nonorientable.

side, as shown in Figure 4.22. This path reverses orientation. Since there is an orientation-reversing path, the manifold is nonorientable. ❄

There is a surprise when we look at the orientability and sidedness of surfaces in 3-manifolds. Recall from Section 3.3 that a surface in $\mathbb{R}^3$ is two-sided if it is orientable, and it is one-sided if it is nonorientable. But consider the surface $z = 0$ in the 3-manifold $K^2 \times S^1$ of Figure 4.22. This surface is a Klein bottle, hence nonorientable. However, a bug walking on top of this surface and going off the right side or the left side or the front or the back of the glued cube always returns on the top of the surface. There is no way for it to get from the top of the surface to the bottom of the surface by walking on the surface. This surface is a two-sided Klein bottle.

On the other hand, consider the surface $x = y$ in this 3-manifold, shown in Figure 4.23. The edge gluings show that this surface is a torus. A bug walking on the front of this sur-

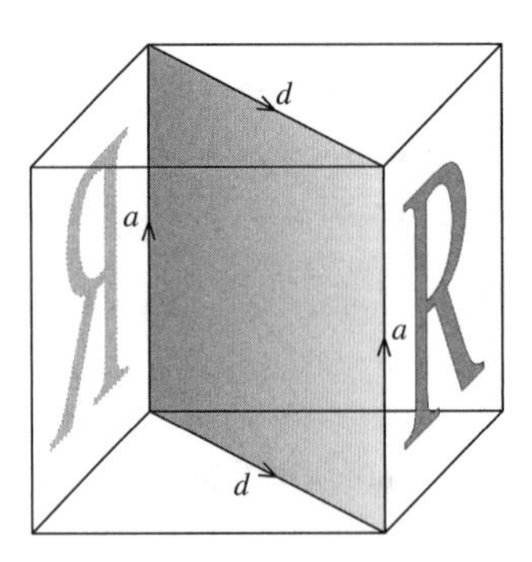

FIGURE 4.23
A one-sided torus in $K^2 \times S^1$.

face, going off the left side of the cube, and reappearing on the right side, will reappear on the back of the surface. This surface is a one-sided torus. Exercise 9 asks you to investigate a connection between these strange phenomena and the fact that the 3-manifold $K^2 \times S^1$ is nonorientable.

Three-manifolds are of interest to cosmologists because it is probable that the universe is a 3-manifold. Certainly our local neighborhood of the universe looks like $\mathbb{R}^3$, and it seems reasonable to believe that this should be the case everywhere in the universe. As a 3-manifold, the universe could be infinite in size. But our examples of closed 3-manifolds show that the universe could also be finite in size while still being without boundary. If the universe is a closed 3-manifold, it is interesting to ask which 3-manifold it might be. Two conditions impose limits on our choices. First, most cosmologists believe that the universe is orientable: it should not be possible for astronauts to return from a voyage through space with their hearts on the right sides of their bodies and with reversed orientations of all their molecules. The second limitation has to do with the curvature of the universe. Curvature is not a topological concept: it is determined by measuring distances and angles, so it requires a geometric structure on a space. However, it turns out that the topological structure of a manifold limits what kind of geometric structure it can have. For example, S^3 and P^3 can only have uniform geometries with positive curvature, T^3 and $K^2 \times S^1$ can only have uniform geometries with zero curvature, and many other 3-manifolds can only have uniform geometries with negative curvature. Recent experiments and theoretical developments support the belief that our universe has zero curvature.

Compact 3-manifolds have not been classified. We will see one of the main difficulties in Section 6.6. However, in the 1930s, German topologists Nowacki, Hantschze, and Wendt proved that there are exactly six closed, orientable, path-connected 3-manifolds that can support a geometry with zero curvature. One of these is T^3; another is the half-twist cube of Figure 4.18. The other four are listed in Exercise 11. It may turn out that our universe is one of these six 3-manifolds. For more details on possible shapes for the universe, see the article "The Shape of the Universe: Ten Possibilities" [43] by Colin Adams and Joey Shapiro and the book *The Shape of Space* [46] by Jeffrey Weeks. In the article by Adams and Shapiro, the last four possibilities are noncompact 3-manifolds corresponding to infinite universes. The last two chapters of Weeks' book describe an ingenious experimental program to determine the shape of the universe by analyzing patterns in cosmic microwave background radiation.

Exercises 4.3

1. Glue together triangles that span across the corners of the solid cube in Figure 4.13 to verify that the boundary of a small neighborhood of vertex A in Example 4.12 is S^2.

2. Glue together triangles that span across the corners of the solid cube in Figure 4.15 to verify that the boundary of a small neighborhood of vertex A in Example 4.14 is the projective plane P^2.

3. Try to describe what might happen to an astronaut who flies through a point that has arbitrarily small neighborhoods bounded by projective planes.

4. Consider the pseudo-manifold obtained from gluing the faces of a solid cube as indicated in Figure 4.19. Show that the boundary of a small neighborhood of the point A is the connected sum of four projective planes.

5. Glue the faces of solid cubes as shown in the two diagrams of Figure 4.24. In each case, decide whether or not the result is a 3-manifold. If it is a 3-manifold, is it orientable?

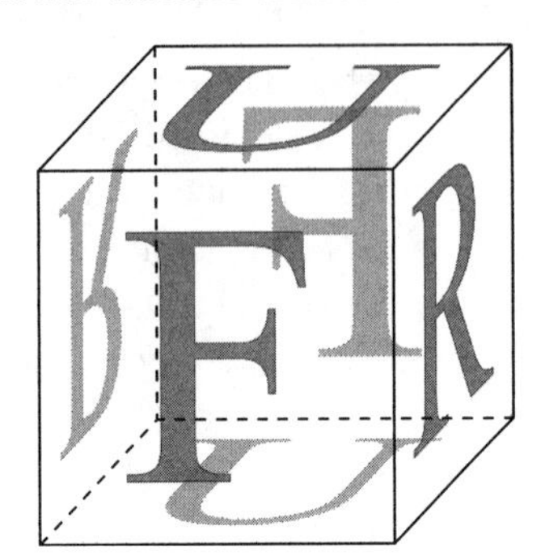

FIGURE 4.24
Are these 3-manifolds?

6. **(a)** Use the model of S^3 as two 3-dimensional balls glued together at corresponding points of their boundaries to show that S^3 is orientable.
   (b) Use the model of projective 3-space P^3 as a 3-dimensional ball with antipodal points on its boundary glued together to show that, perhaps surprisingly, P^3 is orientable.
   (c) How is the orientability of a surface S related to orientability of the 3-manifold $S \times S^1$?

7. Sketch a two-sided Möbius band in $K^2 \times S^1$.

8. In the space formed by gluing a solid cube as illustrated in the diagram on the left of Figure 4.24, consider the surfaces $x = 0$, $y = 0$, and $z = 0$. Determine what surfaces these become under the gluings, and whether they are one-sided or two-sided.

9. Show that a 3-manifold is nonorientable if and only if it contains a two-sided Klein bottle. It might be useful to think of the tubular region traced out by the body of a 3-dimensional bug as it travels around an orientation-reversing path.

10. Investigate the different ways of gluing the faces of a tetrahedron together in pairs. Which ways give 3-manifolds?

11. The other four closed, orientable, path-connected 3-manifolds that support a geometry with zero curvature are listed below. Verify that each one is a 3-manifold, and that it is orientable.
    (a) The quarter-twist cube formed by gluing a solid cube as illustrated in the diagram at the top left of Figure 4.25.
    (b) The Hantschze-Wendt manifold, or double cube space, formed by gluing together the faces of two solid cubes as illustrated in the diagram at the top right of Figure 4.25.

(c) A hexagonal prism in which each side face is glued directly across to the opposite face and the top is glued to the bottom with a $\frac{1}{6}$ twist, as illustrated in the diagram at the bottom left of Figure 4.25.

(d) A hexagonal prism in which each side face is glued directly across to the opposite face and the top is glued to the bottom with a $\frac{1}{3}$ twist, as illustrated in the diagram at the bottom right of Figure 4.25.

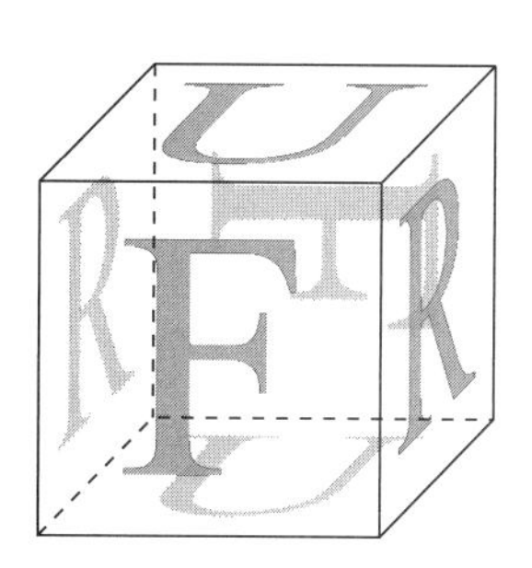

The quarter-twist cube.

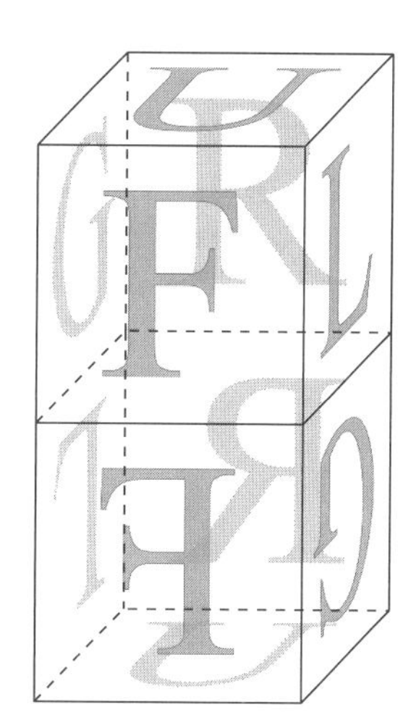

The Hantschze-Wendt manifold.

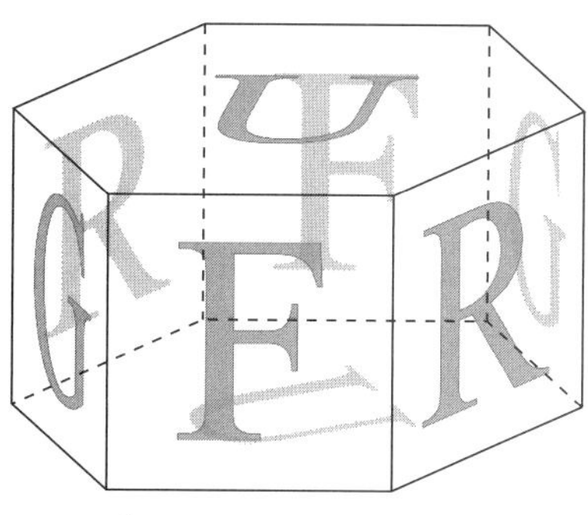

The $\frac{1}{6}$-twist hexagonal prism.

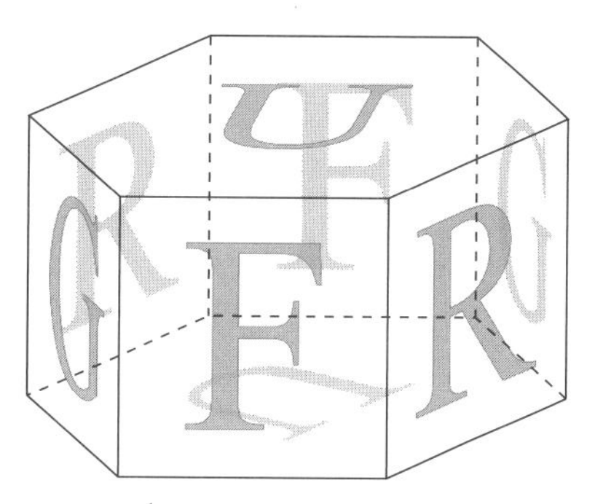

The $\frac{1}{3}$-twist hexagonal prism.

FIGURE 4.25
The four additional compact orientable 3-manifolds that have geometries with zero curvature.

4.4 Heegaard Splittings

In his dissertation in 1898 at the University of Copenhagen, Poul Heegaard put forth a method for describing any closed orientable 3-manifold. We will examine this method in terms of a decomposition of the 3-manifold into 3-cells attached together in various ways. In this section we will assume without further mention that all spaces are triangulated and that subsets are piecewise linear with respect to the space that contains them. We will also restrict our attention to orientable 3-manifolds.

Start with a 3-cell embedded in $\mathbb{R}^3$. Attach the two endpoints of an arc to the boundary of the 3-cell. If we thicken up the arc, we will find that we have attached a second 3-cell to the original 3-cell by identifying a pair of disks in the boundary of the first 3-cell with a pair of disks in the second 3-cell. The resulting 3-manifold is obviously a solid torus (see Example 4.3). The following definition extends this operation to attaching a handle to the boundary of any 3-manifold.

Definition 4.26 *A* ***handle of index*** **1** *(or a* **1*-handle****) is a* 3*-cell that is attached to the boundary of a* 3*-manifold by gluing a pair of disks in the boundary of the* 3*-cell to a pair of disks in the boundary of the* 3*-manifold. A* ***handle of index*** **0** *(or a* **0*-handle****) is a* 3*-cell to which handles of higher indices can be attached.*

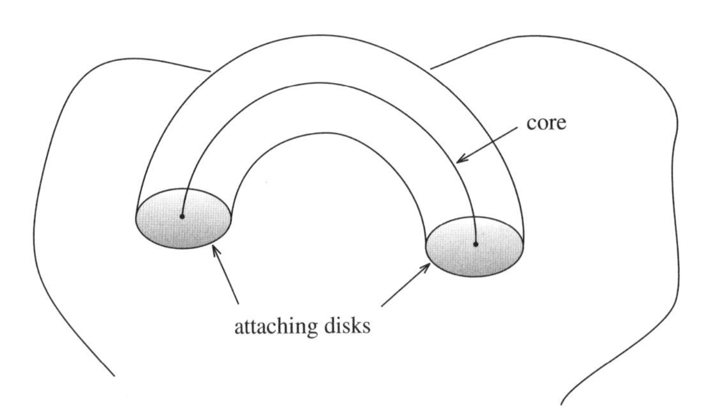

FIGURE 4.27
Attaching a 1-handle.

The index of a 1-handle refers to the 1-dimensional arc at the core of the handle. The endpoints of the arc are interior points of the disks along which we glue. The points on the boundary of the manifold to which we attach the endpoints of this arc specify the essential geometry of the 1-handle. The index of a 0-handle refers to the single point at the center of the cell. It is not attached to anything, but the cell containing it acts like a vertex in a triangulation with handles of higher index being attached to its boundary.

Starting with a single 3-cell as a 0-handle, we can attach any number of 1-handles to its boundary. Each 1-handle will increase the genus of the boundary of the resulting 3-manifold by 1. Because the boundary is connected, the choice of the pair of attachment points for the core of the 1-handles does not affect the topological type of the resulting 3-manifold (see Exercise 12 of Section 3.1).

Definition 4.28 *A **handlebody** is a* 3*-manifold formed by attaching a finite number of* 1*-handles to a* 3*-cell. The **genus** of the handlebody is the number of* 1*-handles attached.*

On the boundary of your favorite 3-manifold find a simple closed curve. Attach a disk to the 3-manifold by gluing the boundary of the disk along the curve. Now thicken up the disk so that it is a 3-cell with the original disk slicing through the middle. The 3-cell should intersect the original 3-manifold in an annulus with the simple closed curve between the two boundary curves of the annulus. Even if the 3-manifold is embedded in $\mathbb{R}^3$, depending on the choice of the simple closed curve, it may not be possible to embed the disk in $\mathbb{R}^3$. You can work in a higher-dimensional Euclidean space or, better yet, just think of adjoining the disk and 3-cell abstractly to form a new 3-manifold as a topological space with its own intrinsic existence.

Definition 4.29 *A **handle of index 2** (or a **2-handle**) is a* 3*-cell that is attached to the boundary of a* 3*-manifold by gluing an annulus in the boundary of the* 3*-cell to an annulus in the boundary of the* 3*-manifold.*

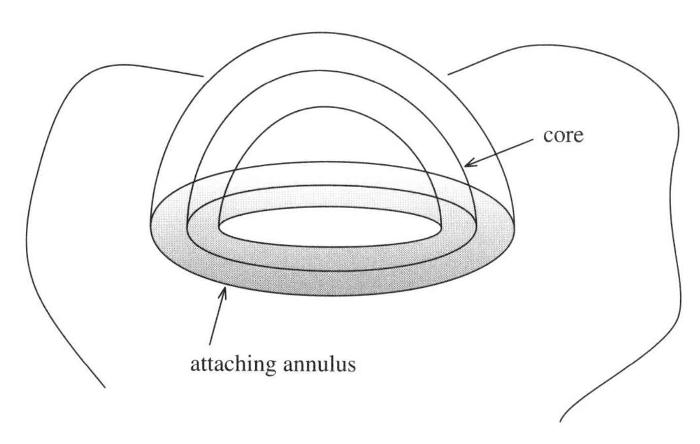

FIGURE 4.30
Attaching a 2-handle.

The index of a 2-handle refers to the 2-dimensional disk at the core of the handle. The boundary of the disk is the centerline of the annulus along which we glue. This curve on the boundary of the manifold specifies the essential geometry of the 2-handle.

The boundary components of a compact orientable 3-manifold are closed orientable surfaces. On any boundary component of positive genus, we can find a simple closed curve that does not separate the surface. If we attach a 2-handle along such a curve, we will reduce the genus of the boundary component by 1. By repeating this process, the boundary components of the manifold will eventually become 2-spheres. We can then attach a 3-cell to each of the boundary components of the manifold to create a closed 3-manifold.

Definition 4.31 *A* ***handle of index*** **3** *(or a* **3*-handle****) is a* 3*-cell that is glued along its* 2*-sphere boundary to a* 2*-sphere boundary component of a* 3*-manifold.*

The boundary of a handlebody of genus n is of course a surface of genus n. Thus, we can find n disjoint simple closed curves that do not separate the surface and attach 2-handles along their curves. Then fill in the resulting 2-sphere boundary component with a 3-cell to produce a closed 3-manifold. By choosing curves that wind around the boundary of the handlebody in intricate patterns, we can use this process to construct some quite complicated closed 3-manifolds.

The question now arises as to whether every closed orientable 3-manifold can be constructed as a sequence of handles. Exercises 1 to 6 at the end of this section show that we can modify any handle decomposition of a 3-manifold so each handle is attached to handles of strictly lower indices. The key to answering the question is to observe how, after we perform these modifications, the 2-handles intersect the 3-handles. A 2-handle is attached along an annulus in its boundary. Nothing else is attached to the two remaining boundary disks until it comes time to attach the 3-handles. Thus, the 2-handles are attached to 3-handles along two disks in their boundaries. That is, we can view the 3-handles as 0-handles with the 2-handles attached as 1-handles. Together they form a handlebody! And this handlebody intersects the handlebody formed from the original 0-handles and 1-handles along a common surface. The following theorem guarantees that any closed orientable 3-manifold has such a handle decomposition. In fact, a triangulation of the 3-manifold automatically provides a handle decomposition.

Theorem 4.32 *Any closed orientable* 3*-manifold is the union of two handlebodies that intersect along their common boundaries.*

Proof. Consider a triangulation of the 3-manifold. Consider a small 3-cell around each vertex. These balls can be chosen so they are disjoint and have boundaries composed of disks that span across the corners of the tetrahedral solids. The vertices of the triangulation are the cores of these 0-handles. Each edge of the triangulation connects two of these 0-handles. Run a thin beam along each edge to attach a 1-handle to the 0-handles. Each face of the triangulation spans three 0-handles and three 1-handles. Thicken up each of these disks into a plate to yield a 2-handle determined by each face. Now all that remains of the 3-manifold are 3-cells, one in the interior of each tetrahedral solid in the triangulation. These can be attached as 3-handles to complete the handle decomposition of the 3-manifold.

The 0-handles and 1-handles form one handlebody in the 3-manifold; the 3-handles and the 2-handles form another. These two handlebodies intersect along their common boundary as required. ❄

Definition 4.33 *A **Heegaard splitting** of a 3-manifold is a decomposition of the 3-manifold as the union of two handlebodies that intersect along their common boundaries. The **genus** of a closed orientable 3-manifold is the minimum genus for the handlebodies in a Heegaard splitting of the 3-manifold.*

Of course the genus of a 3-manifold is a topological invariant. From the discussion in Section 4.1, we know the 3-sphere is of genus 0. By Exercise 10 of that section, we know that the 3-sphere is the only 3-manifold of genus 0.

The 3-manifolds of genus 1 are obtained by gluing together two solid tori along their boundaries. We can think of a solid torus as a 1-handle attached to a 0-handle. Gluing the second torus amounts to changing our perspective on the 1-handle of the second torus and thinking of it as a 2-handle attached to the first solid torus. Thus, the essential information is the curve in the boundary of the solid torus along which we will attach the core of the 2-handle. The remainder of the gluing is just filling in the spherical hole with a 3-handle. The core of this 2-handle is known as a **meridional disk** of the second solid torus. The boundary of a meridional disk is a **meridional curve**. A simple closed curve in the boundary of a solid torus that intersects a meridional curve transversely at a single point is a **longitudinal curve**.

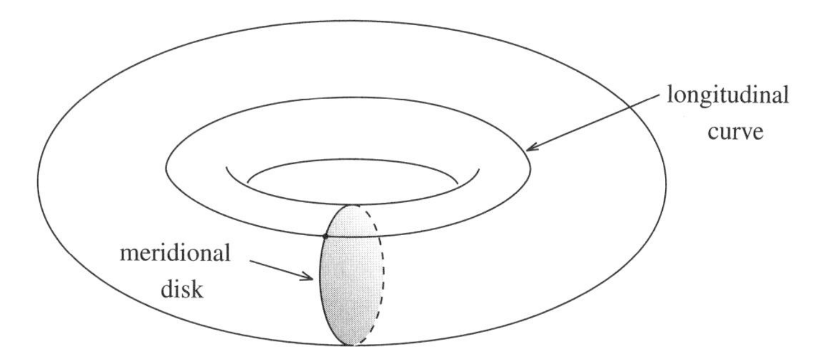

FIGURE 4.34
A longitudinal curve and a meridional disk of a solid torus.

Exercise 9 asks you to show that if a meridional curve is glued to a longitudinal curve, the result is a 3-sphere. Exercise 10 asks you to show that if a meridional curve is glued to a meridional curve, the result is $S^2 \times S^1$.

Other 3-manifolds of genus 1 are obtained by gluing together the solid tori so a meridional curve goes p times longitudinally and q times meridionally for some relatively prime integers p and q. The resulting 3-manifold is called a lens space and is denoted $L(p, q)$. Exercise 11 gives an alternative description of lens spaces that is more suggestive of their name.

Exercises 4.4

1. Consider the process of attaching a 1-handle to the boundary of a 3-manifold.

(a) Draw pictures to show that you can shrink the attaching disks to be as small as desired.

(b) Draw pictures to show that you can slide the attaching disks around to any position on the boundary components without affecting the topological type of the resulting 3-manifold.

2. (a) Draw pictures to show that attaching a 1-handle to two distinct 0-handles is the same as merging the two 0-handles into one.

(b) Consider a connected 3-manifold constructed from 0-handles, 1-handles, and 2-handles. Show that the handles can be rearranged so that the number of 0-handles required is reduced to one.

3. Consider a 3-manifold constructed by attaching 1-handles to a collection of 0-handles. A new 1-handle is now attached to this manifold. Draw pictures to show how to slide the attaching disks off the old 1-handles so that the new 1-handle is attached to the 0-handles. This can be done with an isotopy of the 3-manifold, so the two ways of attaching the new 1-handle yield topologically identical 3-manifolds.

4. Consider a 3-manifold constructed from 0-handles, 1-handles, and 2-handles. Suppose all the 1-handles have been attached to 0-handles and all the 2-handles have been attached to 0-handles and 1-handles.

(a) Suppose a new 2-handle is attached to this manifold. Draw pictures to show how to slide the attaching annulus off the old 2-handles so that the new 2-handle is attached to the 0-handles and 1-handles. This can be done with an isotopy of the 3-manifold, so the two ways of attaching the new 2-handle yield topologically identical 3-manifolds.

(b) Suppose a new 2-handle is attached to the 0-handles and 1-handles of the 3-manifold. Suppose the core disk of the new 2-handle intersects a 1-handle in a simple closed curve. Notice that there are two essentially different ways this can happen. Draw pictures to show how to slide the attaching annulus off the 1-handles so that the new 2-handle is attached to the 0-handles. This can be done with an isotopy of the 3-manifold, so the two ways of attaching the new 2-handle yield topologically identical 3-manifolds.

5. Consider a 3-manifold constructed from 0-handles, 1-handles, and 2-handles. Suppose all the 1-handles have been attached to 0-handles. Suppose all the 2-handles have been attached to 0-handles and 1-handles so that none of the core disks intersect the 1-handles in simple closed curves.

(a) Suppose a new 1-handle is attached to this manifold. Draw pictures to show how to slide the attaching disks off the old 2-handles so that the new 1-handle is attached to the 0-handles and 1-handles. This can be done with an isotopy of the 3-manifold, so the two ways of attaching the new 1-handle yield topologically identical 3-manifolds.

(b) Suppose a new 1-handle is attached to the 0-handles and 1-handles of the 3-manifold. Draw pictures to show how to slide the attaching disks off the old 1-handles so that the new 1-handle is attached to the 0-handles. This can be done with an isotopy of the 3-manifold, so the two ways of attaching the new 1-handle yield topologically identical 3-manifolds.

6. Based on the results of the previous exercises, show that a handle decomposition of a 3-manifold can be rearranged so that there is one 0-handle, all 1-handles are attached to this 0-handle, and all 2-handles are attached to the 0-handles and 1-handles.

7. Show that the Euler characteristic of a handlebody of genus n is $1 - n$. What is the Euler characteristic of its boundary?

8. **(a)** When we adjoin a 0-handle to a 3-manifold, what change occurs in the Euler characteristic of the manifold and in the Euler characteristic of its boundary?

(b) When we adjoin a 1-handle to a 3-manifold, what change occurs in the Euler characteristic of the manifold and in the Euler characteristic of its boundary?

(c) When we adjoin a 2-handle to a 3-manifold, what change occurs in the Euler characteristic of the manifold and in the Euler characteristic of its boundary?

(d) When we adjoin a 3-handle to a 3-manifold, what change occurs in the Euler characteristic of the manifold and in the Euler characteristic of its boundary?

9. Suppose two solid tori are glued together along their boundaries so that a meridional curve in the boundary of one is a longitudinal curve in the boundary of the other. Show that the resulting 3-manifold is a 3-sphere.

10. Suppose two solid tori are glued together along their boundaries so that a meridional curve in the boundary of one is a meridional curve in the boundary of the other. Show that the resulting 3-manifold is topological equivalent to $S^2 \times S^1$.

11. Here is an alternative method for constructing the lens spaces. Choose a positive integer $p \geq 2$ and a positive integer q relatively prime to p. Consider a 3-cell in the shape

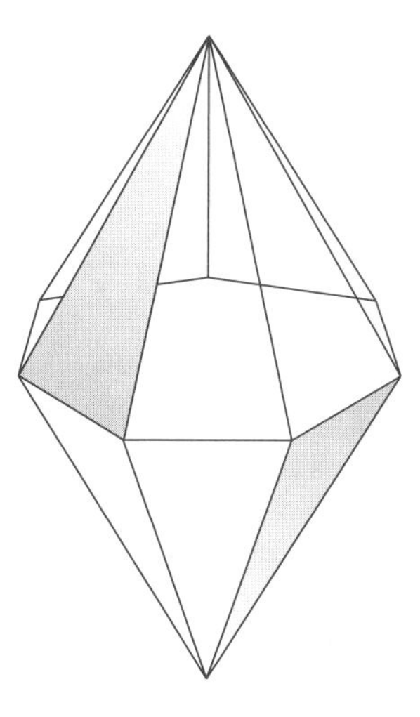

FIGURE 4.35
A pair of faces to be identified in forming the lens space $L(7, 2)$.

of a double pyramid based on a p-sided polygon. Identify each of the triangles in the boundary of the upper pyramid with a triangle q sides around in the boundary of the lower pyramid. Figure 4.35 illustrates this construction for $p = 7$ and $q = 2$.

(a) Drill out a cylindrical core from the top of this lens to the bottom. Show that when the identifications are made, the drilled-out portion of the lens is a solid torus.

(b) Show that when the identifications are made, the remaining portion of the lens is also a solid torus.

(c) See Chapter 9, Section B, of *Knots and Links* [25] by Dale Rolfsen for a description of assembling the outer portion of the lens as a big cheese and assembling the whole thing as the lens space $L(p, q)$.

References and Suggested Readings for Chapter 4

43. Colin Adams and Joey Shapiro, "The shape of the universe: ten possibilities," *American Scientist*, 89 (2001), 443–453.

One of these ten 3-dimensional manifolds is likely to be a model for the global structure of the physical universe.

44. Donald Blackett, *Elementary Topology: A Combinatorial and Algebraic Approach*, Academic Press, New York, 1982.

In addition to material on surfaces, vector fields, and the topology of networks, this book also discusses the gluing of polyhedral solids.

45. Robert Osserman, *Poetry of the Universe: A Mathematical Exploration of the Cosmos*, Anchor Books, New York, 1995.

The unfolding discovery of the shape of the universe is placed in a historical and social context parallel to attemps to map the surface of the earth.

46. Jeffrey Weeks, *The Shape of Space* (second edition), Marcel Dekker, New York, 2002.

An intuitive presentation of the geometry and topology of surfaces and 3-dimensional manifolds culminating in the description of an ongoing project to determine the global structure of the universe from cosmic microwave background radiation.

5

Fixed Points

When the range of a function is the same as its domain, it is possible for the function to map certain points back to themselves. In this chapter we will examine properties that guarantee the existence of such points that are fixed under the operation of the function. Continuity of the function plays an important role, as does the topological nature of the domain.

5.1 Continuous Functions on Closed Bounded Intervals

A function is determined by the operation it performs on the points of its domain. The simplest operation is to do nothing: the identity function on a set X does this to all the points. But a function $f : X \to X$ other than the identity function might still map a point $x \in X$ to x. In algebraic terms, such a point is a solution to the equation $f(x) = x$. The following definition captures the essence of this situation.

Definition 5.1 *A **fixed point** of a function $f : X \to X$ is a point $x \in X$ such that $f(x) = x$.*

When X is a subset of the real numbers, there is a nice graphical interpretation of a fixed point. The equation $f(x) = x$ is comparing the function f with the identity function. A fixed point, as a solution of this equation, occurs where the graph of f intersects the graph of id_X. Figure 5.2 illustrates a typical situation of a function $f : [0, 1] \to [0, 1]$ with two fixed points.

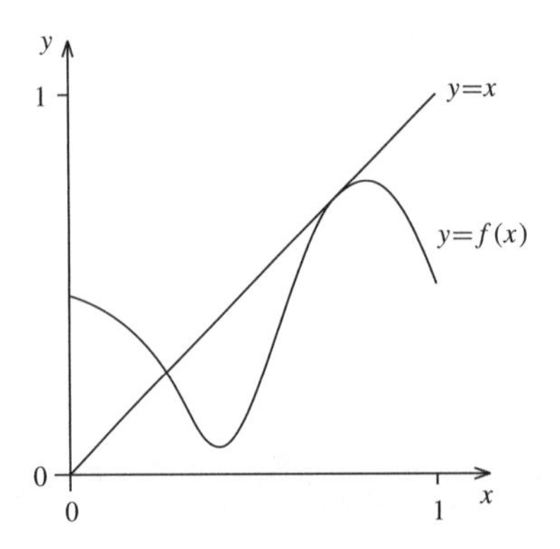

FIGURE 5.2
A function with two fixed points.

We can also visualize the action of a function $f : X \to X$ in terms of how it moves the points of X. Imagine a function from the interval $(0, 1)$ to itself that pushes each point left toward 0. The function defined by the formula $f(x) = x^2$ is a simple case where this happens. Despite the way things are bunching up near 0, each point moves, and so there are no fixed points in the domain $(0, 1)$. Notice how the situation changes if we include the endpoint 0 in the domain of such a function. Since there are no points to the left of 0, the only reasonable way to extend the definition of f is to let $f(0)$ be 0. By adjoining 0 to the domain, we have introduced a fixed point.

A similar situation occurs if the domain is unbounded. On the domain $[0, \infty)$ for example, a function can add a positive quantity to each point and push it to the right. Again the situation changes if the domain is bounded. A function that tries to increase each point of a bounded domain will have to make smaller and smaller increments for points near the upper end of the domain. The limiting endpoint, if it is included in the domain, will thus be a fixed point.

There is a third way for a function to try to sneak away without fixed points. The domain might be missing points that we would normally be inclined to include. For example, let $X = [0, 1] \cup \mathbb{Q}$, the rational numbers between 0 and 1, and let $f : X \to X$ be defined by $f(x) = \frac{1}{3}(x^2 + 1)$. To find fixed points of f, we solve the equation $\frac{1}{3}(x^2 + 1) = x$. The quadratic formula quickly reveals two roots: $(3 \pm \sqrt{5})/2$. But since these irrational numbers are not in the domain of f, they are not fixed points. This situation is related to the previous examples. Instead of a missing endpoint, however, the missing point is in the interior of the interval.

These examples suggest that to guarantee a function has a fixed point, we might begin by requiring the domain to be a closed bounded interval. Of course, we do not want the graph of the function to jump over line $y = x$. A discontinuous function such as $f : [0, 1] \to [0, 1]$ defined by

$$f(x) = \begin{cases} 1 & \text{if } x < \frac{1}{2}, \\ 0 & \text{if } x \geq \frac{1}{2} \end{cases}$$

shows the importance of continuity in the fixed-point business.

In the proof of the Fixed-Point Theorem for an Interval (Theorem 5.4 below), we will construct a sequence of points and show this sequence converges. For this, we need the following theorem, which we will find useful several times in later sections as well as in the proof of the Fixed-Point Theorem for an Interval. Exercise 2 provides some guidance in applying the Completeness Property of the Real Numbers (Property 1.42) to prove this result.

Monotone Convergence Theorem 5.3 *A nondecreasing sequence of real numbers that is bounded above converges. A nonincreasing sequence of real numbers that is bounded below converges.*

The following theorem should have an intuitive obviousness to it. Many calculus students would gladly accept it after examining the graphs of a few typical examples. Exercise 7 indicates how this result follows from the Intermediate Value Theorem. However, the Intermediate Value Theorem is another deep theorem whose proof is based on the Completeness Property of the Real Numbers. Our proof of Theorem 5.4 is built directly on the Completeness Property. It is a case study in the use of mathematics to put intuition on firm logical ground. It points out the crucial roles played by the properties of the domain and by the continuity of the function. And it provides a constructive algorithm for determining lower and upper approximations for the fixed point along with an error bound for these approximations.

The Fixed-Point Theorem for an Interval also hints at a surprising generalization to higher dimensions that we will consider in Section 5.4.

Fixed-Point Theorem for an Interval 5.4 *Any continuous function $f : [a, b] \to [a, b]$ has a fixed point.*

Proof. This proof relies on repeated bisection of intervals in the domain $[a, b]$. Figure 5.5 illustrates this for a typical function. Begin by letting $a_0 = a$ and $b_0 = b$. Consider the midpoint $(a_0 + b_0)/2$. If

$$f\left(\frac{a_0 + b_0}{2}\right) > \frac{a_0 + b_0}{2},$$

let

$$a_1 = \frac{a_0 + b_0}{2}$$

and $b_1 = b_0$. Otherwise, let $a_1 = a_0$ and

$$b_1 = \frac{a_0 + b_0}{2}.$$

Use the principle of mathematical induction to continue this bisection process. After n bisections, we will have sequences $a_0 \leq a_1 \leq \cdots \leq a_n$ and $b_0 \geq b_1 \geq \cdots \geq b_n$ of points in $[a, b]$ with $f(a_k) \geq a_k$ and $f(b_k) \leq b_k$. These points also satisfy $b_k - a_k = (b - a)2^{-k}$.

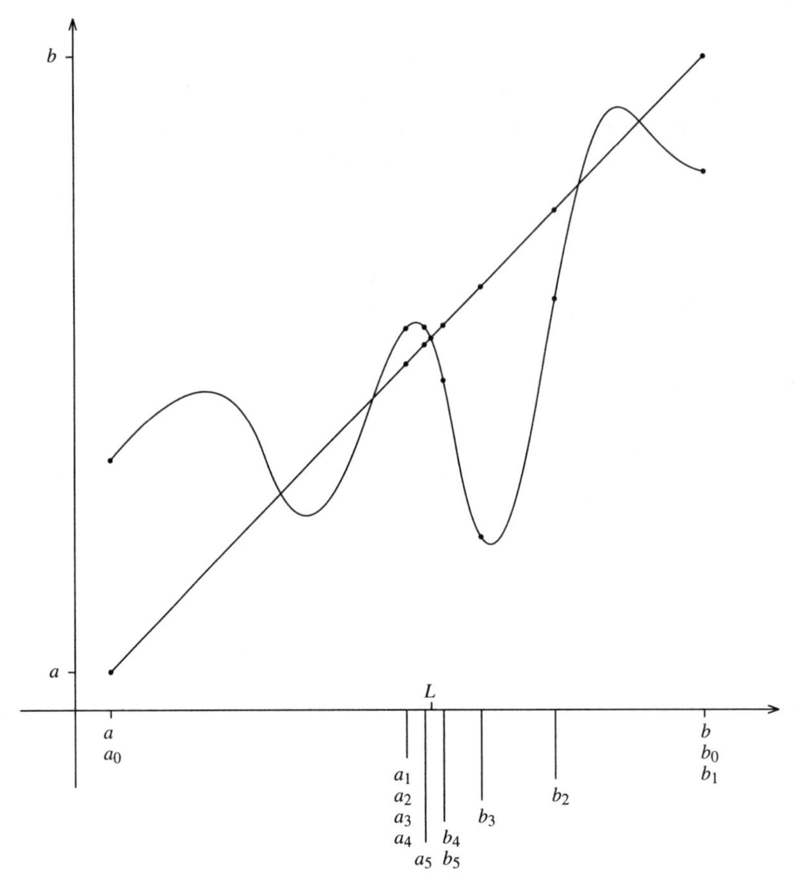

FIGURE 5.5
Finding a fixed point by the bisection method.

Since the sequence $a_0, a_1, \ldots$ is nondecreasing and bounded above, the Monotone Convergence Theorem guarantees that it converges to a limit L. Likewise, the sequence $b_0, b_1, \ldots$ is nonincreasing and bounded below, so it converges to a limit. Also observe that

$$\lim_{n\to\infty} b_n = \lim_{n\to\infty} (a_n + (b-a)2^{-n}) = \lim_{n\to\infty} a_n = L.$$

Using the continuity of f and the fact that $f(a_k) \geq a_k$, we have

$$f(L) = f\left(\lim_{n\to\infty} a_n\right) = \lim_{n\to\infty} f(a_n) \geq \lim_{n\to\infty} a_n = L.$$

Likewise,

$$f(L) = f\left(\lim_{n\to\infty} b_n\right) = \lim_{n\to\infty} f(b_n) \leq \lim_{n\to\infty} b_n = L.$$

It follows that $f(L) = L$. That is, L is a fixed-point of f. ❄

We conclude this section with a definition to describe the property of closed bounded intervals given in Theorem 5.4. Exercise 5 asks you to verify that this is a topological property.

Definition 5.6 *A space X has the* ***fixed-point property*** *if and only if every continuous function $f : X \to X$ has a fixed point.*

Exercises 5.1

1. The first proposition of Euclid's *Elements* is the construction of an equilateral triangle with a given line segment as one of its sides. With A and B as the endpoints of the given segment, Euclid constructs a circle with center at A passing through B and a circle with center at B passing through A. He designates as C one of the points where the circles cross and argues that the three segments AB, BC, and CA are of equal length, thus forming an equilateral triangle. But what justifies the assumption that the two circles do in fact intersect? What if the plane consisted only of points whose coordinates were rational multiples of the length of AB?

2. Carry out the following steps in proving the Monotone Convergence Theorem (Theorem 5.3).

(a) Suppose $x_1, x_2, \ldots$ is a nondecreasing sequence of real numbers that is bounded above. Apply the Completeness Property of the Real Numbers (Property 1.42) to obtain the least upper bound L for the set $\{x_k \mid k = 1, 2, \ldots\}$.

(b) Let $\varepsilon > 0$ be given. Show that there must be a point x_N in the sequence with $x_N > L - \varepsilon$. Suggestion: Derive a contradiction to a property of L if $x_n \leq L - \varepsilon$ for $n = 1, 2, \ldots$.

(c) Use the fact that the sequence is nondecreasing to show that $n \geq N \Longrightarrow L - \varepsilon < x_n \leq L$.

(d) Conclude that $n \geq N \Longrightarrow |L - x_n| < \varepsilon$, so that the sequence satisfies the definition of $\lim_{n\to\infty} x_n = L$.

(e) Carry out a similar proof in the case that the sequence is nonincreasing and bounded below.

3. Sketch the graph of $\cos : [0, 1] \to [0, 1]$. Add to your sketch the sequences $a_0 \leq a_1 \leq \cdots \leq a_n$ and $b_0 \geq b_1 \geq \cdots \geq b_n$ of points as defined by the bisection algorithm given in the proof of Theorem 5.4.

4. Sketch the graph of an interesting function $f : [a, b] \to [a, b]$. Add to your sketch the sequences $a_0 \leq a_1 \leq \cdots \leq a_n$ and $b_0 \geq b_1 \geq \cdots \geq b_n$ of points as defined by the bisection algorithm given in the proof of Theorem 5.4.

5. (a) Suppose a function $f : X \to X$ has a fixed point and $h : X \to Y$ is a bijection. Show that the composition $h \circ f \circ h^{-1} : Y \to Y$ has a fixed point.

(b) Suppose a space X has the fixed-point property and $h : X \to Y$ is a homeomorphism. Show that Y has the fixed-point property.

6. The Fixed-Point Theorem for an Interval (Theorem 5.4) is related to the Intermediate-Value Theorem, which states that if a continuous function $f : [a, b] \to \mathbb{R}$ satisfies $f(a) < c$ and $f(b) > c$ for some $c \in \mathbb{R}$, then there is $L \in [a, b]$ with $f(L) = c$. One proof of the Intermediate-Value Theorem uses a bisection method similar to that of the proof of Theorem 5.4. Fill in the details in the following outline to prove the Intermediate-Value Theorem.

(a) Apply the bisection technique to obtain a sequence of nested intervals $[a_n, b_n]$ of length $(b - a)2^{-n}$ where $f(a_n) < c$ and $f(b_n) \geq c$.

(b) Show that $\lim_{n\to\infty} a_n$ and $\lim_{n\to\infty} b_n$ exist and are equal.

(c) Use the continuity of f to conclude that

$$f\Big(\lim_{n\to\infty} a_n\Big) \leq c \quad \text{and} \quad f\Big(\lim_{n\to\infty} b_n\Big) \geq c.$$

(d) Conclude that $L = \lim_{n\to\infty} a_n = \lim_{n\to\infty} b_n$ satisfies $f(L) = c$.

7. Prove Theorem 5.4 by applying the Intermediate-Value Theorem to the function defined by $f(x) - x$.

8. Suppose $f : S^1 \to \mathbb{R}$ is a continuous function defined on the circle $S^1 = \{(x, y) \mid x^2 + y^2 = 1\}$. Show that there are antipodal points (x, y) and $(-x, -y)$ such that $f(x, y) = f(-x, -y)$. Suggestion: Apply the Intermediate-Value Theorem to the function $g : [0, \pi] \to \mathbb{R}$ defined by $g(t) = f(\cos t, \sin t) - f(-\cos t, -\sin t)$.

9. (a) Show that on any great circle of the earth, there always exists a pair of antipodal points on the circle with exactly the same temperature.

(b) This result requires the assumption that temperature is a continuous function of position. Check with your friends who are chemistry or physics majors as to whether this is a reasonable assumption at the molecular level. A more basic question is whether temperature is even defined for points between molecules.

10. Let $p : \mathbb{R} \to \mathbb{R}$ be a polynomial of odd degree. Assume the coefficient of the highest-power term is positive.

(a) Show that there is a number $a \in \mathbb{R}$ sufficiently far from 0 in the negative direction such that $p(a) < 0$.

(b) Show that there is a number $b \in \mathbb{R}$ sufficiently far from 0 in the positive direction such that $p(b) > 0$.

(c) Conclude that there is a number $x \in \mathbb{R}$ such that $p(x) = 0$.

5.2 Contraction Mapping Theorem

Your study of trigonometry very likely included methods for solving equations involving trigonometric functions. But an equation as simple as $\cos x = x$ defies all analytical techniques for writing a solution in closed form. On the other hand, you can grab your calculator, enter any number, repeatedly compute the cosine (in radian mode), and obtain results that very quickly converge to a number equal to its cosine, at least to the accuracy of your calculator.

Many other functions exhibit the behavior that iterating the function produces a sequence of number converging to a fixed point of the function. The crucial property is that the function maps pairs of points closer together than they were originally.

Definition 5.7 *Let X be a subset of $\mathbb{R}$. A function $f : X \to X$ is a* **contraction** *if and only if there is a scaling factor $s < 1$ such that $|f(x) - f(y)| \leq s|x - y|$ for all $x, y \in X$.*

Example 5.8 Show that $\cos : [0, \frac{3}{2}] \to [0, \frac{3}{2}]$ is a contraction.

Solution. The Mean Value Theorem of calculus gives a way to relate the distance between a pair of points to the distance between their images. For a differentiable function $f : [a, b] \to [a, b]$, this theorem says that for any $x, y \in [a, b]$ there is a point c between x and y such that

$$f(x) - f(y) = f'(c)(x - y).$$

For $f(x) = \cos(x)$, we have $f'(x) = -\sin x$. So on the interval $[0, \frac{3}{2}]$, the maximum of the derivative in absolute value is $\sin 1.5$. Thus, for $s = .9975$ (a value slightly greater than $\sin 1.5$, but definitely less than 1), we have

$$\begin{aligned} |\cos x - \cos y| &= |-(\sin c)(x - y)| \\ &= |\sin c||x - y| \\ &\leq s|x - y| \end{aligned}$$

for all $x, y \in [0, \frac{3}{2}]$. ❄

The main result of this section is Theorem 5.10, the Contraction Mapping Theorem. The proof involves iterating a function to obtain a convergent sequence of points. The main difficulty inherent in this proof is that we need to show that the sequence converges prior to knowing the limit of the sequence. There are several ways to solve this problem. In a metaphor suggested by Victor Bryant in his text *Yet Another Introduction to Analysis*, here is a result that meets our needs and is useful in other contexts.

Spanish Hotel Theorem 5.9 *Every sequence $x_1, x_2, \ldots$ of points in $\mathbb{R}$ has a monotone subsequence. That is, there is a sequence $x_{n_1}, x_{n_2}, \ldots$ selected from the original sequence with $n_1 < n_2 < \cdots$, such that either $x_{n_1} \leq x_{n_2} \leq \cdots$ or $x_{n_1} \geq x_{n_2} \geq \cdots$.*

Proof. A coastal town in Spain has an infinite row of hotels along a road leading down to the beach. The tourist bureau that rates these hotels has a special designation for any hotel having a view of the sea. Any hotel that is at least as tall as the rest of the hotels on the road to the sea receives this view-rating. Starting with the hotel farthest from the sea, let $x_1, x_2, \ldots$ denote the heights of the hotels in order as we travel along the road to the beach.

One possibility is that an infinite number of hotels are view-rated. In this case the heights of these hotels form a nonincreasing subsequence of the original sequence of heights.

The other possibility is that only a finite number of hotels are view-rated. In this case we can obtain an increasing subsequence as follows. Walk along the road to the beach past all the view-rated hotels. Let n_1 be the index of next hotel. Since it does not have a view of the sea, there is a taller hotel closer to the shore that blocks its view. Let n_2 denote the index of that hotel. Since it is not view-rated, there is an even taller hotel closer to the shore. Continue obtaining indices $n_1 < n_2 < \cdots$ of taller and taller hotels. This gives an increasing subsequence $x_{n_1}, x_{n_2}, \ldots$ of the original sequence. ❄

The discussion at the beginning of this section described how, starting with any real number x_0, we can iterate the cosine function to obtain a sequence of points that converges to a fixed point of the function. Example 5.8 showed that the cosine function is a contraction. It was no accident that iterating the cosine function gives a convergent sequence. This is quite a general phenomenon. The following theorem states this result for the case of a real-valued function of a real variable.

The Contraction Mapping Theorem 5.10 *Suppose $f : \mathbb{R} \to \mathbb{R}$ is a contraction. Then f has a unique fixed point. For any $x_0 \in \mathbb{R}$, inductively define $x_{n+1} = f(x_n)$ for $n = 0, 1, \ldots$. Then the sequence $x_0, x_1, \ldots$ converges to the fixed point.*

Proof. This proof has a peculiar sense of pulling itself up by its bootstraps. Begin by letting s be a scaling factor for f. For $n > m$ we have

$$\begin{aligned} |x_n - x_m| &= |f(x_{n-1}) - f(x_{m-1})| \le s|x_{n-1} - x_{m-1}| \\ &= s|f(x_{n-2}) - f(x_{m-2})| \le s^2|x_{n-2} - x_{m-2}| \\ &= \qquad \cdots \qquad \le s^m|x_{n-m} - x_0|. \end{aligned}$$

Now for any $k > 0$ we have

$$\begin{aligned} |x_k - x_0| &\le |x_k - x_{k-1}| + |x_{k-1} - x_{k-2}| + \cdots + |x_1 - x_0| \\ &\le s^{k-1}|x_1 - x_0| + s^{k-2}|x_1 - x_0| + \cdots + |x_1 - x_0| \\ &= (s^{k-1} + s^{k-2} + \cdots + 1)|x_1 - x_0| \\ &= \frac{1 - s^k}{1 - s}|x_1 - x_0| \\ &\le \frac{1}{1 - s}|x_1 - x_0|. \end{aligned}$$

The first consequence of this is that the sequence $x_0, x_1, \ldots$ is bounded. By the Spanish Hotel Theorem, the sequence has a monotone subsequence. Since the subsequence is also bounded, we know by the Monotone Convergence Theorem (Theorem 5.3) that it will converge. Let x denote the limit of the subsequence.

We can also use the inequality established in the second paragraph of this proof to continue the string of inequalities begun in the first paragraph.

$$\begin{aligned} |x_n - x_m| &\leq s^m |x_{n-m} - x_0| \\ &\leq \frac{s^m}{1-s} |x_1 - x_0|. \end{aligned}$$

Consider x_m to be a term in the subsequence and x_n to be a term in the original sequence with $n > m$. By choosing sufficiently large values for m, we can make the distances $|x_n - x_m|$ arbitrarily small. Since the subsequence converges to x, the original sequence must also converge to x.

Since the contraction f is continuous (see Exercise 4), we have

$$f(x) = f\Big(\lim_{n\to\infty} x_n\Big) = \lim_{n\to\infty} f(x_n) = \lim_{n\to\infty} x_{n+1} = x.$$

That is, the limit x of the sequence $x_0, x_1, \ldots$ is a fixed point of f.

Finally, it is easy to show that a contraction can have at most one fixed point. Suppose x and y are fixed points for f. Then

$$|x - y| = |f(x) - f(y)| \leq s|x - y|.$$

Since $s < 1$, this can only happen if $|x - y| = 0$. That is, we must have $x = y$. ❄

In the middle of the proof of the Contraction Mapping Theorem, we used the Spanish Hotel Theorem to obtain a convergent subsequence of a given sequence. We record this useful result for future reference.

Bolzano-Weierstrass Theorem 5.11 *A bounded sequence of real numbers has a convergent subsequence.*

Exercises 5.2

1. Investigate the behavior obtained by iterating the cosine function. How does the limiting behavior of the sequences of iterates depend on the choice of the initial point?

2. **(a)** Show that on $[\frac{1}{2}, \infty)$ the square root function is a contraction.

   (b) Investigate the behavior obtained by iterating the square root function on $[0, \infty)$. How does the limiting behavior of the sequence of iterates depend on the choice of the initial point?

   (c) What features of the cosine and the square root functions account for the qualitatively different behaviors of the iterates of these functions (oscillations about the limit for the cosine function and monotone convergence for the square root function)?

3. Give an example of a function $f : \mathbb{R} \to \mathbb{R}$ such that $|f(x) - f(y)| < |x - y|$ for all $x, y \in \mathbb{R}$ with $x \neq y$, but which does not have a fixed point.

4. Suppose a function $f : X \to X$ is a contraction. Show that f is continuous.

5. **(a)** Find a function you can iterate to solve the equation $\sin x = x + \frac{1}{2}$. On what intervals is your function a contraction?

(b) Find a function you can iterate to solve the equation $\sin x = \frac{1}{2}(x + \frac{1}{2})$. On what intervals is your function a contraction? What happens when you iterate your function near fixed points outside these intervals?

6. How many solutions does the equation $e^x = 4x$ have? Find a function you can iterate to solve this equation. Investigate the different behaviors of iterating this function when you start near the different solutions.

7. Consider the function $f : [1, 2] \to [1, 2]$ defined by $f(x) = \frac{x}{2} + \frac{1}{x}$.

(a) Show that f is a contraction.

(b) Compute enough iterates of f to guess the value of a fixed point.

(c) Confirm algebraically that your guess is indeed a fixed point of f.

8. Newton's method for finding the root of a function f involves starting with an initial approximation x_0 to the root and computing iterates:

$$x_{n+1} = x_n - \frac{f(x_n)}{f'(x)} \quad \text{for} \quad n = 0, 1, \ldots.$$

Investigate conditions on f under which

$$F(x) = x - \frac{f(x)}{f'(x)}$$

is a contraction.

9. Extract a clear proof of the Bolzano-Weierstrass Theorem from the proof of the Contraction Mapping Theorem.

10. Extend the Bolzano-Weierstrass Theorem to sequences in Euclidean space.

(a) Prove that a bounded sequence of points in the plane has a convergent subsequence. Hint: As Gertrude Stein might have said, "A subsequence of a subsequence is a subsequence."

(b) Prove that a bounded sequence of points in $\mathbb{R}^n$ has a convergent subsequence.

5.3 Sperner's Lemma

The proof of Theorem 5.4 illustrated a way of labeling the points of an interval $[a, b]$ according to the direction they were moved by a function. The subdivision of the interval then led to a fixed point of the function. The main result of this section is a similar combinatorial result that was first proved by Emanuel Sperner in 1928. Within a year, other mathematicians noted that Sperner's result leads to a simple proof that an n-cell has the fixed-point property. In this section, we will prove the 2-dimensional version of Sperner's Lemma. Then in the next section, we will apply this result to prove that the 2-dimensional disk $D^2 = \{ (x, y) \in \mathbb{R}^2 \mid x^2 + y^2 \leq 1 \}$ has the fixed-point property.

We begin with a result that boils down to a simple observation about a cat that goes in and out of the house through one of those little pet doors. If the cat is inside when the people go to bed and is outside when they wake up in the morning, the cat must have gone through the door an odd number of times during the night. The proof given below introduces the key idea that we will exploit in the proof of Sperner's Lemma. The following definitions provide some convenient terminology.

Definition 5.12 *A **partition** of an interval $[a, b]$ is a finite set $\{x_0, x_1, \ldots, x_n\}$ of points of the interval such that $a = x_0 < x_1 < \cdots < x_n = b$. A **proper labeling** of a partition is an assignment of either the letter a or the letter b to each point of the partition with each endpoint receiving its own letter as its label. The label of each subinterval $[x_{k-1}, x_k]$ of the partition is the unordered pair of labels of its endpoints.*

Thus, if a subinterval has vertices labeled a and b (in either order), the label of the subinterval itself will be ab. If the vertices are both labeled a, the subinterval will be labeled aa; if they are both labeled b, the subinterval will be labeled bb.

Cat-Door Lemma 5.13 *A properly labeled partition of an interval $[a, b]$ contains an odd number of ab subintervals.*

Proof. Cut the interval at each of the interior partition points. Count the number of times a occurs as the label of an endpoint of these subintervals. This number is clearly odd since a occurs as a label once as the left endpoint of the leftmost subinterval, whereas each interior partition point labeled a yields two endpoints of subintervals with this label.

On the other hand, the aa and bb subintervals have an even number of endpoints labeled a, and each ab interval contributes one a label. Figure 5.14 illustrates a typical situation. In order for the total to be odd, the number of ab intervals must be odd. ❄

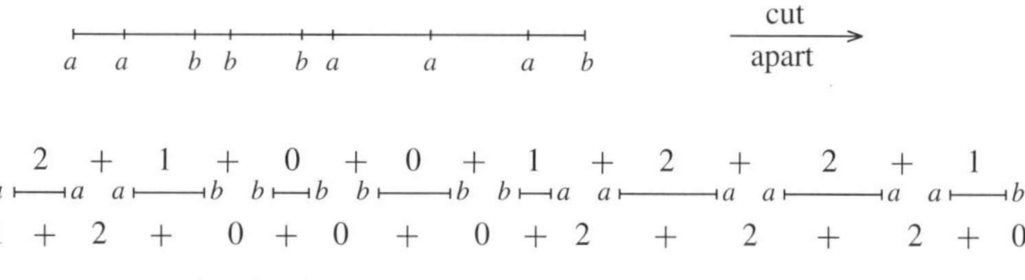

FIGURE 5.14
Two ways to count the number of endpoints labeled a.

Recall from Definition 3.18 that a triangulation of the disk is a subdivision of the disk into triangular regions such that the intersection of any pair of distinct triangles is either empty, a vertex of both triangles, or an edge of both triangles. Now we extend the terminology of Definition 5.12 to a triangular region.

Definition 5.15 *Consider a triangular region with vertices labeled a, b, and c. A* ***proper labeling*** *of a triangulation of the region is an assignment of either a, b, or c to each vertex of the triangulation in such a way that the three edges of the region are properly labeled as intervals. The label of each triangle of the triangulation is the unordered triple of the labels of its vertices.*

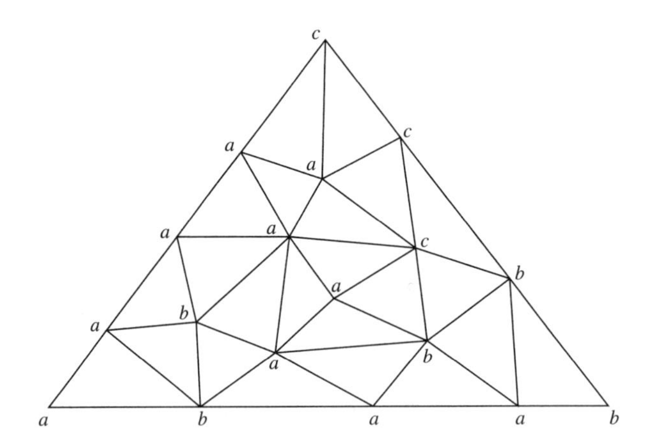

FIGURE 5.16
A properly labeled triangulation.

We are finally ready for Sperner's Lemma. This combinatorial result does for triangles what the Cat-Door Lemma does for intervals. The proofs are strikingly similar. Just as the proof of the Cat-Door Lemma relies on the fact that the interval $[a, b]$ has an odd number of endpoints (namely one) labeled a, the proof of Sperner's Lemma uses the fact that a properly labeled partition of an interval has an odd number of ab subintervals.

Sperner's Lemma 5.17 *Consider a triangular region with vertices labeled a, b, and c. A properly labeled triangulation contains an odd number of abc triangles.*

Proof. Cut up the triangular region along the edges of the triangulation. Count the number of ab edges that occurs. This number is clearly odd. Indeed, the original triangle has no ab edges along two of its sides and (by the Cat-Door Lemma) an odd number along the third side. And every ab edge in the interior of the triangulation will be an edge of two of the cut-apart triangles.

On the other hand, the only cut-apart triangles with ab edges are those labeled aab, abb, or abc. The first two kinds have two ab edges, while the abc triangles contribute exactly one ab edge each. In order for the total number of ab edges to be odd, there must be an odd number of abc triangles. ❄

Exercises 5.3

1. Think of cutting apart the region in Figure 5.16 along the edges of the triangulation.

(a) Count the number of ab edges that come from the perimeter of the region.

(b) Count the number of ab edges that come from the interior of the region.

(c) Count the number of ab edges that come from aab triangles.

(d) Count the number of ab edges that come from abb triangles.

(e) Count the number of ab edges that come from abc triangles.

(f) Confirm that the two ways of counting the number of ab edges give the same total.

2. In Figure 5.16 find an ab edge along the base of the region. Draw an arc to connect the midpoint of this edge to the midpoint of the other ab edge of this triangle. Continue extending this arc through adjacent triangles so that it joins the midpoints of ab edges. Describe the two possible ways this arc can end. Use this arc trick to give a proof of Sperner's Lemma that actually produces an abc triangle.

3. Fill in the outline given below to extend Sperner's Lemma to a 3-dimensional solid.

(a) Use Definition 4.5 to define a triangulation of a tetrahedral solid.

(b) Extend Definition 5.15 to define a proper labeling of a triangulation of a tetrahedral solid.

(c) Cut apart the tetrahedral region along the faces of the triangulation. Apply the 2-dimensional version of Sperner's Lemma to prove that the number of abc faces is odd.

(d) Enumerate the possible labelings of the tetrahedron in the triangulation and verify that only $abcd$ tetrahedra yield an odd number of abc faces.

(e) Conclude that the number of $abcd$ tetrahedra in the triangulation is odd.

4. Extend the argument of Exercise 3 to prove a version of Sperner's Lemma for cells of any dimension.

5.4 Brouwer Fixed-Point Theorem for a Disk

We saw in Section 5.1 that a closed bounded interval has the fixed-point property: any continuous function mapping the interval back to itself will have a fixed point. What happens with a disk? Rotating a disk leaves the center fixed. Squeezing the disk toward the top fixes the uppermost point. But perhaps some combination of rotating, squeezing, folding, and other continuous transformations will move all the points to new locations. Does the second dimension give enough room to maneuver?

The Brouwer Fixed-Point Theorem (first proved by L. E. J. Brouwer in 1910) says no, the disk also has the fixed-point property. Thus, if you pick up a sheet of paper, crumple it up any way you like (but be careful not to tear it), and flatten it out inside the rectangular region where it began, there will always be a point in exactly the same place it started.

The Brouwer Fixed-Point Theorem also applies to cells of higher dimensions. For example, in three dimensions, no matter how you stir a cup of coffee (as long as it doesn't splash discontinuously) there will always be one point that returns to its starting position.

In this section we will consider the 2-dimensional version of the Brouwer Fixed-Point Theorem. We begin by using Sperner's Lemma to prove that a disk cannot be retracted (as defined precisely below) onto its boundary, and then proceed to the theorem itself. The proof for a cell of higher dimension is entirely analogous, although harder to visualize.

Definition 5.18 *A retraction of a set X onto a subset A is a continuous function $r : X \to A$ such that $r(x) = x$ for all $x \in A$.*

Example 5.19 Find a retraction of the punctured disk

$$X = \{(x, y) \in \mathbb{R}^2 \mid 0 < x^2 + y^2 \leq 1\}$$

onto its boundary

$$A = \{(x, y) \in \mathbb{R}^2 \mid x^2 + y^2 = 1\}.$$

Solution. Move each point of X radially out to the boundary. The function $r : X \to A$ defined by

$$r(x, y) = \left(\frac{x}{\sqrt{x^2 + y^2}}, \frac{y}{\sqrt{x^2 + y^2}} \right)$$

does the job. It maps each point of X to a point at a unit distance from the origin while leaving fixed any point that is already a unit distance from the origin. Since the origin is not an element of X, we do not have to worry about division by zero in the definition of r spoiling its continuity. ❄

Example 5.20 Prove that there is no retraction of an interval $[a, b]$ onto the set $\{a, b\}$ of its two endpoints.

Solution. By the Intermediate-Value Theorem, a continuous function $r : [a, b] \to \{a, b\}$ that maps a to a and b to b must map some points of $[a, b]$ to intermediate values between a and b. Thus, no such function can have $\{a, b\}$ as its image. ❄

The following theorem has an intuitively physical appeal. Think of a balloon stretched over the mouth of a jar. We wouldn't expect to be able to deform the balloon onto the rim of the jar unless we cut the balloon, poked a hole it, or performed some other discontinuous operation.

First, here is a definition that extends the idea of the mesh of a partition of an interval (useful in studying the convergence of Riemann sums to the corresponding integral) to triangulations. By Exercise 2, any triangular region can be triangulated with a mesh that is arbitrarily small.

Definition 5.21 *The **mesh** of a triangulation is the maximum distance between two points in a single triangle of the triangulation.*

No-Retraction Theorem 5.22 *There does not exist a retraction of a disk onto its boundary.*

Proof. Suppose there is a retraction r of a disk onto its boundary. By Exercise 1, the existence of such a retraction is a topological property of the space. Thus, we can think of the disk as a triangular region.

In preparation for the use of Sperner's Lemma, we want to label every point of the region so that any triangulation will have a proper labeling. Label the vertices a, b, and c. Label the points along the edge from a to b by the closer of the two endpoints of this edge; the midpoint can be labeled arbitrarily. Similarly label the points along the other two edges of the region. Since the retraction maps points of the triangular region to points of the boundary, we can label any point of the region with the label of its image under the retraction. The retraction does not move points of the boundary, so of course their labels do not change. In particular, the labels of the vertices of any triangulation will be a proper labeling.

Consider a sequence of triangulations of the region such that the meshes of these triangulations converge to zero. By Sperner's Lemma, each triangulation will have an odd number of triangles with vertices labeled abc. In particular, since zero is not an odd number, we can select from each triangulation one abc triangle for special consideration.

By the Bolzano-Weierstrass Theorem (Theorem 5.11 and Exercise 10 of Section 5.2), a subsequence of a vertices of these abc triangles will converge. Let x denote the limit point. Notice first that x must be in the triangular region. Otherwise, there would be a positive distance such that all points of the region would be farther away from x than this distance. In this case, the sequence of a vertices would not converge to x. Notice also that since the meshes of the triangulations converge to zero, the corresponding sequences of b and c vertices converge to x.

The retraction r now faces a conflict of interest in assigning a boundary point to x. Since r is continuous and since x is a limit of a points, $r(x)$ must be the limit of the images of the a points. Thus, $r(x)$ must be in one of the two edges containing the vertex a and not past the midpoint away from a along either of these edges. But $r(x)$ must be likewise near the vertex b and near the vertex c. Since no such boundary point exists, we conclude that the proposed retraction is impossible. ❄

We now have all the preliminary work in place to prove the Brouwer Fixed-Point Theorem for a disk.

Brouwer Fixed-Point Theorem 5.23 *A continuous function $f : D^2 \to D^2$ has a fixed point.*

Proof. This is again a proof by contradiction. Let $f : D^2 \to D^2$ be a continuous function. Suppose that f has no fixed points. We will proceed to construct a retraction of the disk onto its boundary. Since we have proved that no such retraction exists, we must conclude that f must have a fixed point.

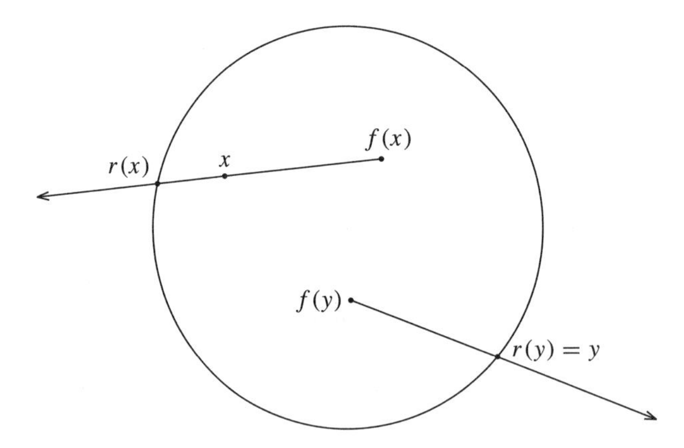

FIGURE 5.24
The ray from $f(x)$ to x crosses the boundary of D^2 at $r(x)$. The ray from $f(y)$ to a boundary point y crosses at $r(y) = y$.

For each point $x \in D^2$ there is a well-defined ray from $f(x)$ through x. Define $r(x)$ to be the point where this ray crosses the boundary of D^2 on its way out of the disk. Exercises 3 and 4 give two ways to see that r is continuous. Notice that if a point y is on the boundary of D^2, then the ray from $f(y)$ through y will cross through the boundary at y. Hence r does not move points on the boundary of D^2. That is, r is a retraction. By Theorem 5.22 we know this cannot exist. Thus, f must have a fixed point. ❄

Exercises 5.4

1. Show that the existence of a retraction $r : X \to A$ is a topological property of the space X and the subset A. That is, if $h : X \to Y$ is a homeomorphism, then there is a retraction of X onto A if and only if there is a retraction of Y onto $h(A)$.

2. **(a)** Show that the maximum distance between two points in a triangular region is the distance between two of the vertices of the triangle.

(b) Show that the three medians of a triangle subdivide the triangle into smaller triangles for which the mesh is less than or equal to two-thirds of the mesh of the original triangle.

(c) Given any positive number ε, show that any triangular region can be triangulated with mesh less than ε.

3. Let D^2 denote the unit disk $\{(x, y) \mid x^2 + y^2 \leq 1\}$ in $\mathbb{R}^2$, and let $f : D^2 \to D^2$ be the continuous function that we assume has no fixed points. Let $r : D^2 \to S^1$ be the retraction as described in the proof of the Brouwer Fixed-Point Theorem. Follow these steps to derive an algebraic formula for r thus confirming its continuity.

(a) Let (u, v) denote the image of $(x, y) \in D^2$ under the fixed-point free function f. Show that $r(x, y) = (u, v) + t((x, y) - (u, v))$ for the scalar $t > 0$ such that the $r(x, y)$ is a unit vector.

(b) Write the condition $||r(x, y)||^2 = 1$ as a quadratic equation in the variable t.

(c) Use the quadratic formula to write t as a continuous function of x, y, u, and v. Notice that you will need to choose the positive square root so that $t > 0$. Notice also how the assumption that f has no fixed points ensures that you are not dividing by zero.

4. Let D^2 denote the unit disk $\{(x, y) \mid x^2 + y^2 \leq 1\}$ in $\mathbb{R}^2$, and let $f : D^2 \to D^2$ be the continuous function that we assume has no fixed points. Let $r : D^2 \to S^1$ be the retraction as described in the proof of the Brouwer Fixed-Point Theorem. Follow these steps to produce a geometrical argument for the continuity of r.

(a) Consider a small arc of S^1 around the image $r(x)$ of a point $x \in D^2$. Draw lines through the midpoint of the line segment between x and $f(x)$ and each of the endpoints of the arc.

(b) Choose a disk around $f(x)$ that does not cross the lines. Use the continuity of f to get a disk around x that does not cross the lines and that f maps into the disk around $f(x)$.

(c) Show that r maps the points of the disk around x to the arc around $r(x)$.

References and Suggested Readings for Chapter 5

47. William G. Chinn and Norman Steenrod, *First Concepts of Topology*, Random House, New York, 1966.

    Continuity, compactness, connectedness, and other basic concepts of topology are applied to prove the pancake theorem, the ham-sandwich theorem, the fixed-point theorem for a disk, and the fundamental theorem of algebra.

48. Yu. A. Shashkin, *Fixed Points*, American Mathematical Society, Providence, RI, 1991.

    A self-contained development of basic results of fixed-point theory.

49. Francis Su, "Rental harmony: Sperner's lemma in fair division," *American Mathematical Monthly*, 106 (1999), 930–942.

    Sperner's lemma is generalized to an n-dimensional simplex and applied to fair division problems such as devising a fair way for apartment mates to choose rooms and split the rent.

6

The Fundamental Group

Numerical invariants and invariant properties enable us to distinguish certain topological spaces. We can go further and associate with a topological space a set having an algebraic structure. The fundamental group is the most basic of such possibilities. It not only provides a useful invariant for topological spaces, but the algebraic operation of multiplication defined for this group reflects the global structure of the space.

6.1 Deformations with Singularities

A circle, an annulus, a Möbius band, and a solid torus all have the basic shape of a ring going around a hole. Even though these four spaces are topologically distinct, we would like to characterize this ring-like property. We would like to distinguish these spaces from the 2-sphere, for example. The 2-sphere has a hole all right, but the hole of a sphere seems quite a bit different from the hole of a lifesaver. While the 2-dimensional torus has a ring-like shape, it has a hole where the dough of a doughnut would be as well as a hole where you grab to dunk the doughnut into your coffee. We will use loops in the space to get a handle on these kinds of topological properties. Amazing as it may seem, we will be able to detect the ring-like property intrinsically from the space itself, with no need to see how the space is embedded in some ambient Euclidean space.

Recall from Definition 1.40 that a path is a continuous function defined on the closed unit interval [0, 1]. The following definition imposes the condition that the function maps the two endpoints to the same point.

Definition 6.1 *A* ***loop*** *in a space X is a path $\alpha : [0, 1] \to X$ such that $\alpha(0) = \alpha(1)$. The element $\alpha(0) = \alpha(1)$ of X is the* ***base point*** *of the loop α.*

You may recall from Exercise 1 of Section 1.5 the idea of combining two paths into a single path. You simply travel along the first path at twice the normal speed. Then, as long as the final point of the first path coincides with the initial point of the second path, you can continue along the second path at twice the normal speed. For loops at a common base point, we can always perform this kind of splicing.

Definition 6.2 *Suppose the loops α and β in a space X have a common base point x_0. The* ***concatenation*** *of α with β is the path denoted $\alpha \cdot \beta$ and defined by*

$$(\alpha \cdot \beta)(s) = \begin{cases} \alpha(2s) & \text{for } 0 \le s \le \frac{1}{2}, \\ \beta(2s-1) & \text{for } \frac{1}{2} \le s \le 1. \end{cases}$$

Since paths α and β with base point x_0 are continuous, we see from the formulas that $\alpha \cdot \beta$ is continuous on the intervals $[0, \frac{1}{2}]$ and $[\frac{1}{2}, 1]$. Since

$$\alpha\big(2 \cdot \tfrac{1}{2}\big) = \alpha(1) = x_0 = \beta(0) = \beta\big(2 \cdot \tfrac{1}{2} - 1\big),$$

the two formulas for $\alpha \cdot \beta$ agree at $s = \frac{1}{2}$. By Theorem 1.27, $\alpha \cdot \beta$ is continuous. Of course, $(\alpha \cdot \beta)(0) = \alpha(2 \cdot 0) = \alpha(0) = x_0$ and $(\alpha \cdot \beta)(1) = \beta(2 \cdot 1 - 1) = \beta(1) = x_0$. Hence, the concatenation of two loops based at x_0 is likewise a loop based at x_0.

Figure 6.3 illustrates three typical loops in an annulus. The arrow indicates which way the image is traced as the domain parameter increases from 0 to 1. The loop α does not do any significant traveling around the annulus. In particular, if we lay a string along the image of α, we could wind in the string without letting go of the ends. Even though the loop β ventures forth more boldly, it eventually doubles back. Thus, β can also be deformed back to a constant loop at the base point. Notice that the loop β is not a one-to-one function. Thus, the stages of the deformation will necessarily involve loops that cross themselves. The loop γ goes around the annulus in an essential way. It is clear, at least intuitively, that there is no way to deform γ staying within the annulus, keeping the ends fixed, and finishing with a constant loop.

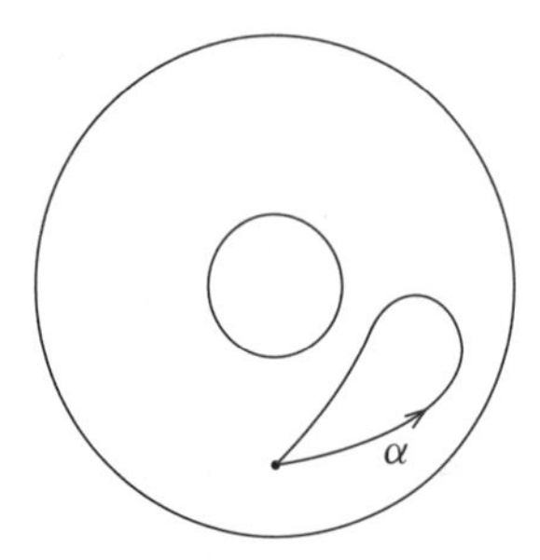

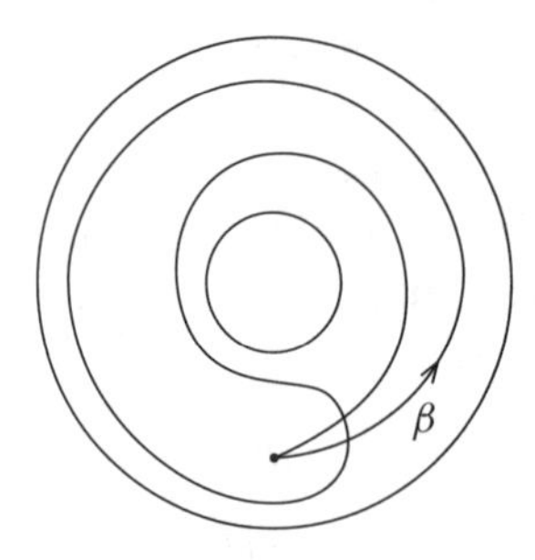

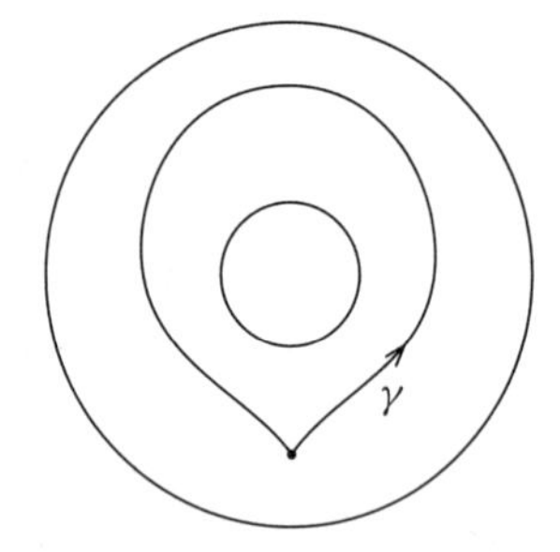

FIGURE 6.3
Typical loops in an annulus.

These deformations of the loops in the annulus illustrate a concept that yields an equivalence relation among loops with a common base point. The following definition makes precise the idea of a continuously parameterized family of loops representing the various stages of a deformation of one loop to another loop. Although continuity is required, the functions involved do not have to be one-to-one. Thus, the loops may cross themselves and they may cross the loops at other stages of the deformation. The loops at all stages of the deformation must map into the space X. Also, the loops at all stages of the deformation must map the end points of the interval to the common base point.

Definition 6.4 *Suppose the loops α and β in a space X have a common base point x_0. A* ***homotopy*** *from α to β is a continuous function $H : [0, 1] \times [0, 1] \to X$ such that the loops $H_t : [0, 1] \to X$ defined by $H_t(s) = H(s, t)$ all have the base point x_0 and such that $H_0 = \alpha$ and $H_1 = \beta$. We say the loop α is* ***homotopic*** *to the loop β if and only if there is such a homotopy from α to β. This relation is denoted $\alpha \sim \beta$.*

Picture a homotopy from a loop $\alpha : [0, 1] \to X$ to a loop $\beta : [0, 1] \to X$ as mapping a square disk into X. For the ordered pair $(s, t) \in [0, 1] \times [0, 1]$, the t-coordinate is the parameter determining which path $H_t : [0, 1] \to X$ to use and the s-coordinate is the distance parameter along this path. Thus, the bottom of the square maps to X via α, the top of the square maps via β, and the two vertical sides map to the base point x_0.

Figure 6.5 shows how these pieces fit together. The labels in both the domain and range show which pieces correspond to the functions $\alpha = H_0$, $\beta = H_1$, and a typical intermediate loop H_t.

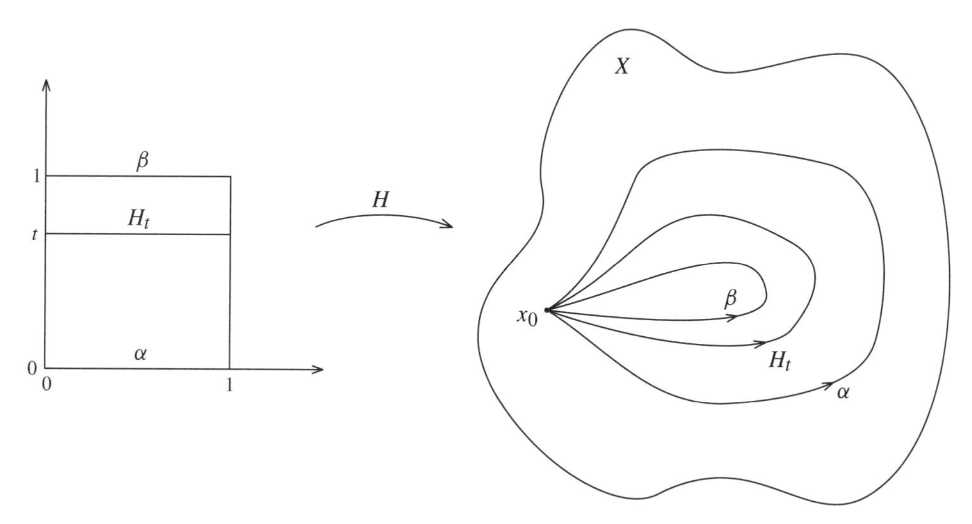

FIGURE 6.5
A homotopy H from a loop α to a loop β.

Example 6.6 Show that any loop in the disk $D^2 = \{(x, y) \mid x^2 + y^2 \leq 1\}$ is homotopic to the constant loop at the base point.

Solution. Let $\alpha : [0, 1] \to D^2$ denote a loop with base point x_0. For any $s \in [0, 1]$ there is a line segment in D^2 from $\alpha(s)$ to x_0. In fact for $t \in [0, 1]$, $\alpha(s) + t(x_0 - \alpha(s))$ parameterizes this segment as a path. The formula $H(s, t) = \alpha(s) + t(x_0 - \alpha(s))$ defines a continuous function $H : [0, 1] \times [0, 1] \to D^2$. We easily check that H is a homotopy from α to the constant loop at x_0. Indeed,

$$H_0(s) = H(s, 0) = \alpha(s) + 0(x_0 - \alpha(s)) = \alpha(s),$$
$$H_1(s) = H(s, 1) = \alpha(s) + 1(x_0 - \alpha(s)) = x_0,$$

and for all $t \in [0, 1]$,

$$H_t(0) = H(0, t) = \alpha(0) + t(x_0 - \alpha(0)) = x_0 + t(x_0 - x_0) = x_0,$$
$$H_t(1) = H(1, t) = \alpha(1) + t(x_0 - \alpha(1)) = x_0 + t(x_0 - x_0) = x_0.$$

❄

Theorem 6.7 *The relation of homotopy among loops with base point x_0 in a space X is an equivalence relation on the set of all such loops.*

Proof. We simply cook up homotopies to verify the reflexive, symmetric, and transitive properties as required by Definition 1.2. Exercise 7 at the end of this section asks you to check that the given formulas do indeed define the required homotopies. ❄

We are more interested in the net amount of winding a loop does in a space than in the details of how fast it travels or any backtracking it does. Thus, we will be more interested in the equivalence class of a loop than in the loop itself. Recall from Exercise 2 of Section 1.1 that the set of all loops in a space at a given base point is the disjoint union of these equivalence classes.

Definition 6.8 *For a loop α with base point x_0 in a space X, the set of all loops homotopic to α is the* **homotopy class** *of α. This set is denoted $\langle\alpha\rangle$.*

In the next section, we will see how concatenation of loops leads to a kind of multiplication on the homotopy classes of loops. The resulting set with its algebraic operation will be a topological invariant of the underlying space.

Exercises 6.1

1. Draw the images of two interesting loops α and β with a common base point.
   (a) Label the points $\alpha(\frac{k}{10})$ and $\beta(\frac{k}{10})$ for $k = 0, 1, \ldots, 10$.

(b) Draw the loop $\alpha \cdot \beta$. Label the points $(\alpha \cdot \beta)(\frac{k}{10})$ for $k = 0, 1, \ldots, 10$.

(c) Draw the loop $\beta \cdot \alpha$. Label the points $(\beta \cdot \alpha)(\frac{k}{10})$ for $k = 0, 1, \ldots, 10$.

2. Draw several intermediate loops to show how the loop β in Figure 6.3 can be deformed to a constant loop at the base point. Remember to stay within the annulus at all times and to keep the end points of the loops at the base point.

3. If we do not require the paths at all stages of a deformation to start and end at a common base point, show that any loop $\alpha : [0, 1] \to X$ can be deformed to a constant loop.

4. Extend the idea of Example 6.6 to show that in a disk, any two loops with a common base point are homotopic.

5. **(a)** A subset X of a Euclidean space is **convex** if and only if for every pair of points $x, y \in X$ the line segment between x and y is also in X. Extend the idea of Example 6.6 to show that in a convex set, any loop is homotopic to a constant loop.

(b) In a **star-shaped** set there is a special point from which you can see all other points of the set without looking outside the set. Formulate a precise definition of this concept of a star-shaped set. Prove that any loop based at the special point of a star-shaped set is homotopic to a constant loop.

(c) What can you say about loops based at other points of a star-shaped set?

6. **(a)** Show that any loop in the annulus $A = \{ (x, y) \in \mathbb{R}^2 \mid 1 \leq x^2 + y^2 \leq 9 \}$ with base point at $(2, 0)$ is homotopic to a loop whose image lies in the circle $\{ (x, y) \in \mathbb{R}^2 \mid x^2 + y^2 = 4 \}$ in A^2.

(b) Show that any loop in a Möbius band M^2 with base point on the centerline of M^2 is homotopic to a loop whose image lies in the centerline of M^2.

7. Suppose the point x_0 is designated as a base point for loops in a space X. Verify the conditions for the functions defined below to be homotopies as required to prove Theorem 6.7.

(a) Reflexivity: Given a loop α, show that $H(s, t) = \alpha(s)$ defines a homotopy from α to α.

(b) Symmetry: Given a homotopy G from the loop α to the loop β, show that the reverse parameterization of G defined by $H(s, t) = G(s, 1 - t)$ is a homotopy from β to α.

(c) Transitivity: Given a homotopy F from the loop α to the loop β and a homotopy G from the loop β to the loop γ, we run through the parameterization of F at twice the normal rate and follow that by running through the parameterization of G at twice the normal rate. Check that the resulting function defined by

$$H(s, t) = \begin{cases} F(s, 2t) & \text{for } 0 \leq t \leq \frac{1}{2}, \\ G(s, 2t - 1) & \text{for } \frac{1}{2} \leq t \leq 1 \end{cases}$$

is a homotopy from α to γ.

6.2 Algebraic Properties

In the previous section we saw how to concatenate two loops to produce a new loop. The goal of this section is to use concatenation to define an algebraic structure that will give us some information about the shape of a topological space. We run into several immediate difficulties if we try to work with the loops themselves. For example, we would like a constant path ε to act as the algebraic identity element. However, for any reasonably interesting path $\alpha : [0, 1] \to X$, the concatenation $\alpha \cdot \varepsilon$ will not equal α. Similarly, we would like to be able to cancel a loop by concatenating it with a loop that traces the path of the original loop in the reverse direction. But again, for a nonconstant loop, there is no way to concatenate it with any loop to produce a constant loop.

The solution to this problem is to consider homotopy classes of loops rather than the loops themselves. This also allows us to concentrate on the essential shape of the space without being distracted by the meandering of a loop through the space. Here then is the definition of the product of two homotopy classes of loops.

Definition 6.9 *Let $\alpha : [0, 1] \to X$ and $\beta : [0, 1] \to X$ be two loops with base point x_0 in a space X. The* ***product*** *of the homotopy classes $\langle\alpha\rangle$ and $\langle\beta\rangle$ of these two loops is denoted $\langle\alpha\rangle\langle\beta\rangle$ and is defined to be the homotopy class $\langle\alpha \cdot \beta\rangle$ of the concatenation of α and β.*

The above definition uses two loops α and β to represent their homotopy classes. This raises the question as to whether the product of the classes depends on the choice of the representatives. That is, if $\alpha' \sim \alpha$ and $\beta' \sim \beta$, then $\langle\alpha\rangle = \langle\alpha'\rangle$ and $\langle\beta\rangle = \langle\beta'\rangle$. We need to verify in this situation that $\langle\alpha\rangle\langle\beta\rangle = \langle\alpha'\rangle\langle\beta'\rangle$. This is a typical example of showing that an operation is well-defined.

Theorem 6.10 *Suppose $\alpha \sim \alpha'$ and $\beta \sim \beta'$ as loops based at x_0 in a space X. Then $\langle\alpha\rangle\langle\beta\rangle = \langle\alpha'\rangle\langle\beta'\rangle$.*

Proof. Since $\langle\alpha\rangle\langle\beta\rangle = \langle\alpha \cdot \beta\rangle$ and $\langle\alpha'\rangle\langle\beta'\rangle = \langle\alpha' \cdot \beta'\rangle$, we need to show that $\alpha \cdot \beta \sim \alpha' \cdot \beta'$. We use a homotopy F from α to α' and a homotopy G from β to β' to paste together a homotopy H from $\alpha \cdot \beta$ to $\alpha' \cdot \beta'$. Figure 6.11 provides a guide for pasting together these homotopies. The domain $[0, 1] \times [0, 1]$ is labeled with the functions used on the various pieces.

Define H by

$$H(s,t) = \begin{cases} F(2s, t) & \text{for } 0 \le s \le \frac{1}{2}, \\ G(2s-1, t) & \text{for } \frac{1}{2} \le s \le 1. \end{cases}$$

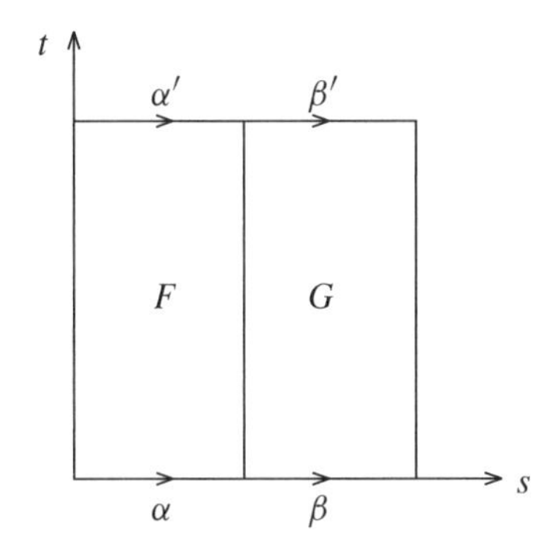

FIGURE 6.11
A guide to defining a homotopy from $\alpha \cdot \beta$ to $\alpha' \cdot \beta'$.

Notice that domains of the two portions of the definition overlap along the line where $s = \frac{1}{2}$. For these points we have $F(2s, t) = F(1, t) = x_0 = G(0, t) = G(2s - 1, t)$. Therefore H is well-defined, and by Theorem 1.27 it is continuous. Now at any stage $t \in [0, 1]$, we have that $H_t = F_t \cdot G_t$. Therefore, H_t is a loop with base point x_0. It follows that H is a homotopy from $H_0 = F_0 \cdot G_0 = \alpha \cdot \beta$ to $H_1 = F_1 \cdot G_1 = \alpha' \cdot \beta'$. ❄

Having verified that concatenation of loops gives a well-defined product of homotopy classes of loops, we are now ready to consider the algebraic properties of this multiplication.

Theorem 6.12 *Let x_0 be a common base point for all loops in a space X. The product of homotopy classes of loops in X satisfies the following three properties:*

Associativity: *For any loops α, β, and γ in X, we have $(\langle\alpha\rangle\langle\beta\rangle)\langle\gamma\rangle = \langle\alpha\rangle(\langle\beta\rangle\langle\gamma\rangle)$.*

Existence of an identity element: *The constant path ε defined by $\varepsilon(s) = x_0$ for all $s \in [0, 1]$ determines a homotopy class $\langle\varepsilon\rangle$ that satisfies $\langle\alpha\rangle\langle\varepsilon\rangle = \langle\alpha\rangle = \langle\varepsilon\rangle\langle\alpha\rangle$ for any loop α.*

Existence of inverses: *For any loop α, the reverse loop α^{-1} defined by $\alpha^{-1}(s) = \alpha(1 - s)$ determines a homotopy class $\langle\alpha^{-1}\rangle$ that satisfies $\langle\alpha\rangle\langle\alpha^{-1}\rangle = \langle\varepsilon\rangle$ and $\langle\alpha^{-1}\rangle\langle\alpha\rangle = \langle\varepsilon\rangle$.*

Proof. The proofs of these three properties involve deforming the parameterization of the loops involved to form the required homotopies. The ideas are quite simple, although the details of writing down the formulas is more of a technical exercise in analytic geometry.

Associativity: From the definition of multiplication in terms of concatenation, $(\langle\alpha\rangle\langle\beta\rangle)\langle\gamma\rangle = \langle\alpha \cdot \beta\rangle\langle\gamma\rangle = \langle(\alpha \cdot \beta) \cdot \gamma\rangle$ and $\langle\alpha\rangle(\langle\beta\rangle\langle\gamma\rangle) = \langle\alpha\rangle\langle\beta \cdot \gamma\rangle = \langle\alpha \cdot (\beta \cdot \gamma)\rangle$. Thus, we need a homotopy from the loop $(\alpha \cdot \beta) \cdot \gamma$ to the loop $\alpha \cdot (\beta \cdot \gamma)$. Figure 6.13 provides a guide to defining the required homotopy

$$F(s,t) = \begin{cases} \alpha\left(\dfrac{4s}{1+t}\right) & \text{for } 0 \le s \le \dfrac{1+t}{4}, \\ \beta(4s-1-t) & \text{for } \dfrac{1+t}{4} \le s \le \dfrac{2+t}{4}, \\ \gamma\left(\dfrac{4s-2-t}{2-t}\right) & \text{for } \dfrac{2+t}{4} \le s \le 1. \end{cases}$$

Exercise 2 at the end of this section asks you to verify that F does the job required of a homotopy from $(\alpha \cdot \beta) \cdot \gamma$ to $\alpha \cdot (\beta \cdot \gamma)$.

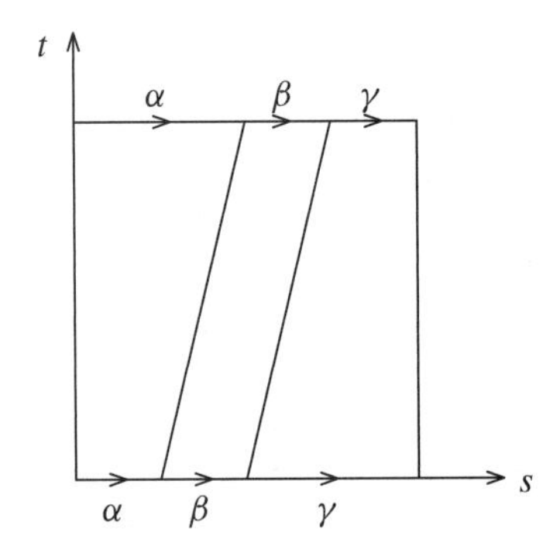

FIGURE 6.13
A guide to defining a homotopy from $(\alpha \cdot \beta) \cdot \gamma$ to $\alpha \cdot (\beta \cdot \gamma)$.

Existence of an identity element: Since $\langle \varepsilon \rangle \langle \alpha \rangle = \langle \varepsilon \cdot \alpha \rangle$, we need to find a homotopy from the loop $\varepsilon \cdot \alpha$ to the loop α. Exercise 3 at the end of this section asks you to draw the guide to construction of such a homotopy G and to verify the formulas in the following definition:

$$G(s,t) = \begin{cases} x_0 & \text{for } 0 \le s \le \dfrac{1}{2} - \dfrac{1}{2}t, \\ \alpha\left(\dfrac{2s-1+t}{1+t}\right) & \text{for } \dfrac{1}{2} - \dfrac{1}{2}t \le s \le 1. \end{cases}$$

Exercise 4 at the end of this section asks you to construct a similar homotopy from $\alpha \cdot \varepsilon$ to α to show that $\langle \alpha \rangle \langle \varepsilon \rangle = \langle \alpha \rangle$.

Existence of inverses: Since $\langle \alpha \rangle \langle \alpha^{-1} \rangle = \langle \alpha \cdot \alpha^{-1} \rangle$ we need to find a homotopy from the loop $\alpha \cdot \alpha^{-1}$ to the constant loop ε. The easiest way to construct the desired homotopy is to have $H_t(s)$ travel along α as far as $\alpha(1-t)$ and stay there until there is just enough time left to return to the base point along α^{-1}. The following definition gives the desired homotopy:

$$H(s,t) = \begin{cases} \alpha(2s) & \text{for } 0 \le s \le \dfrac{1-t}{2}, \\ \alpha(1-t) & \text{for } \dfrac{1-t}{2} \le s \le \dfrac{1+t}{2}, \\ \alpha(2-2s) & \text{for } \dfrac{1+t}{2} \le s \le 1. \end{cases}$$

Since the reverse of α^{-1} is α, we can interchange the roles of α and α^{-1} in the definition of this homotopy to obtain a homotopy from $\alpha^{-1} \cdot \alpha$ to ε. This shows that $\langle \alpha^{-1} \rangle \langle \alpha \rangle = \langle \varepsilon \rangle$.

❄

Associativity, the existence of an identity, and the existence of inverses are the three properties that define the algebraic structure known as a **group**. Thus, the previous theorem can be summarized as stating that the collection of homotopy classes of loops forms a group under the multiplication. Theorem 6.12 justifies the terminology given in the following definition.

Definition 6.14 *Let X be a topological space with base point x_0. The set of homotopy classes of loops in X based at x_0 along with the operation of forming products of homotopy classes is the* ***fundamental group*** *of the space X based at x_0. This group is denoted $\pi_1(X, x_0)$.*

We have defined the fundamental group in terms of homotopy classes of functions defined on an interval. Topologists have considered the analogous situation with functions defined on the n-fold Cartesian product of intervals. This leads to the higher homotopy groups $\pi_n(X, x_0)$ of which the fundamental group is the first. Many of our results about the fundamental group extend to the higher homotopy groups.

Exercises 6.2

1. Decide which of the following give functions that are well-defined on the set of homotopy classes of loops in $\mathbb{R}^2 - \{(0,0)\}$ based at $(1,0)$. If the rule does not yield a well-defined function, give two homotopic loops that have different values under the rule.

(a) The maximum distance you are from the origin as you travel around the loop.

(b) The number of quadrants you visit as you travel around the loop.

(c) The angle measured counterclockwise from the positive x-axis to the ray from the origin through the image of $\frac{1}{2}$.

(d) The net number of times you wind around the origin as you travel along the loop.

2. Suppose α, β, and γ are three loops based at a point x_0 in a space X. The associative property of multiplication of the homotopy classes of these loops requires a homotopy from $(\alpha \cdot \beta) \cdot \gamma$ to $\alpha \cdot (\beta \cdot \gamma)$. In the first of these concatenations, α is traced out for $s \in [0, \frac{1}{4}]$, followed by β for $s \in [\frac{1}{4}, \frac{1}{2}]$, and then by γ for $s \in [\frac{1}{2}, 1]$. In the second, α is traced out for $s \in [0, \frac{1}{2}]$, followed by β for $s \in [\frac{1}{2}, \frac{3}{4}]$, and then by γ for $s \in [\frac{3}{4}, 1]$. The homotopy used in the proof of Theorem 6.12 changes the parameterizations of the three loops as t varies from 0 to 1. Here are some general formulas for performing this kind of reparameterization.

(a) For $a < b$, verify that $(s - a)/(b - a)$ defines a bijection that stretches or shrinks the interval $[a, b]$ to the interval $[0, 1]$ with a mapping to 0 and b to 1.

(b) Verify that as t increases from 0 to 1, the formulas $a + (a' - a)t$ and $b + (b' - b)t$ will change in value from a to a' and from b to b'.

(c) Combine the results of parts (a) and (b) to obtain the formula

$$\frac{s - (a + (a' - a)t)}{(b + (b - b')t) - (a + (a - a')t)}$$

that for any $t \in [0, 1]$ defines a bijection from the interval $[a+(a'-a)t, b+(b'-b)t]$ to the interval $[0, 1]$.

(d) Apply the result of part (c) to the guide given in Figure 6.13. Show that the resulting three formulas simplify to the three parts of the definition of the homotopy F in the proof of Theorem 6.12.

3. Suppose α is any loop based at a point x_0 in a space X. Let ε denote the constant loop at the base point x_0.

(a) Draw a unit square as the domain of a homotopy G from $\varepsilon \cdot \alpha$ to α. Construct a guide to defining G by labeling the axes and the sides of the square with the loops used on the various pieces of the square.

(b) Apply the principles of Exercise 2(d) to verify the formulas given for G in the proof of Theorem 6.12.

4. Suppose α is any loop based at a point x_0 in a space X. Let ε denote the constant loop at the base point x_0.

(a) Draw a guide to constructing a homotopy from $\alpha \cdot \varepsilon$ to α.

(b) Adapt the formulas given in the proof of Theorem 6.12 to determine formulas for a homotopy from $\alpha \cdot \varepsilon$ to α.

5. Suppose α is any loop based at a point x_0 in a space X. Let ε denote the constant loop at the base point x_0. Try to construct a homotopy from $\alpha \cdot \alpha^{-1}$ to ε by speeding up the parameterization of the entire loop α, staying at the base point, and then following the loop α^{-1} at the increased speed. Point out the discontinuities that tend to appear as you approach the constant loop at the final stage.

6. Suppose α is any loop based at a point x_0 in a space X. Let ε denote the constant loop at the base point x_0. Construct a homotopy from $\alpha \cdot \alpha^{-1}$ to ε so that there is no waiting at position $1 - s$. Gradually slow down the parameterization of the portion of α defined on the interval $[0, 1 - s]$ so that you travel to position $1 - s$ along α, then immediately turn around, and return to the base point.

6.3 Invariance of the Fundamental Group

With any topological space and designated base point in the space, we have associated an algebraic structure known as the fundamental group. We want to obtain some geometric information about the space from the fundamental group. In particular, we want to use the fundamental group as a topological invariant of the space. In this section we will examine the mathematical concepts that make the fundamental group a valuable tool in studying a topological space.

The first question is the algebraic version of the problem of identity: what does it mean for two groups to be the same? As with geometric figures, we will give different answers to the question depending on the information we are after. We typically do not want to insist that the elements are literally identical. No two distinct objects would be identified under this relation. However, we usually want more than just a bijection between the set of elements that comprise the group. This would tell us nothing about the algebraic structure of the underlying sets. The most fruitful concept of two groups being essentially the same is captured in the following definition.

Definition 6.15 *Let G and H be groups. Denote the group operation in both by juxtaposition of the elements. A* ***homomorphism*** *from G to H is a function $h : G \to H$ such that $h(ab) = h(a)h(b)$ for any two elements $a, b \in G$. An* ***isomorphism*** *is a homomorphism that is a bijection. A group G is* ***isomorphic*** *to a group H if and only if there is an isomorphism from G to H. We write $G \cong H$ to denote this relation.*

Thus, an isomorphism respects the set theoretical structure of the groups as well as the algebraic structure. When we try to determine the fundamental group of a space, we will be satisfied to know it is isomorphic to some group that we can describe in terms of familiar mathematical objects (the integers, reflections and rotations of polygons, strings of characters with certain cancellation rules, and so forth).

Recall that a path component of a space is the set of all points that can be joined to a given point by a path. Because loops are paths, the loops in a space all lie in the path component containing the base point. Thus, it is reasonable to consider the fundamental group only for spaces that are path-connected. Once we adopt this convention, the following theorem says that we get the same group (up to isomorphism) no matter what base point we designate in a space.

Theorem 6.16 *Suppose X is a path-connected space. For any two points x_0 and x_1 in X, the group $\pi_1(X, x_0)$ is isomorphic to the group $\pi_1(X, x_1)$.*

Proof. The proof of this theorem consists of defining an isomorphism from $\pi_1(X, x_0)$ to $\pi_1(X, x_1)$ and verifying that the function satisfies the various properties required to confirm that it is indeed an isomorphism. The significant steps in this proof are outlined below. The details are left to you to supply as the solution to Exercise 6 at the end of this section. This is an excellent opportunity for you to demonstrate your ability to construct homotopies.

Let γ be a path from x_0 to x_1. A loop based at x_0 can have its base point switched to x_1 by traveling along γ in the reverse direction (from x_1 to x_0), then traversing the loop, and returning to x_1 by traveling along γ in the forward direction. More formally, we define a function $\Gamma : \pi_1(X, x_0) \to \pi_1(X, x_1)$ by $\Gamma(\langle\alpha\rangle) = \langle\gamma^{-1} \cdot \alpha \cdot \gamma\rangle$ for any loop α based at x_0. There are several details for you to check about this definition of Γ. These are spelled out in parts (a) through (e) of Exercise 6.

For any loops α, β based at x_0, part (g) of Exercise 6 gives an easy way to prove that $\Gamma(\langle\alpha\rangle\langle\beta\rangle) = \Gamma(\langle\alpha\rangle)\Gamma(\langle\beta\rangle)$. That is, Γ is a homomorphism.

The way we used the path γ to construct Γ can be applied using the reverse path γ^{-1} to construct a function $\Gamma' : \pi_1(X, x_1) \to \pi_1(X, x_0)$. As you can check in part (h) of Exercise 6, this function is the inverse of Γ. Hence Γ is a bijection. ❄

The next theorem is the key to proving that the fundamental group is a topological invariant. For a continuous function between topological spaces, we can compose this function with any loop in the domain to get a loop in the range. When we view this as a transformation of homotopy classes of loops, this correspondence is a homomorphism between the fundamental groups of the spaces.

Theorem 6.17 *Suppose $f : X \to Y$ is a continuous function and x_0 is designated as the base point in X. Then f induces a homomorphism $f_* : \pi_1(X, x_0) \to \pi_1(Y, f(x_0))$ defined by $f_*(\langle\alpha\rangle) = \langle f \circ \alpha\rangle$ for all $\langle\alpha\rangle \in \pi_1(X, x_0)$.*

Proof. You will find the proof of this theorem is more interesting to work out on your own than it would be to read. There are three main steps: check that $f \circ \alpha$ is indeed a loop in Y based at $f(x_0)$, show that f_* is well-defined, and confirm that f_* is a homomorphism. Exercise 7 at the end of this section asks you to fill in the details. ❄

Now all the pieces are in place to show that the fundamental group is a topological invariant. This is most easily done as a consequence of the properties stated in the following theorem.

Theorem 6.18 *Suppose X, Y, and Z are topological spaces. Let x_0 be designated as the base point for X.*

1. *The identity function $\mathrm{id}_X : X \to X$ induces the identity homomorphism $\mathrm{id}_{\pi_1(X,x_0)} : \pi_1(X, x_0) \to \pi_1(X, x_0)$.*
2. *If $f : X \to Y$ and $g : Y \to Z$ are continuous functions, then $(f \circ g)_* = f_* \circ g_*$.*

Proof. For any element $\langle\alpha\rangle \in \pi_1(X, x_0)$, we have

$$\begin{aligned}(\mathrm{id}_X)_*(\langle\alpha\rangle) &= \langle \mathrm{id}_X \circ \alpha\rangle \\ &= \langle\alpha\rangle \\ &= \mathrm{id}_{\pi_1(X,x_0)}(\langle\alpha\rangle).\end{aligned}$$

Hence, $(\mathrm{id}_X)_* = \mathrm{id}_{\pi_1(X,x_0)}$.

We also have

$$\begin{aligned}(f \circ g)_*(\langle\alpha\rangle) &= \langle (f \circ g) \circ \alpha\rangle \\ &= \langle f \circ (g \circ \alpha)\rangle \\ &= f_*(\langle g \circ \alpha\rangle) \\ &= f_*(g_*(\langle\alpha\rangle)) \\ &= (f_* \circ g_*)(\langle\alpha\rangle).\end{aligned}$$

Hence, $(f \circ g)_* = f_* \circ g_*$. ❄

In the language of category theory, the previous theorem says that the assignment of the fundamental group to a topological space and the assignment of the induced homomorphism to a continuous function is a **covariant functor**. Watch how easily the topological invariance follows from the two conditions in Theorem 6.18.

Theorem 6.19 *The fundamental group is a topological invariant for path-connected topological spaces.*

Proof. Suppose $h : X \to Y$ is a homeomorphism between path-connected spaces X and Y. Let x_0 be designated as the base point of X. Because h is a homeomorphism, h^{-1} exists, and both h and h^{-1} are continuous. Therefore, $(h^{-1})_* \circ h_* = (h^{-1} \circ h)_* = (\mathrm{id}_X)_* = \mathrm{id}_{\pi_1(X,x_0)}$, and $h_* \circ (h^{-1})_* = (h \circ h^{-1})_* = (\mathrm{id}_Y)_* = \mathrm{id}_{\pi_1(Y,f(x_0))}$. It follows that the homomorphism $h_* : \pi_1(X, x_0) \to \pi_1(Y, h(x_0))$ induced by the homeomorphism $h : X \to Y$ has an inverse $(h^{-1})_*$. Thus, h_* is an isomorphism from $\pi_1(X, x_0)$ to $\pi_1(Y, h(x_0))$.

By Theorem 6.16, any choice of base points for X and Y will give groups isomorphic to $\pi_1(X, x_0)$ and $\pi_1(Y, h(x_0))$. Therefore, the fundamental group depends only on the topological type of the space and not on the choice of base point. ❄

Exercises 6.3

1. **(a)** Show that the composition of two homomorphisms is a homomorphism.
   (b) Show that the composition of two isomorphisms is an isomorphism.
   (c) Show that the identity function on a group is an isomorphism from the group to itself.

(d) Show that the inverse of an isomorphism is an isomorphism.

(e) Show that the relation of isomorphism is an equivalence relation among groups.

2. This exercise asks you to prove some basic facts about homomorphisms between groups. If you have not studied group theory, you are encouraged to track down a textbook on abstract algebra for a few hints. Suppose $h : G \to H$ is a homomorphism from a group G to a group H. Let e denote the identity element of G.

(a) Show that $h(e)$ is the identity element of H.

(b) For any element $a \in G$, show that $h(a^{-1}) = (h(a))^{-1}$.

(c) Show that h is one-to-one if and only if e is the only element of G that h maps to the identity element of H.

3. Show that the group of integers with the operation of addition is isomorphic to the group of integer powers of 2 with the operation of multiplication.

4. Show that the group of complex numbers with the operation of addition is isomorphic to the group of $\mathbb{R}^2$ with the operation of addition of vectors.

5. Consider the set

$$X = \left\{ \begin{bmatrix} a & b \\ -b & a \end{bmatrix} \middle| \, a, b \in \mathbb{R}, (a, b) \neq (0, 0) \right\}$$

of nonzero 2×2 matrices.

(a) Show that if A and B are elements of X, then the matrix product AB is also an element of X.

(b) Show that if $A \in X$, then A is invertible and $A^{-1} \in X$.

(c) Show that the identity matrix I is an element of X.

(d) Show that X with the operation of matrix multiplication is isomorphic to the group of nonzero complex numbers with the operation of multiplication.

6. Fill in the details of the proof of Theorem 6.16.

(a) Use the path $\gamma : [0, 1] \to X$ from x_0 to x_1 to define a path γ^{-1} from x_1 to x_0.

(b) Define the operation of concatenation of paths. Be sure to impose appropriate conditions on the endpoints of the paths so that the second path begins where the first one leaves off.

(c) For any loop α based at $\gamma(0)$, show that $(\gamma^{-1} \cdot \alpha) \cdot \gamma$ is homotopic to $\gamma^{-1} \cdot (\alpha \cdot \gamma)$ as loops based at $\gamma(1)$.

(d) Extend the concept of homotopy of loops to a definition of homotopy of paths. Show this is an equivalence relation. State and prove a general associative law concerning homotopy classes of paths.

(e) Show that Γ is well-defined. That is, suppose α and α' are two loops based at x_0 with $\langle \alpha \rangle = \langle \alpha' \rangle$. Use the homotopy from α to α' to construct a homotopy from $\gamma^{-1} \cdot \alpha \cdot \gamma$ to $\gamma^{-1} \cdot \alpha' \cdot \gamma$. Conclude that $\Gamma(\langle \alpha \rangle) = \Gamma(\langle \alpha' \rangle)$.

(f) Show that $\gamma \cdot \gamma^{-1}$ is a loop based at x_0. Show that this loop is homotopic to ε, the constant loop at x_0.

(g) Show that Γ is a homomorphism. That is, suppose α and β are loops based at x_0. Use parts (e) and (f) of this exercise along with the fact that $\langle\varepsilon\rangle$ is the identity element of the group $\pi_1(X, x_0)$ to justify the following steps:

$$\begin{aligned}\Gamma(\langle\alpha\rangle\langle\beta\rangle) &= \langle\gamma^{-1}\cdot\alpha\cdot\beta\cdot\gamma\rangle\\ &= \langle\gamma^{-1}\cdot\alpha\cdot\gamma\cdot\gamma^{-1}\cdot\beta\cdot\gamma\rangle\\ &= \langle\gamma^{-1}\cdot\alpha\cdot\gamma\rangle\langle\gamma^{-1}\cdot\beta\cdot\gamma\rangle\\ &= \Gamma(\langle\alpha\rangle)\Gamma(\langle\beta\rangle).\end{aligned}$$

(h) Show that $\Gamma' : \pi_1(X, x_1) \to \pi_1(X, x_0)$ as defined in the proof of Theorem 6.16 is the inverse of Γ. That is, for any $\langle\alpha\rangle \in \pi_1(X, x_0)$, show that $(\Gamma' \circ \Gamma)(\langle\alpha\rangle) = \langle\alpha\rangle$; and for any $\langle\alpha'\rangle \in \pi_1(X, x_1)$, show that $(\Gamma \circ \Gamma')(\langle\alpha'\rangle) = \langle\alpha'\rangle$.

7. Suppose $f : X \to Y$ is a continuous function and x_0 is designated as the base point in X.

(a) Suppose $\alpha : [0, 1] \to X$ is a loop based at x_0. Quote a general theorem on continuity to justify the claim that $f \circ \alpha : [0, 1] \to Y$ is continuous. Check that the endpoints of $f \circ \alpha$ are both $f(x_0)$. Conclude that $f \circ \alpha$ is a loop in Y based at $f(x_0)$.

(b) Suppose α and α' are loops in X based at x_0 with $\langle\alpha\rangle = \langle\alpha'\rangle$. Use the homotopy from α to α' to define a homotopy from $f\circ\alpha$ to $f\circ\alpha'$. Be sure to check the starting and ending loops of your homotopy. Also check that all stages of your homotopy keep the end points of the loops at $f(x_0)$. Conclude that $f_*(\langle\alpha\rangle) = \langle f \circ \alpha\rangle$ is equal to $f_*(\langle\alpha'\rangle) = \langle f \circ \alpha'\rangle$, and therefore f_* is well-defined.

(c) Suppose α and β are loops in X based at x_0. Give a justification for each link in the following chain of equalities:

$$\begin{aligned}f_*(\langle\alpha\rangle\langle\beta\rangle) &= f_*(\langle\alpha\cdot\beta\rangle)\\ &= \langle f\circ(\alpha\cdot\beta)\rangle\\ &= \langle(f\circ\alpha)\cdot(f\circ\beta)\rangle\\ &= \langle f\circ\alpha\rangle\langle f\circ\beta\rangle\rangle\\ &= f_*(\langle\alpha\rangle)f_*(\langle\beta\rangle).\end{aligned}$$

Conclude that f_* is a homomorphism.

8. Suppose we have a way to associate to any topological space X a group $H(X)$. Suppose for any continuous function $f : X \to Y$ we have a homomorphism $f_* : H(X) \to H(Y)$. Suppose that $(\mathrm{id}_X)_* = \mathrm{id}_{H(X)}$ and if $f : X \to Y$ and $g : Y \to Z$, then $(g \circ f)_* = g_* \circ f_*$. In the language of category theory, H is a covariant functor from the category of topological spaces and continuous functions to the category of groups and homomorphisms.

Consider $S^1 = \{(x, y) \in \mathbb{R}^2 \mid x^2 + y^2 = 1\}$ as the boundary of $D^2 = \{(x, y) \in \mathbb{R}^2 \mid x^2 + y^2 \le 1\}$. Suppose $H(D^2) = \{\mathbf{0}\}$ and $H(S^1) = \mathbb{Z}$.

(a) Suppose there is continuous function $r : D^2 \to S^1$ with $r(z) = z$ for all $z \in S^1$. Describe the homomorphism r_*.

(b) Consider the inclusion function $i : S^1 \to D^2$. Describe the homomorphism i_*.

(c) What is $r \circ i$?

(d) Show that this leads to a contradiction, and conclude that no continuous function can retract a disk onto its boundary leaving the points of the boundary fixed.

6.4 The Sphere and the Circle

The first three sections of this chapter have laid a solid foundation for the theory of the fundamental group. Now we are ready to dig in and develop techniques for computing the fundamental groups of some interesting topological spaces.

Let us begin with a simple example based on Exercise 5 of Section 6.1.

Example 6.20 Let X be a convex subset of some Euclidean space. Let x_0 be designated as the base point of X. Show that $\pi_1(X, x_0)$ is isomorphic to the trivial group $\{\mathbf{0}\}$, the group whose only element is the multiplicative identity.

Solution. By Exercise 5 of Section 6.1, every loop in X is homotopic to the constant loop ε at the base point. Hence $\pi_1(X, x_0) = \{\langle\varepsilon\rangle\}$. Since $\langle\varepsilon\rangle$ is the multiplicative identity in $\pi_1(X, x_0)$, the function that maps $\langle\varepsilon\rangle$ to the identity element in $\{\mathbf{0}\}$ is an isomorphism. ❄

From the point of view of the algebraic structures, a space with a trivial fundamental group is in a sense the simplest. This is reflected in the following definition.

Definition 6.21 *A path-connected space X with a trivial fundamental group is called* ***simply connected.***

As an immediate application of the previous example, we see that a Euclidean space of any dimension is simply connected. So also is a ball $B^n = \{(x_1, \ldots, x_n) \mid x_1^2 + \cdots + x_n^2 \leq 1\}$ of any dimension. Exercises 9 through 12 at the end of this section illustrate some simply connected spaces that are not convex.

The next result illustrates several important topological principles.

Theorem 6.22 *The* 2*-sphere* S^2 *is simply connected.*

Proof. Take any loop α in S^2. For the sake of definiteness, we can think of the base point as the south pole of the sphere. We want to show that α is homotopic to the constant loop at the base point. The trick is to push α off the north pole so that its image lies entirely in a topological disk south of some sufficiently northerly latitude. Since a loop in a disk is homotopic to a constant loop, we will then be able to shrink the original loop to a constant loop.

This is simple enough for well-behaved loops. But what if α crosses through the north pole an infinite number of times (see Figure 6.23)? What if it acts like the curve in Example 7.31? To handle the general case, we need to invoke topological principles involving open sets, continuous functions, and the concept of compactness. The outline is given in the following paragraphs. The material in Chapter 7 will supply the missing justifications.

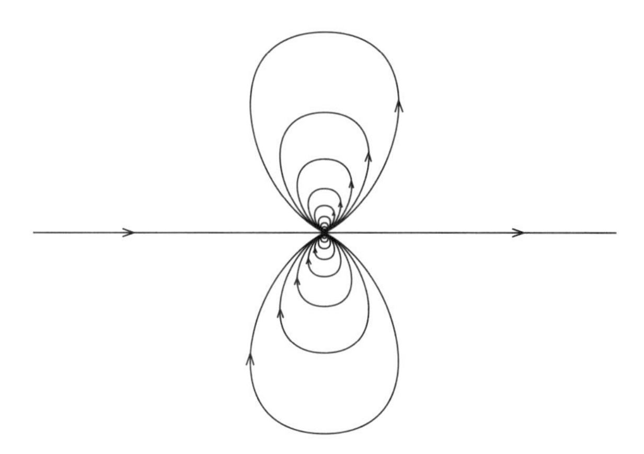

FIGURE 6.23
A loop passing through the north pole an infinite number of times.

Adjoin a strip south of the equator onto the northern hemisphere and a strip north of the equator onto the southern hemisphere to obtain two overlapping disks whose union is S^2. For instance, let D_+ be all points north of the Tropic of Capricorn and let D_- be all points south of the Tropic of Cancer.

Because α is continuous, we can find for any point $t \in [0, 1]$ an open interval $(t - \delta_t, t + \delta_t)$ such that α maps $(t - \delta_t, t + \delta_t) \cap [0, 1]$ entirely into D_+ or entirely into D_- (possibly into both). By the Heine-Borel Theorem (Theorem 7.34 of Section 7.34), the unit interval $[0, 1]$ has the crucial property known as compactness. This says that from among the infinite number of intervals $(t - \delta_t, t + \delta_t)$, we can choose a finite number whose union still contains all of $[0, 1]$.

Now use this finite collection of intervals to define a partition of $[0, 1]$ as follows. Let $c_0 = 0$ and let I_1 be one of these intervals that contains c_0. As you move to the right there will be a region where the interval I_1 overlaps an interval I_2 containing the right end point of I_1. Let c_1 be a point in this overlap region. Proceeding inductively, we obtain partition points $c_0, \ldots, c_m$ so that for $k = 1, \ldots, m$, the interval $[c_{k-1}, c_k]$ is contained in one of the intervals I_k in the finite collection. In particular, α maps each of the intervals $[c_{k-1}, c_k]$ entirely into D_+ or entirely into D_-.

For an interval $[c_{k-1}, c_k]$ that α maps into D_+, find $i \leq k - 1$ and $j \geq k$ so that α maps $[c_i, c_j]$ into D_+ and the endpoints into the Tropical region $D_+ \cap D_-$. Perform a homotopy of paths that leaves the endpoints fixed throughout the homotopy and deforms the restriction $\alpha|_{[c_i, c_j]}$ to a path in $D_+ \cap D_-$. Repeat such deformations as necessary to obtain a loop that lies entirely in the disk D_-. This loop is homotopic to the constant loop ε at the base point. It follows that the original loop is itself homotopic to ε. ❄

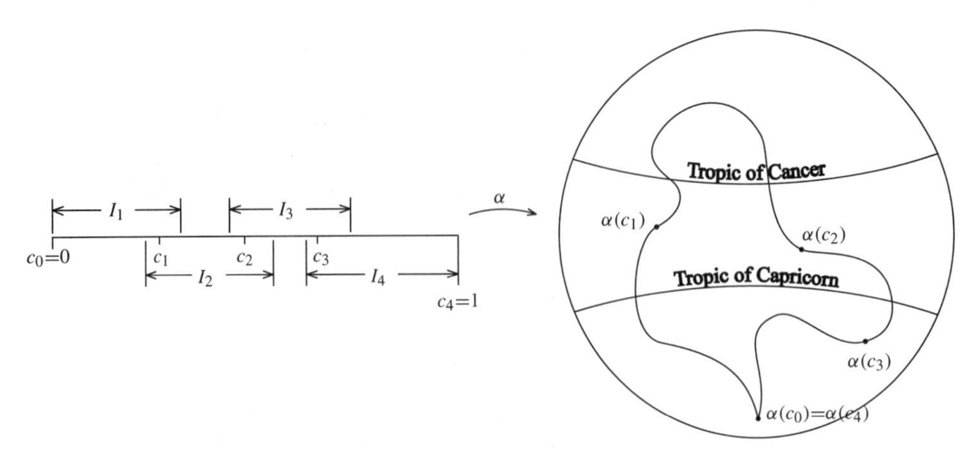

FIGURE 6.24
A loop α traveling in and out of the tropics.

The remainder of this section is devoted to computing the fundamental group of the circle S^1. Without further mention, we will use the point $(1, 0)$ on the x-axis as the base point. The following definition provides the key to determining $\pi_1(S^1)$.

Definition 6.25 *For a loop $\alpha : [0, 1] \to S^1$, let $\overline{\alpha} : [0, 1] \to \mathbb{R}$ measure the net angle by which α wraps around S^1. That is, $\overline{\alpha}$ is the continuous function from $[0, 1]$ to $\mathbb{R}$ that satisfies $\overline{\alpha}(0) = 0$ and $\alpha(t) = (\cos(\overline{\alpha}(t)), \sin(\overline{\alpha}(t)))$ for all $t \in [0, 1]$.*

Example 6.26 Determine the angle function $\overline{\alpha}$ for the loop $\alpha : [0, 1] \to S^1$ defined by $\alpha(t) = (\cos(2\pi t^2), \sin(2\pi t^2))$.

Solution. Because of the simple formula for α, we can read off the formula $\overline{\alpha}(t) = 2\pi t^2$ for the angle function. ❄

Example 6.27 Consider a loop $\alpha : [0, 1] \to S^1$ that wanders through the four quadrants in the following order: 1, 2, 1, 4, 1, 2, 3, 4, 3, 4, 1, 2, 3, 2, 1, 4, 1, 2, 3, 4. Sketch a graph of a possible angle function $\overline{\alpha} : [0, 1] \to \mathbb{R}$ for such a loop.

Solution. An easy way to sketch such a graph is to divide the range into intervals of length $\frac{\pi}{2}$ and mark them off in groups of four with the labels of the four quadrants. Then just sketch a graph as in Figure 6.28 that increases and decreases to enter into the quadrants in the specified order. ❄

Since any loop $\alpha : [0, 1] \to S^1$ begins and ends at the base point $(1, 0)$ of S^1, the net total angle $\overline{\alpha}(1)$ that α wraps around the circle will be an integral multiple of 2π. In this way we can associate an integer $\frac{1}{2\pi}\overline{\alpha}(1)$ with the loop α.

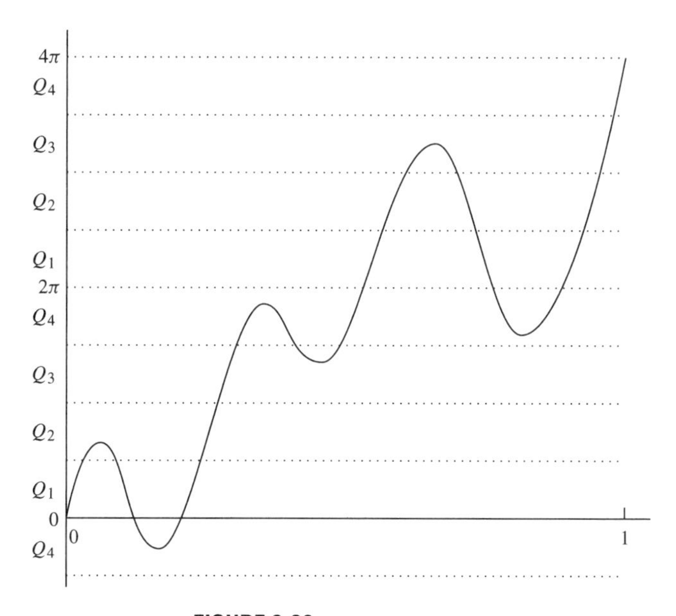

FIGURE 6.28
The angle function for a loop in S^1.

Definition 6.29 *The* ***degree*** *of a loop* $\alpha : [0,1] \to S^1$ *in* S^1 *is the integer denoted* $\deg \alpha$ *and defined by* $\deg \alpha = \frac{1}{2\pi}\overline{\alpha}(1)$.

The following lemma allows us to extend the definition of degree to the homotopy classes of loops in S^1.

Lemma 6.30 *Suppose* α *and* β *are homotopic loops in* S^1. *Then* $\deg \alpha = \deg \beta$.

Proof. Let $H : [0,1] \times [0,1] \to S^1$ be a homotopy from α to β. Then $H_t : [0,1] \to S^1$ defined by $H_t(s) = H(s,t)$ is a family of loops with $H_0 = \alpha$ and $H_1 = \beta$. Now H is continuous, so for any $s \in [0,1]$ the net angle that H_t on $[0,s]$ wraps around S^1 is a continuous function of t. Although the proof of this result actually involves quite a bit of technical fussing, we will accept it as a reasonable consequence of the continuity of the homotopy H. In particular $\overline{H}_t(1)$, the total amount of wrapping, is a continuous function of t. Of course, $\frac{1}{2\pi}\overline{H}_t(1)$ is also continuous. But $\frac{1}{2\pi}\overline{H}_t(1) = \deg(H_t)$ is always an integer. Hence, for any small change in t, there can be no change in $\deg(H_t)$. It follows that $\deg(H_t)$ must be constant for $t \in [0,1]$. Check out Exercises 7 and 8 if you feel the need for a more mathematical justification of the blithe claims of the preceding argument. In any case,

$$\deg \alpha = \deg H_0 = \deg H_1 = \deg \beta.$$

❄

Suppose $\alpha : [0,1] \to S^1$ is a loop in S^1. By the above lemma, any loop β in the homotopy class $\langle\alpha\rangle$, has degree equal to the degree of α. Hence we can define the degree of a homotopy class of loops in terms of the degree of any one of its representatives. With a pardonable reuse of notation, we have the following definition.

Definition 6.31 *The **degree** of an element* $\langle\alpha\rangle \in \pi_1(S^1)$ *is denoted* $\deg\langle\alpha\rangle$ *and defined by* $\deg\langle\alpha\rangle = \deg\alpha$.

Here is another basic fact about the degree function for loops in S^1.

Lemma 6.32 *Suppose α and β are loops in S^1. Then* $\deg(\alpha\cdot\beta) = \deg(\alpha) + \deg(\beta)$.

Proof. Recall that the concatenation $\alpha\cdot\beta$ is the loop that follows along α at twice the normal speed and then follows along β at twice the normal speed. Thus, the net angle that the $\alpha\cdot\beta$ wraps around S^1 when restricted to $[0, \frac{1}{2}]$ equals the net angle that α wraps around S^1. So far, we have a net angle of $\overline{\alpha\cdot\beta}\left(\frac{1}{2}\right) = \overline{\alpha}(1)$. We need to add onto this the net angle $\overline{\beta}(1)$ that β wraps around S^1. Thus,

$$\begin{aligned}
\deg(\alpha\cdot\beta) &= \frac{1}{2\pi}\overline{\alpha\cdot\beta}(1) \\
&= \frac{1}{2\pi}\overline{\alpha\cdot\beta}\left(\frac{1}{2}\right) + \frac{1}{2\pi}\left(\overline{\alpha\cdot\beta}(1) - \overline{\alpha\cdot\beta}\left(\frac{1}{2}\right)\right) \\
&= \frac{1}{2\pi}\overline{\alpha}(1) + \frac{1}{2\pi}(\overline{\beta}(1) - \overline{\beta}(0)) \\
&= \frac{1}{2\pi}\overline{\alpha}(1) + \frac{1}{2\pi}\overline{\beta}(1) \\
&= \deg(\alpha) + \deg(\beta).
\end{aligned}$$

❄

In terms of homotopy classes of loops in S^1, this yields an important result about the degree function.

Theorem 6.33 *The degree function* $\deg : \pi_1(S^1) \to \mathbb{Z}$ *is a homomorphism.*

Proof. We need to show that the degree function converts the multiplication in the group $\pi_1(S^1)$ into ordinary addition in $\mathbb{Z}$. But this is easy, thanks to Definition 6.31 and Lemma 6.32. Let $\langle\alpha\rangle$ and $\langle\beta\rangle$ be two elements of $\pi_1(S^1)$. Then

$$\begin{aligned}\deg(\langle\alpha\rangle\langle\beta\rangle) &= \deg(\langle\alpha\cdot\beta\rangle)\\ &= \deg(\alpha\cdot\beta)\\ &= \deg\alpha+\deg\beta\\ &= \deg\langle\alpha\rangle+\deg\langle\beta\rangle.\end{aligned}$$

❄

We now have all the pieces in place to determine the fundamental group of the circle. Notice that the degree function itself gives the isomorphism from the fundamental group to the additive group of integers.

Theorem 6.34 *The fundamental group of a circle is isomorphic to the integers with the operation of addition.*

Proof. Theorem 6.33 gives that $\deg : \pi_1(S^1) \to \mathbb{Z}$ is a homomorphism. Let us proceed to check the remaining conditions (one-to-one and onto) to confirm that it is an isomorphism.

To show that deg is an injection, let α and β be loops in S^1 with $\deg\langle\alpha\rangle = \deg\langle\beta\rangle$. It follows that $\overline{\alpha}(1) = 2\pi \deg\alpha = 2\pi \deg\beta = \overline{\beta}(1)$. Of course $\overline{\alpha}(0) = 0 = \overline{\beta}(0)$. Thus, we can consider the graphs of $\overline{\alpha}$ and $\overline{\beta}$ as paths in $\mathbb{R}^2$ with common endpoints. For any $s \in [0, 1]$ we can slide $\overline{\alpha}(s)$ linearly along a vertical line to $\overline{\beta}(s)$. And we can put all these slides together to define a homotopy $H(s,t) = \overline{\alpha}(s) + t(\overline{\beta}(s) - \overline{\alpha}(s))$ from the graph of $\overline{\alpha}$ to the graph of $\overline{\beta}$. Now for any $t \in [0, 1]$, this homotopy gives the angle function H_t of a loop in S^1 defined by $(\cos(H_t(s)), \sin(H_t(s)))$. Thus, $(\cos(H(s,t)), \sin(H(s,t)))$ defines a homotopy from the loop α to the loop β. It follows that $\langle\alpha\rangle = \langle\beta\rangle$, and hence that deg is an injection.

To show that deg is a surjection, consider the loop $\alpha : [0, 1] \to S^1$ defined by $\alpha(t) = (\cos 2\pi t, \sin 2\pi t)$. Since α wraps once around the circle, we have that

$$\deg\alpha = \frac{1}{2\pi}\overline{\alpha}(1) = \frac{1}{2\pi}2\pi = 1.$$

Thus, for any $n \in \mathbb{Z}$ we have $\langle\alpha\rangle^n \in \pi_1(S^1)$ with $\deg(\langle\alpha\rangle^n) = n\deg\langle\alpha\rangle = n$. ❄

Exercises 6.4

1. **(a)** Modify the proof that $\pi_1(S^2) \cong \{\mathbf{0}\}$ to show that $\pi_1(S^3) \cong \{\mathbf{0}\}$.
   (b) Generalize your argument to show for any $n \geq 2$ that $\pi_1(S^n) \cong \{\mathbf{0}\}$.
2. Determine the angle function $\overline{\alpha}$ for the loop $\alpha : [0, 1] \to S^1$ defined by $\alpha(t) = (\cos(10\pi t^2 - 4\pi t), \sin(10\pi t^2 - 4\pi t))$. What is the degree of α?
3. Determine the angle function $\overline{\alpha}$ for the loop $\alpha : [0, 1] \to S^1$ defined by $\alpha(t) = (\cos(2\pi 2^t), \sin(2\pi 2^t))$. What is the degree of α?
4. For an arbitrary integer n, give an explicit formula for a loop in S^1 of degree n.

5. Choose two interesting loops α and β in S^1.

(a) Sketch the graphs of $\overline{\alpha}$ and $\overline{\beta}$.

(b) Sketch the graph of $\overline{\alpha \cdot \beta}$.

(c) How does the graph of $\overline{\alpha}$ on $[0, 1]$ compare with the graph of $\overline{\alpha \cdot \beta}$ restricted to $\left[0, \frac{1}{2}\right]$?

(d) How does the graph of $\overline{\beta}$ on $[0, 1]$ compare with the graph of $\overline{\alpha \cdot \beta}$ restricted to $\left[\frac{1}{2}, 1\right]$?

(e) Label your graphs to illustrate that $\overline{\alpha \cdot \beta}\left(\frac{1}{2}\right) = \overline{\alpha}(1)$ and that $\overline{\alpha \cdot \beta}(1) = \overline{\alpha}(1) + \overline{\beta}(1)$.

6. Suppose a continuous function $f : [0, 1] \to \mathbb{R}$ satisfies $f(0) = 0$ and $\frac{1}{2\pi} f(1) \in \mathbb{Z}$.

(a) Show that $\alpha(t) = (\cos(f(t)), \sin(f(t)))$ defines a loop in S^1.

(b) What is the degree of this loop?

(c) What is the angle function of this loop?

7. Suppose $f : [0, 1] \to \mathbb{Z}$ is continuous at $t_0 \in [0, 1]$. Show that there is $\delta > 0$ such that for $t \in (t_0 - \delta, t_0 + \delta) \cap [0, 1]$, we have $f(t) = f(t_0)$. Suggestion: Use the definition of continuity (Definition 1.20) with $\varepsilon = \frac{1}{2}$.

8. Consider a locally constant function $f : [0, 1] \to \mathbb{R}$. That is, for any $t \in [0, 1]$ there is an interval $(t - \delta, t + \delta) \cap [0, 1]$ on which f is constant. Show that f is constant on its entire domain. Suggestion: Consider $\sup\{t \in [0, 1] \mid f(t) = f(0)\}$.

9. We can extend the idea of a homotopy from a parameterization of a family of loops to a parameterization of a family of functions between any two spaces. Let us consider a space X for which there is a **homotopy** from the identity function id_X to a constant function. This means that for some point $x_0 \in X$ there is a continuous function $H : X \times [0, 1] \to X$ such that for all $x \in X$,

$$H_0(x) = H(x, 0) = \mathrm{id}_X(x) = x, \qquad \text{and} \qquad H_1(x) = H(x, 1) = x_0.$$

This homotopy is called a **contraction**. A space X that has a contraction is called a **contractible** space.

(a) Show that $\mathbb{R}$ is contractible.

(b) Show that the interval $[0, 1]$ is contractible.

(c) Show that the interval $(0, 1)$ is contractible. Be careful that the image of your homotopy is contained in the space $(0, 1)$.

(d) Show that a convex subset of any Euclidean space is contractible.

(e) Show that a star-shaped subset of any Euclidean space is contractible.

10. Prove that a contractible space is path-connected.

11. Suppose H is a contraction of a space X to a point x_0. Let α be a loop in X based at x_0.

(a) Show that $H(\alpha(s), t)$ defines a parameterized family of loops from α to the constant loop ε at x_0.

(b) Let γ be the loop traversed by the base point under this deformation. That is, $\gamma(s) = H(\alpha(0), s) = H(\alpha(1), s)$. Show that $\gamma^{-1} \cdot \alpha \cdot \gamma$ is homotopic to ε.

(c) By algebraic manipulations in the group $\pi_1(X, x_0)$, use the equation $\langle \gamma^{-1} \cdot \alpha \cdot \gamma \rangle = \langle \varepsilon \rangle$ to conclude that $\langle \alpha \rangle = \langle \varepsilon \rangle$.

(d) Conclude that a contractible space is simply connected.

12. A **deformation retraction** of a space X to a subset A of X is a continuous function $R : X \times [0, 1] \to X$ such that $R_0(x) = R(x, 0) = x$ for all $x \in X$, $R_1(x) \in A$ for all $x \in X$, and $R_t(x) = R(x, t) = x$ for all $x \in A$ and all $t \in [0, 1]$. The subset A is called a **deformation retract** of the space X.

(a) Show that a deformation retraction of a space to a subset consisting of a single point is a contraction. Give an example of a contraction that is not a deformation retraction to a single-point set.

(b) Suppose R is a deformation retraction of a space X to a subset A. Suppose H is a contraction of X to a point x_0. Show that $R(H(x, s), 1)$ defines a contraction of A.

(c) Bing's house with two rooms is described in Exercise 5 of Section 3.1. Describe how you could deform a solid cube of clay to a model of Bing's house with two rooms. Conclude that Bing's house with two rooms is a deformation retract of the solid cube, and hence it is contractible. This is somewhat surprising since if you try to envision a contraction, you normally think of pushing the set in from some boundary until it is reduced to a point. However, Bing's house with two rooms has no boundary or edge from which to start pushing!

(d) The court jester hat is described in Exercise 5 of Section 3.1. Show that the court jester hat is contractible. Suggestion: Fill in the tip of the hat to get a contractible space that has the hat as a deformation retract.

13. Show that the circle is not contractible.

14. (a) Describe a deformation retraction of an annulus onto a circle.

(b) Describe a deformation retraction of a Möbius band onto a circle.

(c) Describe a deformation retraction of a solid torus onto a circle.

15. Suppose R is a deformation retraction of a space X to a subspace A. Designate a point $x_0 \in A$ as the base point for loops both in A and in X. Consider the homomorphism $i_* : \pi_1(A) \to \pi_1(X)$ induced by the inclusion function $i : A \to X$ defined by $i(a) = a$ for all $a \in A$.

(a) Let α be a loop in X. Show that $R_1 \circ \alpha$ is a loop in A that is homotopic to α. Conclude that i_* is onto.

(b) Let H be a homotopy in X from one loop α in A to another loop β in A. Show that $R_1 \circ H$ is a homotopy in A from α to β. Conclude that i_* is one-to-one.

(c) Conclude that i_* is an isomorphism from $\pi_1(A)$ to $\pi_1(X)$.

16. Show that the fundamental groups of the annulus, the Möbius band, and the solid torus are all isomorphic to the integers. In each case give a generator of the fundamental group (an element whose powers give all elements of the group).

17. Show that the annulus, the Möbius band, and the solid torus are not contractible.

18. Is seems intuitively clear that the disk D^2 cannot be continuously deformed onto its boundary S^1 without moving points of the boundary. The No-Retraction Theorem (Theorem 5.22) justified this intuition. You now have the tools to give a straightforward proof of the impossibility of this task. Start by assuming that a continuous function $r : D^2 \to S^1$ exists with $r(x) = x$ for all $x \in S^1$. Let $i : S^1 \to D^2$ denote the inclusion function. Notice that $r \circ i = \text{id}_{S^1}$. Thus, $r_* \circ i_* = (r \circ i)_* = \text{id}_{\pi_1(S^1)}$. Use the facts that $\pi_1(D^2) \cong \{\mathbf{0}\}$ and $\pi_1(S^1) \cong \mathbb{Z}$ to derive a contradiction.

6.5 Words and Relations

The fundamental group of the complement of a knot or link is a powerful tool for studying knots and links. This section introduces some techniques for computing the fundamental groups of such spaces.

The work in the previous section involved a lot of pushing and adjusting of loops. In proving that the 2-sphere is simply connected, we partitioned the domain and pushed the loop away from the north pole. In our computation of the fundamental group of the circle we saw that any loop α in S^1 can be deformed so it travels at a constant speed around the circle $\deg \alpha$ times.

These kinds of deformations are possible whenever we have a continuous function between two triangulated spaces. We can think of the spaces as the union of the vertices, edges, faces, and higher-dimensional cells in the triangulation. Such a combinatorial decomposition of a space is known as a **simplicial complex**.

The Simplicial Approximation Theorem is a concise statement of the general result for functions between any two triangulated spaces. It says that on a suitable subdivision of the domain, any continuous function can be homotopically deformed by an arbitrarily small amount so that the modified function sends vertices to vertices and is linear on each edge, face, tetrahedron, and higher-dimensional cell of the triangulation. The book *Topology of Surfaces* [42] by L. Christine Kinsey is a nice reference for this marvelously useful result.

In this section we will use the Simplicial Approximation Theorem to replace continuous loops and homotopies with piecewise-linear loops and homotopies. Here is an example of the way this works.

Example 6.35 By the Simplicial Approximation Theorem, we can represent any homotopy class in $\pi_1(S^1)$ by a piecewise-linear loop σ. That is, we can find points $0 = c_0 < c_1 < \cdots < c_m = 1$ to partition $[0, 1]$ and we can consider the circle as a polygon in such a way that σ maps the partition points to vertices of S^1 and is linear on any subinterval $[c_{k-1}, c_k]$. Pick any point $p \in S^1$ that is not a vertex. Show that the degree of σ is the number of subintervals that are mapped so they pass through p going counterclockwise, minus the number that pass through p going clockwise.

Solution. For each $k = 0, \ldots, m$, let γ_k be a path that goes directly from the base point to $\sigma(c_k)$, does not pass through p, and does not collect $200. Let σ_k denote the path obtained by reparameterizing the restriction of σ to the subinterval $[c_{k-1}, c_k]$. Notice that σ is homotopic to the concatenation $\sigma_1 \cdot \sigma_2 \cdot \cdots \cdot \sigma_m$. Replace each σ_k with the loop

$\gamma_{k-1} \cdot \sigma_k \cdot \gamma_k$. Because of all the backtracking along the paths γ_k, we see that σ is homotopic to $(\gamma_0 \cdot \sigma_1 \cdot \gamma_1^{-1}) \cdot (\gamma_1 \cdot \sigma_2 \cdot \gamma_2^{-1}) \cdot \cdots \cdot (\gamma_{k-1} \cdot \sigma_k \cdot \gamma_k^{-1})$. By Lemma 6.30 these two loops have the same degree. By Lemma 6.32, the additive property of the degree function, we have

$$\begin{aligned} \deg \sigma &= \deg((\gamma_0 \cdot \sigma_1 \cdot \gamma_1^{-1}) \cdot (\gamma_1 \cdot \sigma_2 \cdot \gamma_2^{-1}) \cdot \cdots \cdot (\gamma_{m-1} \cdot \sigma_m \cdot \gamma_m^{-1})) \\ &= \sum_{k=1}^{m} \deg(\gamma_{k-1} \cdot \sigma_k \cdot \gamma_k^{-1}). \end{aligned}$$

But

$$\deg(\gamma_{k-1} \cdot \sigma_k \cdot \gamma_k^{-1}) = \begin{cases} +1 & \text{if } \sigma_k \text{ passes counterclockwise through } p, \\ 0 & \text{if } \sigma_k \text{ does not pass through } p, \\ -1 & \text{if } \sigma_k \text{ passes clockwise through } p. \end{cases}$$

Figure 6.36 shows a typical example of how these paths fit together. The paths have been pushed into the interior of the circle for purposes of illustration. The actual paths would travel along the circle. ❄

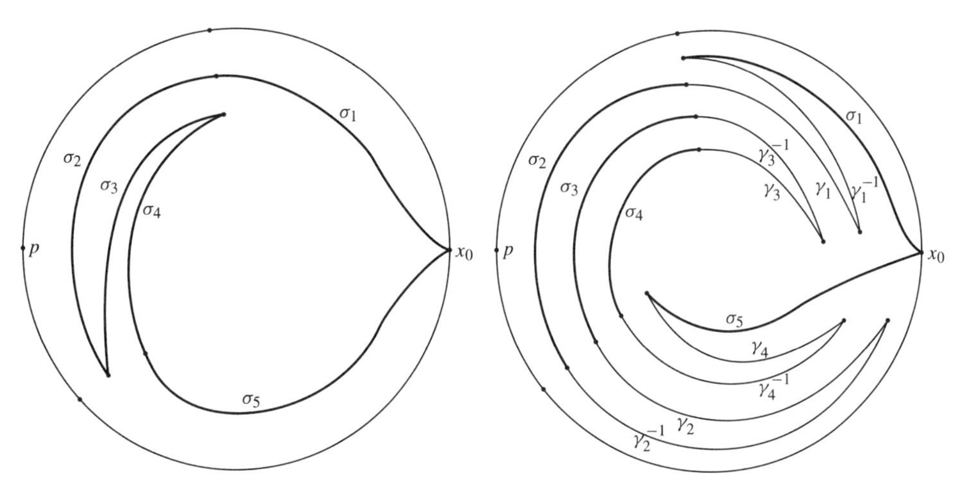

FIGURE 6.36
A loop $\sigma = \sigma_1 \cdot \sigma_2 \cdot \sigma_3 \cdot \sigma_4 \cdot \sigma_5$ and an equivalent concatenation of loops $(\gamma_0 \cdot \sigma_1 \cdot \gamma_1^{-1}) \cdot (\gamma_1 \cdot \sigma_2 \cdot \gamma_2^{-1}) \cdot (\gamma_2 \cdot \sigma_3 \cdot \gamma_3^{-1}) \cdot (\gamma_3 \cdot \sigma_4 \cdot \gamma_4^{-1}) \cdot (\gamma_4 \cdot \sigma_5 \cdot \gamma_5^{-1})$ pushed into the interior of a circle.

We can extend the ideas of Example 6.35 to other spaces. An interesting challenge involves the union of two circles that intersect in a single point. This figure-eight is called the **wedge** of two circles and is denoted $S^1 \vee S^1$. The point at which the two circles meet is designated as the base point.

Let $\langle \alpha \rangle$ denote that homotopy class of a counterclockwise loop around one of the circles, and let $\langle \beta \rangle$ denote the homotopy class of a counterclockwise loop around the other circle. In $\pi_1(S^1 \vee S^1)$ we not only have positive and negative powers of $\langle \alpha \rangle$ and $\langle \beta \rangle$, but we also have elements that involve both circles of this space such as $\langle \alpha \rangle \langle \beta \rangle$, $\langle \beta \rangle \langle \alpha \rangle \langle \beta \rangle^{-1}$, and $\langle \alpha \rangle \langle \beta \rangle \langle \alpha \rangle \langle \beta \rangle \langle \beta \rangle \langle \alpha \rangle \langle \beta \rangle \langle \beta \rangle \langle \beta \rangle \langle \alpha \rangle$.

We can algebraically represent these elements of $\pi_1(S^1 \vee S^1)$ as strings of the four symbols a, a^{-1}, b, and b^{-1} corresponding to $\langle\alpha\rangle$, $\langle\alpha\rangle^{-1}$, $\langle\beta\rangle$, $\langle\beta\rangle^{-1}$, respectively. Such a string is called a **word** in the symbols a and b. From such a word, we delete any occurrence of a letter adjacent to its inverse. We can write these words using exponents to abbreviate a block of repeated letters. For example, $a^2b^5a^{-4}b$ is an abbreviation for $aabbbbba^{-1}a^{-1}a^{-1}a^{-1}b$. Of course something like $a^{-1}a^6a^{-2} = a^{-1}aaaaaaa^{-1}a^{-1}$ reduces to $aaa = a^3$.

We multiply two words by juxtaposing the words and performing any cancellation. For example, the product of a^4b^{-2} times $b^2a^{-1}b$ is $a^4b^{-2}b^2a^{-1}b = a^4a^{-1}b = a^3b$. With this operation, the set of all reduced words forms a group known as the **free group** on the two generators a and b. We denote this group $\langle a, b \mid \; \rangle$. The letters a, b to the left of the vertical bar denote the generators. The blank to the right of the bar indicates that there are no relations among the words in a and b other than the cancellation of the letters adjacent to their inverses.

Consider the function $\theta : \langle a, b \mid \; \rangle \to \pi_1(S^1 \vee S^1)$ that takes a word in a, b to the product formed from the corresponding word in $\langle\alpha\rangle$ and $\langle\beta\rangle$. For example, $\theta(b^{-2}ab^3a^2) = \langle\beta\rangle^{-2}\langle\alpha\rangle\langle\beta\rangle^3\langle\alpha\rangle^2$. The cancellation rules in the free group hold in $\pi_1(S^1 \vee S^1)$, so θ is well-defined. The definition of θ automatically makes it a homomorphism. The following two theorems are the additional conditions we need in order to verify that θ is an isomorphism. These results establish that the fundamental group of $S^1 \vee S^1$ is the free group on two generators.

Theorem 6.37 *The homomorphism $\theta : \langle a, b \mid \; \rangle \to \pi_1(S^1 \vee S^1)$ is a surjection.*

Proof. Let $\langle\sigma\rangle$ be an arbitrary element of $\pi_1(S^1 \vee S^1)$. By the Simplicial Approximation Theorem, we can assume that the loop σ is piecewise linear. That is, we can partition $[0, 1]$ with points $0 = c_0 < c_1 < \cdots < c_m = 1$ and consider the circles as polygons meeting at a common vertex in such a way that σ maps the partition points to vertices of $S^1 \vee S^1$ and is linear on any subinterval $[c_{k-1}, c_k]$.

Choose a point p in the interior of an edge of one of the circles of $S^1 \vee S^1$ and a point q in the interior of an edge of the other. Notice that any loop in $S^1 \vee S^1$ that misses p and q is homotopic to the constant loop ε. Indeed, the complement $S^1 \vee S^1 - \{p, q\}$ is homeomorphic to an $\times$-shaped space. By Exercise 9 of Section 6.4, this convex set is contractible. Hence by Exercise 11 in that section, the set is simply connected.

As in Example 6.35, choose paths from the base point to the vertices of $S^1 \vee S^1$. Use these paths to connect each of the restrictions $\sigma|_{[c_{k-1},c_k]}$ to the base point and convert it into a loop. Any such loop that crosses p or q is homotopic to α, α^{-1}, β, or β^{-1}. All other loops are homotopic to ε. But σ is homotopic to the product of these loops. Thus, the corresponding word in a and b is an element of $\langle a, b \mid \; \rangle$ that θ maps to $\langle\sigma\rangle$. ❄

Theorem 6.38 *The homomorphism $\theta : \langle a, b \mid \; \rangle \to \pi_1(S^1 \vee S^1)$ is an injection.*

Proof. By Exercise 2 of Section 6.3 we need only consider a word in $\langle a, b \mid \ \rangle$ that the homomorphism θ maps to the identity element in $\pi_1(S^1 \vee S^1)$. So suppose some word like $s = a^3ba^{-1}b^2$ maps to the homotopy class $\langle\alpha\rangle^3\langle\beta\rangle\langle\alpha\rangle^{-1}\langle\beta\rangle^2$ that is equal to the identity element. That is, the loop $\sigma = \alpha^3 \cdot \beta \cdot \alpha^{-1} \cdot \beta^2$ is homotopic to the constant loop ε. We will use the homotopy to read off how to insert and delete $\alpha \cdot \alpha^{-1}$, $\alpha^{-1} \cdot \alpha$, $\beta \cdot \beta^{-1}$, and $\beta^{-1} \cdot \beta$ to build σ from ε. This will provide directions for building the word s by the corresponding insertions and deletions of aa^{-1}, $a^{-1}a$, bb^{-1}, and $b^{-1}b$. In particular, we will conclude that s reduces to the identity element in the free group $\langle a, b \mid \ \rangle$.

By the Simplicial Approximation Theorem, we can assume the homotopy $H : [0, 1] \times [0, 1] \to S^1 \vee S^1$ from σ to ε is piecewise-linear. That is, we can triangulate $[0, 1] \times [0, 1]$ and we can consider the circles as polygons meeting at a common vertex in such a way that H maps vertices to vertices, and it maps each edge and triangular face of $[0, 1] \times [0, 1]$ either to a vertex or linearly to an edge of $S^1 \vee S^1$.

In the triangulation of the domain of H, we can collapse the triangles one at a time until all that is left is the interval $[0, 1] \times \{0\}$ on which H corresponds to σ. This is similar to shelling a disk (see Exercise 11 of Section 1.6), but it is even easier since the portion of $[0, 1] \times [0, 1]$ remaining after each triangular move does not need to be a disk. Each collapse deforms one side of a triangle onto the other two sides, or it deforms two sides of a triangle onto the third side, or it deforms three sides of a triangle to a point.

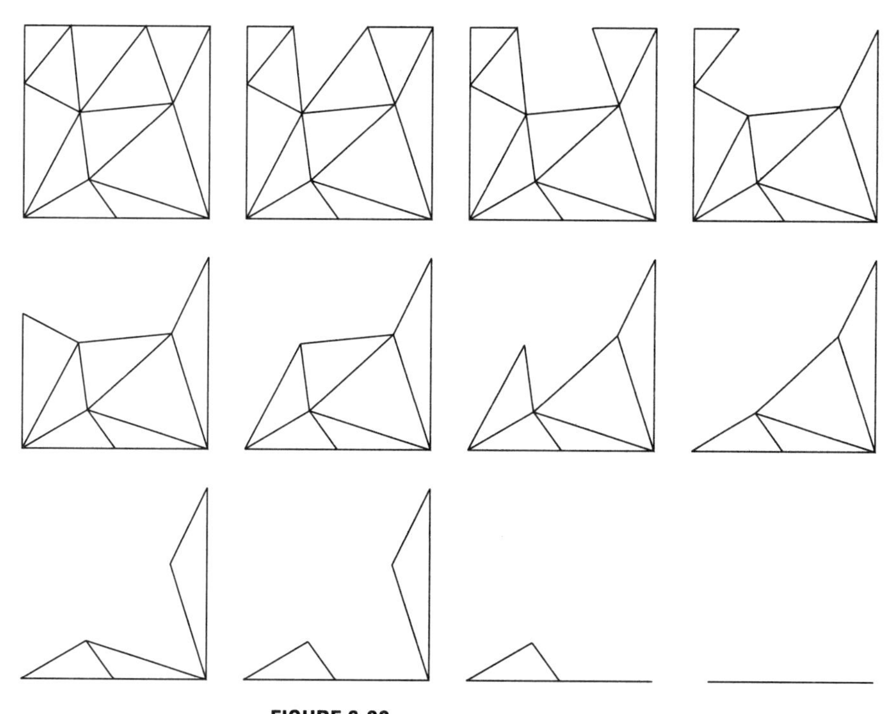

FIGURE 6.39
Collapsing the domain of a homotopy.

Since H maps the left and right sides of $[0, 1] \times [0, 1]$ to the base point of $S^1 \vee S^1$, we can think of starting with ε as the image under H of the path up the left side, across the top, and down the right side of $[0, 1] \times [0, 1]$. The sequence of collapses of the triangles produces a sequence of loops in $S^1 \vee S^1$. Starting with ε and ending with σ, each loop in this sequence is homotopic to the next (see Exercise 6 at the end of this section).

Choose a point p in the interior of an edge of one of the circles of $S^1 \vee S^1$ and a point q in the interior of an edge of the other. As in Example 6.35, choose for each vertex a path that runs directly from the base point to the vertex without passing through p or q. We can write each loop in our sequence as the concatenation of loops formed by a path to a vertex followed by a path along an edge followed by a path back to the base point. Notice that any of these loops through p is homotopic to α or α^{-1}, and any of these loops through q is homotopic to β or β^{-1}. Also recall from the proof of Theorem 6.37 that any of these loops that misses p and q is homotopic to the constant loop. It follows that we can read off the homotopy class of any of our loops as a product of $\langle\alpha\rangle$, $\langle\alpha\rangle^{-1}$, $\langle\beta\rangle$, and $\langle\beta\rangle^{-1}$ by noting the order and direction it crosses through p and q.

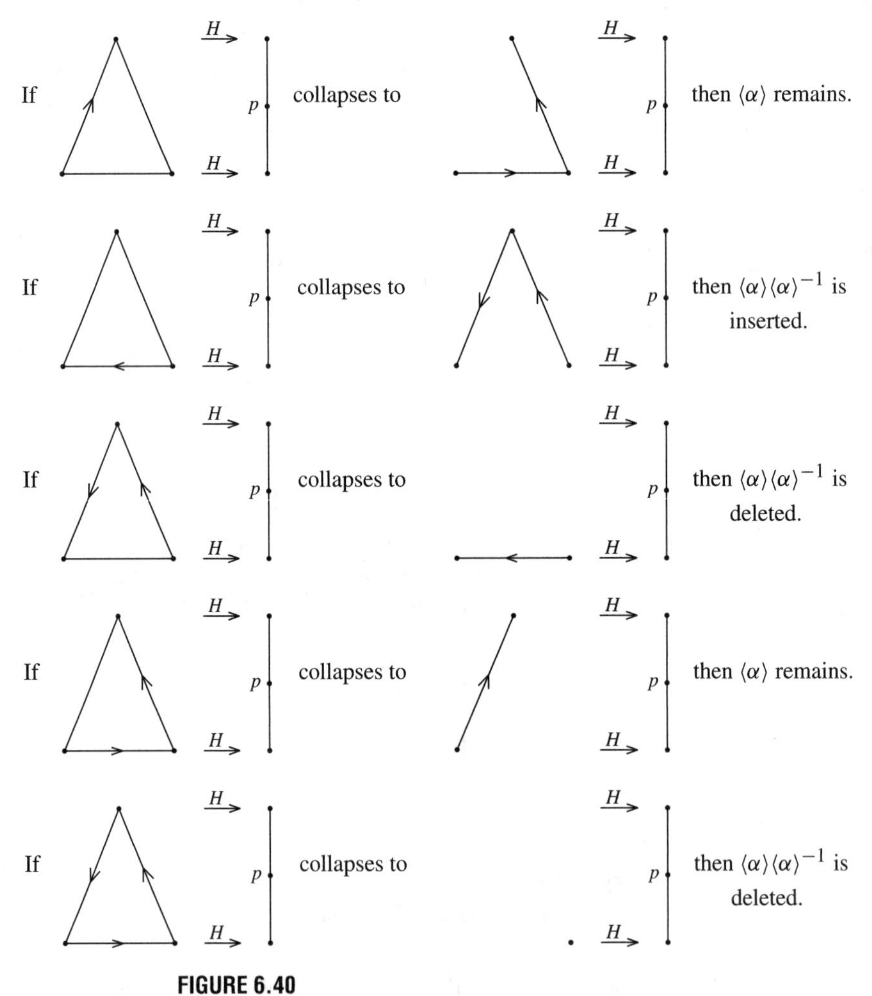

FIGURE 6.40
The collapse of a triangle whose image passes through p.

Aside from reflections and simple variations of orientation, there are only five ways a triangle whose image passes through p can collapse and have a potential effect on the representation of the homotopy class in our sequence of loops. Figure 6.40 illustrates these possibilities and indicates the effect of the collapse on the representation of the homotopy class in terms of $\langle\alpha\rangle$, $\langle\alpha\rangle^{-1}$, $\langle\beta\rangle$, and $\langle\beta\rangle^{-1}$. The only changes are the insertion or deletion of a generator times its inverse. A similar result holds for the collapse of a triangle whose image passes through q.

This shows that the representation of $\langle\sigma\rangle$ as a product of $\langle\alpha\rangle$, $\langle\beta\rangle$, and their inverses can be built up from the null string by inserting and deleting $\langle\alpha\rangle\langle\alpha\rangle^{-1}$, $\langle\alpha\rangle^{-1}\langle\alpha\rangle$, $\langle\beta\rangle\langle\beta\rangle^{-1}$, and $\langle\beta\rangle^{-1}\langle\beta\rangle$. Therefore the corresponding word s in the free group $\langle a, b \mid \ \rangle$ can likewise be built up from the null string by the insertion and deletion of aa^{-1}, $a^{-1}a$, bb^{-1}, and $b^{-1}b$. That is, the reduced form of s is the null string, the identity element in $\langle a, b \mid \ \rangle$. ❄

We have seen how the use of piecewise-linear loops and homotopies permits us to work with the fundamental groups of simple spaces in a reasonable fashion. These techniques work very well when applied to the complements of knots and links in $\mathbb{R}^3$. The general theory is illustrated below in the simple case of the Hopf link.

Example 6.41 The Hopf link consists of two circles A and B embedded in $\mathbb{R}^3$ as indicated in Figure 6.42. Show that the homotopy classes of the two loops α and β in this figure generate the fundamental group of the complement of the Hopf link.

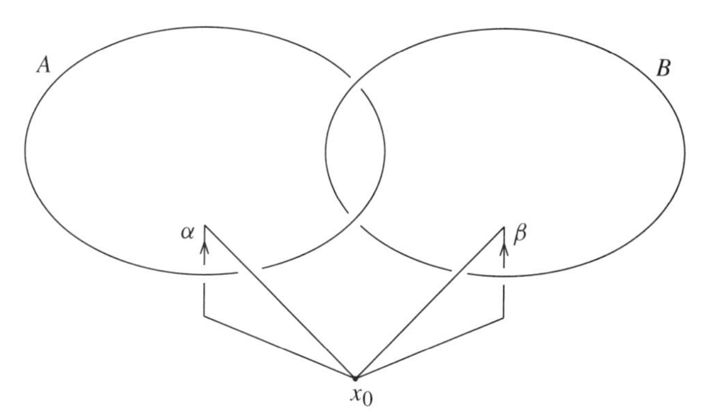

FIGURE 6.42
The Hopf link and generators of the fundamental group of its complement.

Solution. Let σ be a piecewise-linear loop in the complement of the Hopf link. We will identify the function σ with its oriented polygonal image. Assume that the link is also polygonal and that the projections of the link and σ are in general position. In particular, the projection of each edge of σ intersects the projection of the link in at most one point and that point is neither a vertex of σ nor a crossing point of the two components of the link. Run paths from the base point to each of the vertices of σ, making sure they are above the link whenever their projections intersect the projection of the link. Use these paths to convert the edges of σ into loops. As in Example 6.35 and Theorem 6.37, σ is homotopic to the concatenation of these loops. But each of these loops is homotopic to α or α^{-1} if the edge of σ passes under circle A, or it is homotopic to β or β^{-1} if the edge passes under circle B, or it is homotopic to ε if the edge does not pass under the link. In any case, σ is homotopic to a product of the loops α, β, and their inverses. ❄

Example 6.43 Consider the loops α and β indicated in Figure 6.42. Show that $\alpha \cdot \beta$ is homotopic to $\beta \cdot \alpha$ in the complement of the Hopf link.

Solution. Figure 6.44 illustrates steps in a deformation of the loop $\alpha \cdot \beta$ to the loop $\beta \cdot \alpha$. Since this deformation is done with a homotopy, it is legitimate for the loop to pass through itself. Of course, it must not pass through the link. ❄

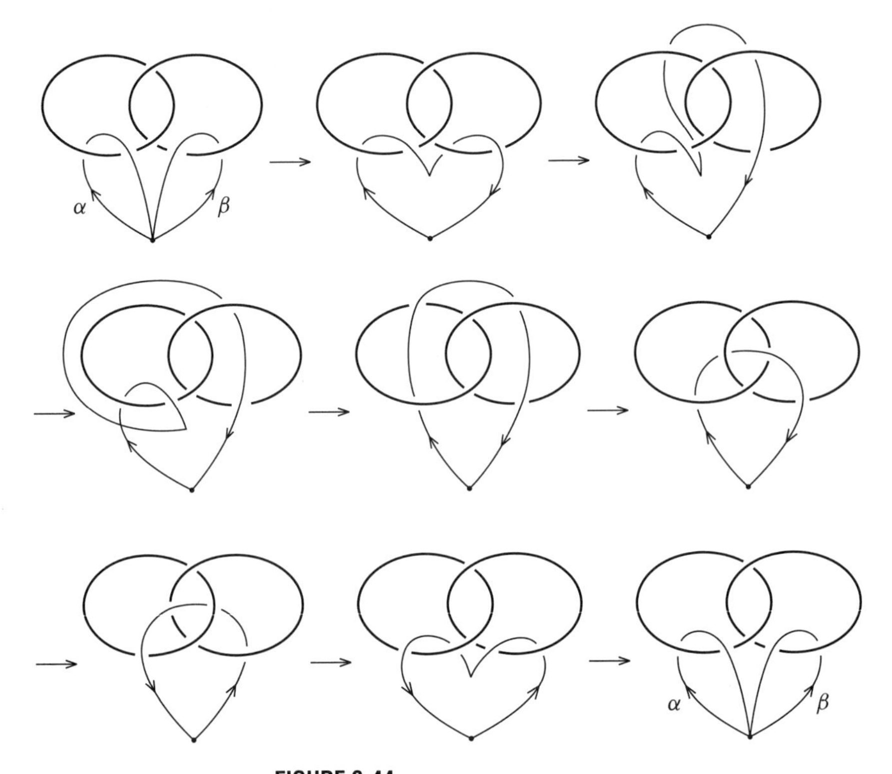

FIGURE 6.44
The generators $\langle\alpha\rangle$ and $\langle\beta\rangle$ commute.

Example 6.45 Consider two homotopic loops in the complement of the Hopf link. Write the loops as the product of the generators α and β as in Example 6.41. Show that the word for one loop can be transformed to the word for the other loop by interchanging powers of α and powers of β and canceling the occurrences of a generator adjacent to its inverse.

Solution. The steps parallel those of the proof of Theorem 6.38. Assume the homotopy is piecewise-linear and that the images of the triangles are in general position under a regular projection of the Hopf link. Then collapse the triangles in the domain of the homotopy leaving the interval corresponding to the final stage of the homotopy. Rewrite the homotopy as a sequence of homotopies obtained from these triangular collapses. As illustrated in Figure 6.46, each triangular disk will have edges that pass under the Hopf link zero, two, or four times. If this number is zero, collapsing the triangle will not change the representation of the loop. If the number is two, collapsing the triangle has the effect of inserting or deleting a generator and its inverse. If the number is four, then collapsing the triangle has the effect of inserting or deleting some permutation of the four paths α, β, α^{-1}, β^{-1}. By Example 6.43 and Exercise 13, the homotopy classes of all four of these paths commute. So this case also amounts to inserting or deleting pairs of a generator times its inverse. ❄

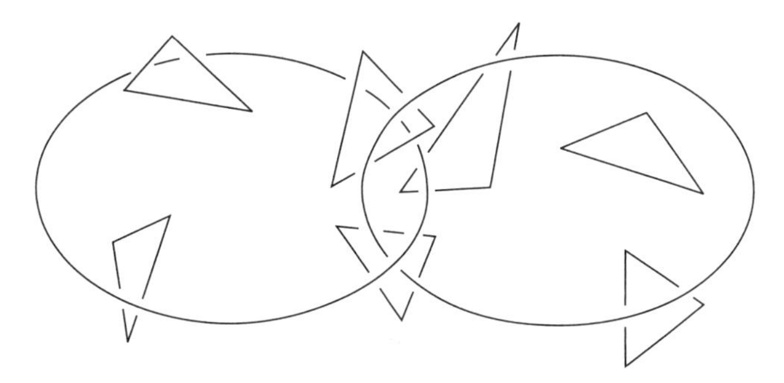

FIGURE 6.46
Typical triangular disks in the complement of the Hopf link.

Exercises 6.5

1. Use the Simplicial Approximation Theorem or the compactness arguement of the proof of Lemma 6.22 to show that the fundamental group of a triangulated space is carried by the edges of the triangulation in the sense that every loop is homotopic to a loop in this one-dimensional skeleton of the space.

2. Use the Simplicial Approximation Theorem to show that the relations among elements of the fundamental group of a triangulated space are carried by the faces of the triangulation in the sense that every homotopy can be represented by one whose image is in the two-dimensional skeleton of the space.

3. **(a)** Describe the elements in the free group $\langle a, b, c \mid \ \rangle$ on three generators. Give some examples of the multiplication of typical elements of this group.

(b) Draw a picture of the wedge of three circles $S^1 \vee S^1 \vee S^1$ with all three circles intersecting at the base point.

(c) Define a homomorphism $\theta : \langle a, b, c \mid \ \rangle \to \pi_1(S^1 \vee S^1 \vee S^1)$ as was done for the wedge of two circles.

(d) Show that θ is a surjection.

(e) Show that θ is an injection.

4. Generalize the previous exercise to compute the fundamental group of the wedge $\bigvee_{k=1}^{n} S^1$ of n circles.

5. Compute the fundamental group of the wedge $S^1 \vee S^2$ of a circle and a sphere.

6. **(a)** Draw a picture to illustrate a continuous function from a square region to a triangular region that maps one side of the square to one edge of the triangle and the opposite side of the square to the other two sides of the triangle. The other two sides of the square should map to vertices of the triangle.

(b) Draw a picture to illustrate a continuous function from a square region to a triangular region that maps one side of the square around the three sides of the triangle and maps the other three sides to a vertex of the triangle.

7. **(a)** Show that $S^1 \vee S^1$ is a deformation retract of a disk with two holes.

(b) Use Exercise 15 of Section 6.4 to show that the fundamental group of the disk with two holes is the free group on two generators.

(c) Draw loops in the disk with two holes that represent ab, ba, bab^{-1}, and one of your favorite words in the free group on a and b.

8. **(a)** Let 1 denote the empty string. This string of no letters is called the **null** string. Show that 1 is the identity element in $\langle a, b \mid \ \rangle$.

(b) Find the inverse of ab in $\langle a, b \mid \ \rangle$. Be sure to check that the inverse you find works on both sides of ab.

(c) Describe the inverse of an arbitrary element of $\langle a, b \mid \ \rangle$.

9. In the free group $\langle a, b \mid \ \rangle$, let $c = ab$. Show that any element of $\langle a, b \mid \ \rangle$ can be written in terms of a and c.

10. It seems obvious that there are no three-dimensional subspaces of the vector space $\mathbb{R}^2$. How could there possibly be a subspace of $\mathbb{R}^2$ that requires three vectors to span it? Think of the many hours you spent in your linear algebra class developing the concepts of spanning and linear independence in order to prove obvious results such as this.

(a) Isn't it equally obvious that a group with two generators cannot have a subgroup that requires more than two generators?

(b) Consider the subset S of $\langle a, b \mid \ \rangle$ consisting of all words in which a and a^{-1} occur equally often. Show this is a subgroup of $\langle a, b \mid \ \rangle$.

(c) Show that any element of S can be written as a product of the elements of the form $a^{-k}ba^k$ and $a^{-k}b^{-1}a^k$ for various integer values of k.

(d) Show that none of the elements described in part (c) can be written as a product of any of the others of these elements.

(e) Conclude that S requires not just two, but an infinite number of generators.

11. Draw a typical polygonal loop σ in the complement of the Hopf link. Run paths above the link from the base point to the vertices of your loop. Use these paths to convert the edges of σ into loops. Convince yourself that each of these loops is homotopic to α or α^{-1} if the edge passes under circle A, homotopic to β or β^{-1} if the edge passes under circle B, and homotopic to ε otherwise.

12. Write down the relations in α and β corresponding to the triangular disks in Figure 6.46. Which one corresponds to a relation that does not hold in the free group on two generators?

13. Suppose a and b are elements of a group that satisfy the condition $ab = ba$.

(a) Show that $ab^{-1} = b^{-1}a$. Suggestion: Write ab^{-1} as $b^{-1}bab^{-1}$.

(b) Show that $a^{-1}b = ba^{-1}$.

(c) Show that $a^{-1}b^{-1} = b^{-1}a^{-1}$.

14. Let X be the space that remains when we remove from $\mathbb{R}^3$ the z-axis and the unit circle in the xy-plane.

(a) Show that $\pi_1(X)$ is generated by two elements.

(b) Show that the two generators commute.

(c) Show that all relations between products of the generators for homotopic loops are consequences of the commutativity and cancellation of a generator with its inverse.

(d) Describe a deformation retract of X onto a torus.

(e) Draw generators of $\pi_1(T^2)$ as loops that go around the two different holes of the torus.

15. (a) Let X be the complement in $\mathbb{R}^3$ of two circles that are unknotted and unlinked. Show that $\pi_1(X)$ is a free group on two generators.

(b) Let X be the complement in $\mathbb{R}^3$ of n circles that are unknotted and unlinked. Show that $\pi_1(X)$ is a free group on n generators.

(c) Prove that the Borromean rings illustrated in Figure 2.8 of Section 2.1 are actually linked.

6.6 The Poincaré Conjecture

Chapter 4 provided many examples of 3-manifolds but not many tools for distinguishing among these examples. Although the Euler characteristic is a powerful invariant for surfaces, it is of little use in distinguishing among 3-manifolds. Indeed, we know from Theorem 4.11 that the Euler characteristic of any closed 3-manifold is 0. In this section we will develop some techniques for computing the fundamental group, and we will see how the fundamental group is much more helpful in distinguishing among 3-manifolds. Alas, however, the fundamental group is not powerful enough to classify 3-manifolds. We do not even know whether there are 3-manifolds other than S^3 with trivial fundamental group. This is one of the most basic problems of geometric topology, a problem that has inspired a great deal of research during its long history, but which is still an open conjecture.

Our first task is to determine the fundamental group of a Cartesian product of two spaces in terms of the fundamental groups of the two factors. We begin with a construction that can be applied to any two groups. (Recall that a group is a set that has an associative multiplication for which there is an identity element, and such that each element has an inverse.)

Example 6.47 Define a group multiplication on the Cartesian product $G \times H$ of two groups G and H.

Solution. For elements (g_1, h_1) and (g_2, h_2) of $G \times H$, we define componentwise multiplication $(g_1, h_1)(g_2, h_2) = (g_1 g_2, h_1 h_2)$. Since the multiplication of elements of the two groups is associative, it is easy to check that this multiplication on $G \times H$ is also associative:

$$\begin{aligned}(g_1, h_1)((g_2, h_2)(g_3, h_3)) &= (g_1, h_1)(g_2 g_3, h_2 h_3)\\ &= (g_1(g_2 g_3), h_1(h_2 h_3))\\ &= ((g_1 g_2)g_3, (h_1 h_2)h_3)\\ &= (g_1 g_2, h_1 h_2)(g_3, h_3)\\ &= ((g_1, h_1)(g_2, h_2))(g_3, h_3).\end{aligned}$$

Let 1_G denote the multiplicative identity in G, and let 1_H denote the multiplicative identity in H. The string of equalities

$$(g, h)(1_G, 1_H) = (g1_G, h1_H) = (g, h) = (1_G g, 1_H h) = (1_G, 1_H)(g, h)$$

shows that $(1_G, 1_H)$ is the multiplicative identity of $G \times H$. Similarly, the string

$$(g, h)(g^{-1}, h^{-1}) = (gg^{-1}, hh^{-1}) = (1_G, 1_H) = (g^{-1}g, h^{-1}h) = (g^{-1}, h^{-1})(g, h),$$

shows that (g^{-1}, h^{-1}) is the inverse of (g, h). ❄

In order to consider loops and homotopies in a Cartesian product of two spaces, we need to know when a function into a Cartesian product is continuous. Recall from your study of multivariable calculus that a function $f : A \to \mathbb{R}^2$ is defined by $f(a) = (f_1(a), f_2(a))$, where $f_1 : A \to \mathbb{R}$ and $f_2 : A \to \mathbb{R}$ give the two coordinates of the point in $\mathbb{R}^2$ that f assigns to any element of its domain A. Furthermore, f is continuous if and only if both functions f_1 and f_2 are continuous. The underlying principle is that points in $\mathbb{R}^2$ are close if and only if the corresponding coordinates of the two points are close. It is reasonable to extend this to the general situation.

Continuity Principle for Cartesian Products 6.48 *A function $f : A \to X \times Y$ from a set A to the Cartesian product of two sets X and Y is continuous if and only if $f_1 : A \to X$ and $f_2 : A \to Y$ are continuous, where $f(a) = (f_1(a), f_2(a))$ for all $a \in A$.*

Let X be a space with base point x_0 and let Y be a space with base point y_0. We designate (x_0, y_0) as the base point for the Cartesian product $X \times Y$. With the use of subscripts to denote the coordinate functions of a function into a Cartesian product, we can associate with any loop $\alpha : [0, 1] \to X \times Y$ a pair of loops (α_1, α_2). Or we can reverse this process and use loops $\alpha_1 : [0, 1] \to X$ and $\alpha_2 : [0, 1] \to Y$ to define a loop $\alpha : [0, 1] \to X \times Y$ such that $\alpha(t) = (\alpha_1(t), \alpha_2(t))$. This bijection between loops in $X \times Y$ and the Cartesian product of the set of loops in X with the set of loops in Y is compatible with the homotopies of loops in these spaces. In fact, this bijection defines an isomorphism as indicated in the following theorem.

Theorem 6.49 *The function* $\theta : \pi_1(X \times Y) \to \pi_1(X) \times \pi_1(Y)$ *defined by* $\theta(\langle\alpha\rangle) = (\langle\alpha_1\rangle, \langle\alpha_2\rangle)$ *is an isomorphism.*

Proof. This is a routine verification of the properties necessary for θ to be an isomorphism. Here is an outline with the details left as Exercise 6. We will continue to use the notation of placing subscripts on a function into a Cartesian product to denote the two coordinate functions that define the given function.

To show that θ is well-defined, suppose H is a homotopy between two loops $\alpha : [0, 1] \to X \times Y$ and $\beta : [0, 1] \to X \times Y$. Then $H_1 : [0, 1] \times [0, 1] \to X$ will be a homotopy from α_1 to β_1, and $H_2 : [0, 1] \times [0, 1] \to Y$ will be a homotopy from α_2 to β_2. In particular, $\theta(\langle\alpha\rangle) = (\langle\alpha_1\rangle, \langle\alpha_2\rangle) = (\langle\beta_1\rangle, \langle\beta_2\rangle) = \theta(\langle\beta\rangle)$.

To show that θ is an injection, suppose $\theta(\langle\alpha\rangle) = \theta(\langle\beta\rangle)$. That is, suppose $(\langle\alpha_1\rangle, \langle\alpha_2\rangle) = (\langle\beta_1\rangle, \langle\beta_2\rangle)$. Use a homotopy $H_1 : [0, 1] \times [0, 1] \to X$ from α_1 to β_1 and a homotopy $H_2 : [0, 1] \times [0, 1] \to Y$ from α_2 to β_2 as the two coordinate functions of a homotopy $H : [0, 1] \times [0, 1] \to X \times Y$ from α to β.

To show that θ is a surjection, let $(\langle\alpha_1\rangle, \langle\alpha_2\rangle)$ be an arbitrary element of $\pi_1(X) \times \pi_1(Y)$. Use the loops $\alpha_1 : [0, 1] \to X$ and $\alpha_2 : [0, 1] \to Y$ as coordinate functions of a loop $\alpha : [0, 1] \to X \times Y$ such that $\theta(\langle\alpha\rangle) = (\langle\alpha_1\rangle, \langle\alpha_2\rangle)$.

The following chain of equalities shows that θ is a homomorphism:

$$\begin{aligned}
\theta(\langle\alpha\rangle\langle\beta\rangle) &= \theta(\langle\alpha \cdot \beta\rangle) \\
&= (\langle(\alpha \cdot \beta)_1\rangle, \langle(\alpha \cdot \beta)_2\rangle) \\
&= (\langle\alpha_1 \cdot \beta_1\rangle, \langle\alpha_2 \cdot \beta_2\rangle) \\
&= (\langle\alpha_1\rangle\langle\beta_1\rangle, \langle\alpha_2\rangle\langle\beta_2\rangle) \\
&= (\langle\alpha_1\rangle, \langle\alpha_2\rangle)(\langle\beta_1\rangle, \langle\beta_2\rangle) \\
&= \theta(\langle\alpha\rangle)\theta(\langle\beta\rangle).
\end{aligned}$$

❄

Example 6.50 Determine the fundamental group of the torus T^2.

Solution. As noted in Exercise 1.12, the torus T^2 is homeomorphic to the Cartesian product $S^1 \times S^1$. By Theorem 6.34, $\pi_1(S^1) \cong \mathbb{Z}$. Thus, by Theorem 6.49,

$$\pi_1(T^2) \cong \pi_1(S^1 \times S^1) \cong \pi_1(S^1) \times \pi_1(S^1) \cong \mathbb{Z} \times \mathbb{Z}.$$

❄

Example 6.51 The 3-torus T^3 is defined to be the three-fold Cartesian product of circles: $T^3 = S^1 \times S^1 \times S^1$. Determine the fundamental group of T^3.

Solution. We can regard the three-fold Cartesian product interchangeably as a set of ordered triples, a set of ordered pairs whose first coordinate is an ordered pair, or as a set of ordered pairs whose second coordinate is an ordered pair. Furthermore, for the triple Cartesian product of groups, the multiplication is compatible from any of these points of view.

Thus, we simply apply Theorem 6.49 twice:

$$\begin{aligned}\pi_1(T^3) &\cong \pi_1((S^1 \times S^1) \times S^1)\\ &\cong \pi_1(S^1 \times S^1) \times \pi_1(S^1)\\ &\cong (\pi_1(S^1) \times \pi_1(S^1)) \times \pi_1(S^1)\\ &\cong \mathbb{Z} \times \mathbb{Z} \times \mathbb{Z}.\end{aligned}$$

❄

Of course not every set can be factored as Cartesian products of simpler sets. Here are two examples of how to use the polyhedral structure of a set in computing the fundamental group.

Example 6.52 Show that the fundamental group of a projective plane P^2 is isomorphic to the group $\mathbb{Z}_2$ of integers modulo 2.

Solution. We will think of the projective plane as the surface obtained by gluing together a Möbius band and a disk along their boundaries. The proof that $\pi_1(P^2) \cong \mathbb{Z}_2$ involves three steps. We first show that a loop α that travels once around the centerline of the Möbius band represents a generator of $\pi_1(P^2)$. Next we will show that this loop concatenated with itself is homotopic to the constant loop. Finally, we will show that the loop itself is not homotopic to the constant loop. It follows that $\pi_1(P^2)$ has exactly two elements, and hence is isomorphic to $\mathbb{Z}_2$, the prototype of a two-element group.

To get started on the first step, take an arbitrary loop in P^2. By the Simplicial Approximation Theorem, there is a polygonal loop that represents the same homotopy class. Now this polygonal loop will miss a point of the disk attached to the Möbius band. So push the loop away from this point onto the boundary of the disk and completely into the Möbius band part of P^2. In the Möbius band, we can further push the loop to lie entirely in the centerline. Since the fundamental group of a circle is generated by a loop that goes once around the circle, we can deform the loop to be a multiple of this generator of the circle sitting as the centerline of the Möbius band. All of this pushing can be done by a homotopy to ensure that the modified loop still represents the same element of $\pi_1(P^2)$ as did the original loop.

Let α denote a loop that goes once around the centerline of the Möbius band in P^2. Push $\alpha \cdot \alpha$ away from the centerline so that it goes out from the base point, follows around the edge of the Möbius band, and returns to the base point as illustrated in Figure 6.53. Now push the loop across the boundary of the Möbius band so that it goes directly from the base point to the boundary, follows around the boundary just inside the disk attached to the Möbius band, and returns directly to the base point. Pull the loop across the disk so that it goes from the base point into the disk and immediately travels out of the disk and back to the base point. One final push will deform this loop to a constant loop at the base point. Thus, $\langle\alpha\rangle\langle\alpha\rangle = \langle\alpha \cdot \alpha\rangle$ is the identity element of $\pi_1(P^2)$.

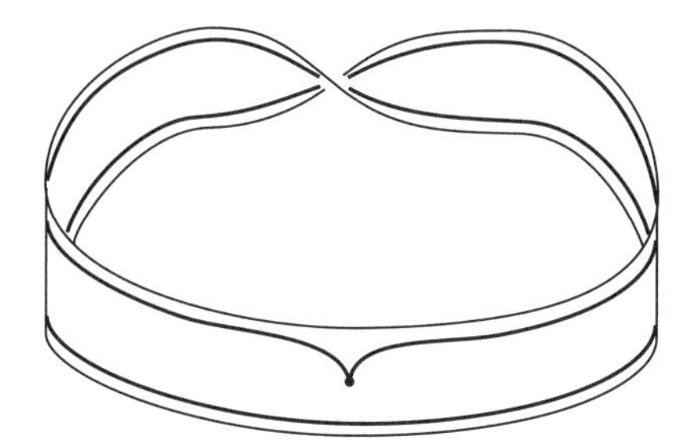

FIGURE 6.53
The loop $\alpha \cdot \alpha$ pushed to the edge of a Möbius band.

Of course, if P^2 turns out to be simply connected, the above two results would be no surprise. The nonending arc trick described below shows that α is not homotopic to a constant loop. See Exercise 10 in Section 3.1 for another situation where this trick is useful.

Begin by pushing α away from the centerline of the Möbius band so that the resulting simple closed curve intersects the centerline only in an arc that cuts across the centerline. If α is homotopic to the constant loop, then so is this homotopic modification. In this case, use the Simplicial Approximation Theorem to get a piecewise linear homotopy that is in general position with respect to the centerline of the Möbius band. That is, the domain $[0, 1] \times [0, 1]$ of the homotopy is subdivided into triangles that map so that if the centerline intersects the image of one of these triangles, it crosses from the image of one edge of the triangle to the image of another edge of the triangle. Thus, the points in $[0, 1] \times [0, 1]$ that map to the centerline will consist of arcs that span across the triangular faces and continue as an arc in the neighboring triangle. Since there are only a finite number of triangles, these arcs must form a family of simple closed curves. The only exception is in one triangle whose edge follows along the modification of α where it crosses the centerline. This arc proceeds into $[0, 1] \times [0, 1]$ and can neither end nor reach another point on the boundary. This contradiction shows that the assumptions about the existence of the homotopy must be fallacious. That is, $\langle\alpha\rangle$ is distinct from the multiplicative identity in $\pi_1(P^2)$. ❄

Here is an extension of Theorem 6.22. Notice how the Simplicial Approximation Theorem allows us to get to the essence of this argument.

Theorem 6.54 *The 3-sphere* $S^3 = \{(w, x, y, z) \in \mathbb{R}^4 \mid w^2 + x^2 + y^2 + z^2 = 1\}$ *is simply connected.*

Proof. Cut the 3-sphere along its equator

$$\{(w, x, y, z) \in \mathbb{R}^4 \mid w^2 + x^2 + y^2 + z^2 = 1, w = 0\}$$

to see S^3 as the union of an upper hemisphere

$$D_+ = \{(w, x, y, z) \in \mathbb{R}^4 \mid w^2 + x^2 + y^2 + z^2 = 1, w \geq 0\}$$

and a lower hemisphere

$$D_- = \{(w, x, y, z) \in \mathbb{R}^4 \mid w^2 + x^2 + y^2 + z^2 = 1, w \leq 0\}.$$

Projection of $\mathbb{R}^4$ onto the last three coordinates shows that D_+ and D_- are homeomorphic to the standard 3-cell $\{(x, y, z) \in \mathbb{R}^3 \mid x^2 + y^2 + z^2 \leq 1\}$.

Now take any loops in S^3. By the Simplicial Approximation Theorem, we can find a polygonal loop that represents the same element of $\pi_1(S^3)$. This loop will miss a point of D_+ so we can homotopically push it away from this point into the boundary of D_+. The resulting curve lies in the 3-cell D_-. Thus, it can be homotopically shrunk to a constant loop. ❄

The fundamental group is a powerful tool both for describing certain algebraic aspects of geometric objects and also as a topological invariant to distinguish among various spaces. From the point of view of the fundamental group, the simplest spaces are those whose groups consist of a single element. These are the simply connected spaces. Among the 3-manifolds without boundary, the 3-sphere stands out as the most basic, and Theorem 6.54 confirms that S^3 is simply connected. You might wonder whether there are other 3-manifolds without boundary and constructed from a finite number of cells that are also simply connected. Henri Poincaré, one of the founders of algebraic topology in the early decades of the twentieth century, conjectured that no others existed.

Poincaré Conjecture 6.55 *The* 3-*sphere is the only closed* 3-*manifold that is simply connected.*

Numerous proofs of the Poincaré Conjecture have appeared, some have even been published, but so far, all have turned up flawed. Surprising as it may seem, generalizations of the Poincaré Conjecture to manifolds of higher dimensions have been settled. In 1960 Steven Smale proved that the conjecture (with suitable strengthened algebraic hypotheses) holds for manifolds of dimension 5 and higher. Then in 1982, Michael Freedman narrowed the gap by proving the Poincaré Conjecture for 4-dimensional manifolds. Despite the many mathematicians who have worked on this problem and the vast amount of research it has spawned, the 3-dimensional Poincaré Conjecture is still an open problem. In 2002, the Russian mathematician Grigori Perelman announced a proof of a general result on the geometric structures of 3-manifolds. This result is known to imply the Poincaré Conjecture. As of 2006, there is considerable optimism that Perelman's proof is correct, but the final word is not in yet.

Exercises 6.6

1. Describe in words or pictures each of the following Cartesian products. Determine the fundamental groups of the space.
   (a) $S^1 \times [0, 1]$
   (b) $(S^1 \vee S^1) \times [0, 1]$
   (c) $(S^1 \vee S^1) \times S^1$
   (d) $S^2 \times [0, 1]$
   (e) $S^2 \times S^1$
   (f) $T^2 \times [0, 1]$
   (g) $T^2 \times S^1$

2. Let D^2 denote a disk. A **solid torus** is defined to be the Cartesian product $S^1 \times D^2$ of a circle and a disk.
   (a) Illustrate a doughnut (with solid dough) as a solid torus.
   (b) Compute the fundamental group of a solid torus.
   (c) Glue two solid tori together by matching up corresponding points in their boundaries. Identify the resulting space as the Cartesian product $S^1 \times S^2$ of a circle and a 2-sphere.

3. Determine $\pi_1(P^2 \times S^1)$.

4. In Exercise 8 of Section 4.1, we described projective three-space P^3 to be the 3-manifold obtained from gluing the boundary of the 3-ball $B^3 = \{(x, y, z) \in \mathbb{R}^3 \mid x^2 + y^2 + z^2 \leq 1\}$ to itself by matching each point (x, y, z) with its antipodal point $(-x, -y, -z)$.
   (a) Show that any plane through the origin of $\mathbb{R}^3$ intersects B^3 in a disk that is glued together to form a projective plane in P^3.
   (b) Show that the projective planes in part a are one-sided surfaces in P^3.
   (c) Show that P^3 is an orientable 3-manifold.
   (d) Show that the z-axis intersects B^3 in an arc that is glued together to form a loop α that represents a generator of $\pi_1(P^3)$.
   (e) Show that $\alpha \cdot \alpha$ is homotopic to the constant loop.
   (f) Show that α is not homotopic to the constant loop. Suggestion: Perform the nonending arc trick on the intersection of the image of the homotopy and the projective plane formed from identifying points on the boundary of the equatorial disk $\{(x, y, z) \in B^3 \mid z = 0\}$.

5. Collect the results of the computation of fundamental groups of 3-manifolds in this section to show that S^3, T^3, $S^2 \times S^1$, $P^2 \times S^1$, and P^3 are distinct 3-manifolds.

6. Fill in the details to show that the function $\theta : \pi_1(X \times Y) \to \pi_1(X) \times \pi_1(Y)$ of Theorem 6.49 is
   (a) well-defined,
   (b) an injection,

(c) a surjection,

(d) a homomorphism.

7. By Example 6.50, $\pi_1(T^2) \cong \mathbb{Z} \times \mathbb{Z}$. With additive terminology for the group operation in $\mathbb{Z} \times \mathbb{Z}$, any element in this group is a multiple of $(1, 0)$ plus a multiple of $(0, 1)$. Find loops in T^2 that correspond to these generators of $\mathbb{Z} \times \mathbb{Z}$.

8. Consider the connected sum $S\#T$ of two surfaces S and T.

(a) Find a function from $S\#T$ onto S that is the identity on the S portion of $S\#T$ and maps T to a disk.

(b) Show that this function induces a homomorphism from $\pi_1(S\#T)$ onto $\pi_1(S)$.

(c) Conclude that if S is not simply connected, then neither is $S\#T$.

9. Let $S^n = \{(x_1, \ldots, x_{n+1}) \in \mathbb{R}^{n+1} \mid x_1^2 + \cdots + x_{n+1}^2 = 1\}$ denote the n-dimensional sphere. Generalize the result of Theorem 6.54 to show S^n is simply connected for any $n \geq 2$.

10. State and prove a version of the Poincaré Conjecture for surfaces.

References and Suggested Readings for Chapter 6

50. William Massey, *Algebraic Topology: An Introduction*, Springer-Verlag, New York, 1989.

The fundamental group is a case study demonstrating the power of algebraic methods applied to surfaces and other topological spaces.

51. Henri Poincaré, “Analysis Situs,” in *Oeuvres*, Vol. VI, 193–288, Gauthier-Villars, Paris, 1953.

Poincaré introduced the notion of the fundamental group in this paper published in 1895.

52. C. T. C. Wall, *A Geometric Introduction to Topology*, Dover, New York, 1993.

Provides the necessary background in topology to develop homology and cohomology for subsets of Euclidean space.

7

Metric and Topological Spaces

It is time to address some questions that we have postponed or treated only informally in previous chapters. Most importantly, we need to define more precisely the idea of a topological space. What exactly are these spaces we have been dealing with? So far, for simplicity, we have usually thought of spaces as subsets of some Euclidean space. Is that necessary? The projective plane P^2 can be represented as a subset of $\mathbb{R}^4$, but why do we need to know this to understand and work with P^2? Indeed, we have analyzed three-dimensional manifolds like T^3 and P^3 without checking that they can be represented as subsets of some Euclidean space. The configuration space of rotations of a sphere described in Exercise 8 of Section 4.1, for instance, is certainly not a subset of a Euclidean space. We need a broader definition that covers these cases.

There are some other ideas we need to clarify. For example, we have been using the idea of connectedness in terms of paths between points and the idea of compactness in terms of triangulations. Can we extend these concepts in a precise way to more general spaces? And what about all the gluing of surfaces and 3-manifolds? What exactly is gluing?

The historical development of topology is similar to that of calculus. Isaac Newton, Gottfried Leibniz, James and John Bernoulli, Leonhard Euler, and others developed the important ideas and techniques of calculus in the 17th and 18th centuries. This was long before Augustin Cauchy and Karl Weierstrass put the subject on a firm logical foundation with the ε-δ definitions of limits and continuity. Likewise, Enrico Betti, Henri Poincaré, and others in the 19th century developed many of the important ideas and techniques of topology before its foundation was laid in the early years of the 20th century.

7.1 Metric Spaces

With his definition of metric spaces in 1906, Maurice Fréchet took a significant first step in establishing a firm foundation for topology. Recall the definition of a continuous function to see the key role metric spaces play in this most important concept in topology. According to Definition 1.20, a function $f : X \to Y$ is continuous at a point $x_0 \in X$ if and only if for every $\varepsilon > 0$ there is a $\delta > 0$ such that $d(x, x_0) < \delta \Longrightarrow d(f(x), f(x_0)) < \varepsilon$. In this definition, the first d refers to distance in the space X and the second d refers to distance in the space Y. Hence to define the concept of a continuous function from a space X to a space Y, we need only to have a way of measuring distances in these two spaces. Fréchet defined a metric space to be a set together with a way of measuring distances between any two elements of the set.

Definition 7.1 *A* ***metric space*** *is a set X together with a function $d : X \times X \to \mathbb{R}$ such that for all $x, y, z \in X$,*

1. *$d(x, y) \geq 0$, with $d(x, y) = 0$ if and only if $x = y$ (positive-definite property)*
2. *$d(x, y) = d(y, x)$ (symmetry)*
3. *$d(x, y) + d(y, z) \geq d(x, z)$ (triangle inequality)*

The function d is called a ***metric*** *on X. A metric space X with a metric d is often designated as an ordered pair (X, d).*

The three properties of a metric abstract those properties of the standard distance function on $\mathbb{R}^n$ that are most important in proving theorems involving distance. As in all abstractions, choices have been made. For instance, the definition requires two distinct points to be a positive distance apart. With a little imagination, we might conceive of situations in which we would want to allow distinct points x and y in X to have $d(x, y) = 0$. If we did allow this, d would not be a metric since it would not satisfy the positive-definite property. Similarly, we sometimes speak of the distance from x to y as the time it takes to get from x to y. If we do this, it may be that $d(x, y) \neq d(y, x)$. For instance, we could be traveling by canoe on a river with y downstream from x. Again, such a d would not be a metric because it would not satisfy the symmetry property. The triangle inequality is perhaps the most important property for proving theorems involving distance. The name is appropriate because the triangle inequality is an abstraction of the property that the sum of the lengths of two sides of a triangle must be at least as large as the length of the third side.

Example 7.2 Here are some of the many ways to define useful metrics on the Euclidean plane $\mathbb{R}^2$. Exercise 1 at the end of this section asks you to verify that these distance functions satisfy the three properties of Definition 7.1. This will confirm that the four spaces $(\mathbb{R}^2, d)$, $(\mathbb{R}^2, d_{TC})$, $(\mathbb{R}^2, d_{RR})$, and $(\mathbb{R}^2, d_D)$ defined below are indeed metric spaces.

(a) The **standard** distance function is given by

$$d((x_1, y_1), (x_2, y_2)) = \sqrt{(x_1 - x_2)^2 + (y_1 - y_2)^2}.$$

(b) The **taxicab** metric is defined by

$$d_{TC}((x_1, y_1), (x_2, y_2)) = |x_1 - x_2| + |y_1 - y_2|.$$

This is an appropriate metric if we are traveling in a city where the streets form a rectangular grid, so that we can only move north-south or east-west.

(c) The **Roman road** metric is defined as follows. Denote the origin (0, 0) by O. If P and Q are any two points in $\mathbb{R}^2$, then

$$d_{RR}(P, Q) = \begin{cases} d(P, Q) & \text{if } P, Q, \text{ and } O \text{ are collinear,} \\ d(P, O) + d(O, Q) & \text{otherwise.} \end{cases}$$

Here d is the standard metric of $\mathbb{R}^2$. In this metric "all roads lead to Rome," which is at the origin. We can only go directly from P to Q if these points are on the same road through O; otherwise, we must go from P to O and then back out to Q. English mathematicians call the metric space $(\mathbb{R}^2, d_{RR})$ "English railway space" because of the perception that if you want to catch a train between any cities in the south of England, you have to go through London.

(d) The **discrete** metric is defined by

$$d_D(P, Q) = \begin{cases} 0 & \text{if } P = Q, \\ 1 & \text{if } P \neq Q. \end{cases}$$

In this strange metric, any two distinct points are at distance 1 from each other. Perhaps you have a machine that will transport you instantly to anywhere in the universe, but it takes an hour to set the machine up! ❄

The previous example provides a sampler of interesting metrics for the Euclidean plane. But the idea of a metric is much broader. We do not need to restrict the underlying set of points to $\mathbb{R}^2$ or even $\mathbb{R}^n$.

Example 7.3 Consider the configuration space X consisting of all chords and tangents to the unit circle, as described in Exercise 8 of Section 3.1. Given two chords or tangents l and m, let c_l and c_m denote their centers or points of tangency and let $\theta(l, m)$ denote the smallest nonnegative angle between the chords (extended if necessary) and tangents. Define

$$d_X(l, m) = d(c_l, c_m) + \theta(l, m),$$

where d is the standard metric on $\mathbb{R}^2$. Check that this is a metric. This formalizes the idea of closeness used informally in identifying this configuration space as a Möbius band.

Solution. We need to verify that d_X satisfies the three conditions of Definition 7.1. Since both functions d and θ are nonnegative, the sum d_X is also nonnegative. Although it is possible for chords to have the same center, this will happen if and only if they are both diameters of the circle. But unless the diameters are equal, there will be a positive angle between them. Thus, d_X is positive-definite. Since both functions d and θ are symmetric and satisfy the triangle inequality, so also does the sum d_X. ❄

Example 7.4 Let $C_{[0,1]}$ be the set of all continuous functions $f : [0,1] \to \mathbb{R}$. The **uniform** metric on the function space $C_{[0,1]}$ is defined by

$$d(f,g) = \sup\{|f(x) - g(x)| \mid x \in [0,1]\}.$$

This says that two functions are close if and only if they are close at every point in [0, 1].

Now Definition 1.20 works unchanged to define a continuous function $f : X \to Y$ between metric spaces X and Y. We simply interpret the first d as the metric on X and the second d as the metric on Y. As an easy consequence, we know what it means for two metric spaces to be homeomorphic. Indeed, we used this idea informally in Exercise 8 of Section 3.1 to show that the configuration space of chords and tangents to a circle is homeomorphic to a Möbius band. By introducing another useful concept of metric spaces, we can reformulate the definition of continuity as a slightly more concise statement.

Definition 7.5 *The **open ball** of radius r about a point x_0 in a metric space (X,d) is the set $B_r(x_0) = \{x \in X \mid d(x,x_0) < r\}$. The **closed ball** of radius r about x_0 is the set $B_r[x_0] = \{x \in X \mid d(x,x_0) \leq r\}$.*

Balls in strange metrics can have strange shapes. Exercise 2 at the end of this section asks you to draw some balls in different metric spaces. Exercise 5 should convince you that the following definition is a direct reformulation of Definition 1.20.

Definition 7.6 *Suppose (X,d_X) and (Y,d_Y) are metric spaces. A function $f : X \to Y$ is **continuous** if and only if for all $x \in X$ and for all $\varepsilon > 0$, there is a $\delta > 0$ such that $f(B_\delta(x)) \subseteq B_\varepsilon(f(x))$.*

The definition of open balls gives a precise way of talking about the set of points close to a given point in a metric space. This leads to the following generalization of the idea of open intervals on the real line.

Definition 7.7 *A set U in a metric space X is **open** if and only if for all $x \in U$, there is $r > 0$ such that $B_r(x) \subseteq U$.*

This definition says that for any point of an open set, all points sufficiently close to the given point are also in the set. Intuitively then, an open set does not contain any of the points on the boundary between the set and its complement.

Exercise 7 asks you to prove the following theorem about the unions and intersections of open sets. These are the essential properties of open sets that will be abstracted in the next section to define a topology on a space. By the principle of mathematical induction, the third property gives that finite intersections of open sets are open. Infinite intersections of open sets may not be open—see Exercise 8.

Theorem 7.8 *Open sets in a metric space X satisfy the following properties:*

1. *$\emptyset$ and X are open sets.*
2. *The union of any collection of open sets is open. That is, if A is any index set and U_α is open for all $\alpha \in A$, then $\bigcup_{\alpha \in A} U_\alpha$ is open.*
3. *If U and V are open, then $U \cap V$ is open.*

The next theorem might seem to be obvious. After all, we certainly wouldn't call an open ball open if it weren't! However, open sets are subsets that have a particular property, and we can't make an open ball have that property just by naming it. Notice the key role the triangle inequality plays in proving that an open ball satisfies the required condition.

Theorem 7.9 *In a metric space, an open ball is an open set.*

Proof. In a metric space (X, d), consider an open ball $B_r(x_0)$. Given any $x \in B_r(x_0)$, we must find an open ball about x that is contained in $B_r(x_0)$. Guided by the picture in Figure 7.10, let $\varepsilon = r - d(x, x_0)$. We will show that $B_\varepsilon(x) \subseteq B_r(x_0)$. To do this, take any $y \in B_\varepsilon(x)$, so that $d(y, x) < \varepsilon$. Then

$$\begin{aligned} d(y, x_0) &\le d(y, x) + d(x, x_0) \qquad \text{(by the triangle inequality)} \\ &< \varepsilon + d(x, x_0) \\ &= r - d(x, x_0) + d(x, x_0) \\ &= r, \end{aligned}$$

so that $y \in B_r(x_0)$. ❄

It is a remarkable fact of importance to the history of topology, that continuous functions can be defined entirely in terms of open sets. In other words, if we know which sets are open in two metric spaces X and Y, we can tell whether a function $f : X \to Y$ is continuous without knowing the precise metrics on X and Y. The statement and proof of the following theorem rely on the notion of the inverse image of a set under a function $f : X \to Y$. For U a subset of Y, you may be familiar with the standard notation

$$f^{-1}(U) = \{x \in X \mid f(x) \in U\}.$$

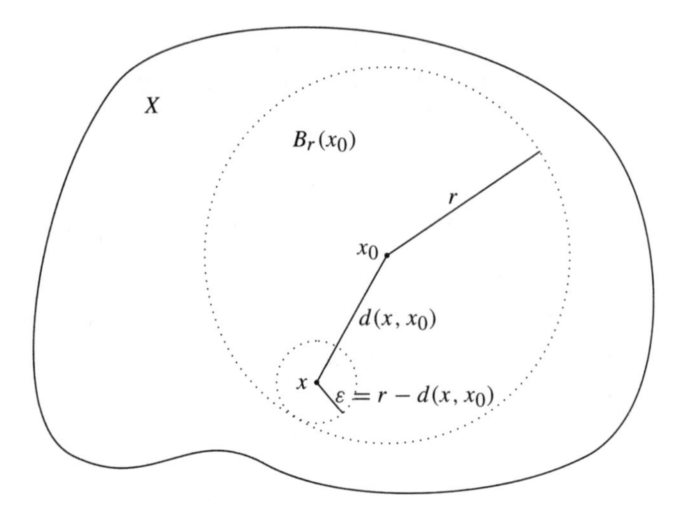

FIGURE 7.10
An open ball is an open set.

In words, $f^{-1}(U)$ is the set of all points in X that f maps into U. Even if f is not a bijection and there is no function f^{-1}, we can still talk about the inverse image of a subset of the range as this subset of the domain of the function. Exercise 12 will make you feel more at home with this notation.

> **Theorem 7.11** *A function $f : X \to Y$ between two metric spaces is continuous if and only if $f^{-1}(U)$ is an open set in X whenever U is an open set in Y.*

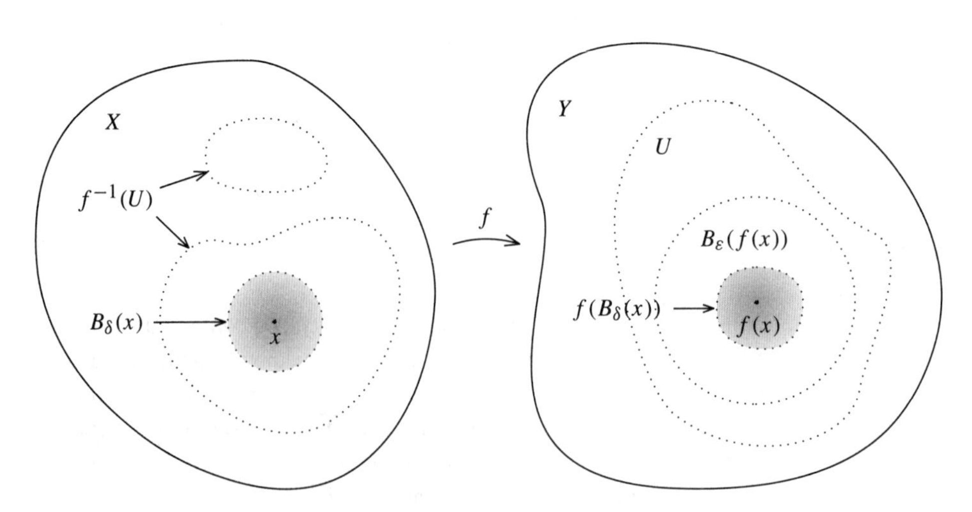

FIGURE 7.12
If f is continuous and U is open, then $f^{-1}(U)$ is open.

Proof. First suppose $f : X \to Y$ is continuous and $U \subseteq Y$ is open. We wish to show that $f^{-1}(U)$ is open in X. So given any $x \in f^{-1}(U)$, we must find a $\delta > 0$ such that $B_\delta(x) \subseteq f^{-1}(U)$. Now since $f(x) \in U$ and U is open, there is an $\varepsilon > 0$ such that $B_\varepsilon(f(x)) \subseteq U$. Since f is continuous, there is a $\delta > 0$ such that $f(B_\delta(x)) \subseteq B_\varepsilon(f(x)) \subseteq U$. Hence $B_\delta(x) \subseteq f^{-1}(U)$.

Conversely, suppose that whenever $U \subseteq Y$ is open, then $f^{-1}(U)$ is open in X. We wish to show that f is continuous. So choose any $x \in X$ and any $\varepsilon > 0$. By Theorem 7.9, the ball $B_\varepsilon(f(x))$ is an open subset of Y. So we know that $f^{-1}(B_\varepsilon(f(x))$ is open in X. Of course x is in $f^{-1}(B_\varepsilon(f(x))$. Hence there is a $\delta > 0$ such that $B_\delta(x) \subseteq f^{-1}(B_\varepsilon(f(x))$. That is, $f(B_\delta(x)) \subseteq B_\varepsilon(f(x))$. ❄

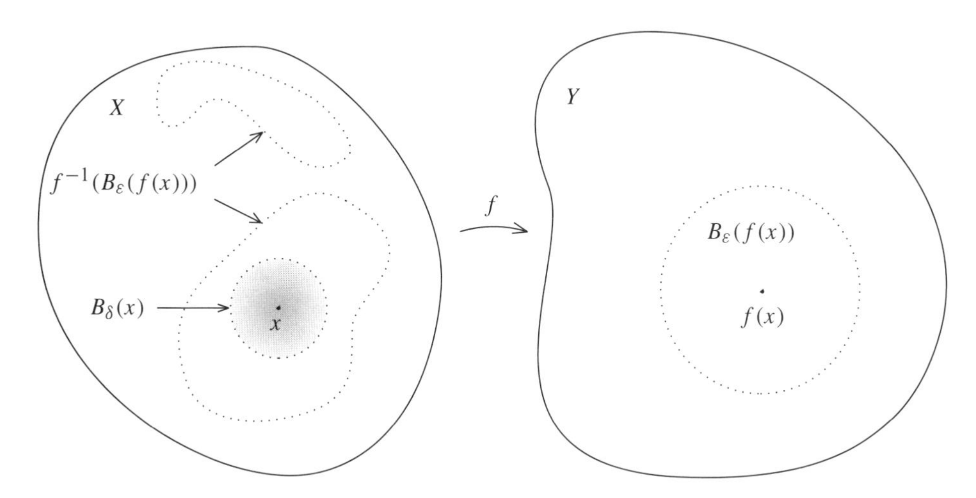

FIGURE 7.13
If $f^{-1}(U)$ is open for any open set U, then f is continuous.

Exercises 7.1

1. Prove that the following functions are metrics on $\mathbb{R}^2$. The hard part is usually proving the triangle inequality.
   - **(a)** The standard metric. You may want to look up this proof in a linear algebra textbook. The Cauchy-Schwarz inequality will be a big help in proving the triangle inequality.
   - **(b)** The taxicab metric.
   - **(c)** The Roman road metric. In proving the triangle inequality, you will need to consider several cases.
   - **(d)** The discrete metric.
2. Draw pictures of the open balls of radii $\frac{1}{2}$, 1, and 2 about the point $(1, 0)$ in $\mathbb{R}^2$ in
   - **(a)** the standard metric,
   - **(b)** the taxicab metric,

(c) the Roman road metric,

(d) the discrete metric.

3. In addition to being called "English railway space," Roman road space is sometimes called "__________ space," where the blank is filled by the name of a candy. Looking at $B_2((1,0))$, can you guess what the candy is?

4. Let $X = \{(x, y) \in \mathbb{R}^2 \mid x \geq 0\}$ with the standard metric inherited from $\mathbb{R}^2$. Draw $B_1((0,0))$.

5. Consider a function $f : X \to Y$. For a subset $A \subseteq X$, let $f(A) = \{f(x) \mid x \in A\}$ denote the **image** under f of the set A.

(a) For subsets $A \subseteq X$ and $B \subseteq Y$, explain why the containment $f(A) \subseteq B$ is a reformulation of the implication $x \in A \Longrightarrow f(x) \in B$.

(b) Suppose X and Y are metric spaces and $x \in X$. For $\varepsilon > 0$ and $\delta > 0$, explain why the containment $f(B_\delta(x)) \subseteq B_\varepsilon(f(x))$ is a reformulation of the implication $d(x', x) < \delta \Longrightarrow d(f(x'), f(x)) < \varepsilon$.

6. Let x and y be any two points in a metric space (X, d).

(a) Let $r = \frac{1}{2}d(x, y)$. Prove that $B_r(x) \cap B_r(y) = \emptyset$.

(b) For any real number r and s with $r + s = d(x, y)$, prove that $B_r(x) \cap B_s(y) = \emptyset$.

7. Prove Theorem 7.8.

8. **(a)** Show that on the real line with the standard metric, an open interval (a, b) is an open set.

(b) Show that on the real line with the standard metric, a single point is not an open set.

(c) Show that an infinite intersection of open sets need not be open by considering

$$\bigcap_{n=1}^{\infty} \left(-\frac{1}{n}, \frac{1}{n}\right).$$

9. **(a)** In $\mathbb{R}^2$, show that every open ball in the standard metric contains an open ball in the taxicab metric with the same center and same radius.

(b) Show that every open ball in the taxicab metric contains an open ball in the standard metric with the same center but a smaller radius.

(c) Deduce that the open sets in $(\mathbb{R}^2, d)$ and $(\mathbb{R}^2, d_{TC})$ are exactly the same.

10. Find a set that is open in $(\mathbb{R}^2, d_{RR})$ but not open in $(\mathbb{R}^2, d)$. Exercise 2 and Theorem 7.9 may be helpful.

11. Prove that with the discrete metric, every subset of $\mathbb{R}^2$ is open.

12. Consider a function $f : X \to Y$. Suppose A and B are subsets of X, and C and D are subsets of Y.

(a) Show that $f(A \cap B) \subseteq f(A) \cap f(B)$.

(b) Find an example in which $f(A \cap B) \neq f(A) \cap f(B)$. Suggestion: Consider a function that is not one-to-one, such as $f : \mathbb{R} \to \mathbb{R}$ given by $f(x) = x^2$.

(c) On the other hand, show that it is always true that $f^{-1}(C \cap D) = f^{-1}(C) \cap f^{-1}(D)$. In this sense, the operator f^{-1} maps sets more nicely than f does.

(d) Investigate what happens when $\cap$ is replaced by $\cup$.

13. Examine the proof of Theorem 1.27 to locate the property of polygonal regions that is crucial for the result of this theorem. Formulate and prove a similar theorem with less restrictive conditions on the sets A and B.

7.2 Topological Spaces

Fréchet's definition of an abstract metric space was an important step, but it leaves something to be desired in terms of topology. After all, topology is supposed to be the study of properties that remain invariant when we deform a space, and deformation changes the distances between pairs of points. Hence topological properties should not depend on a specific metric. One way to make further progress is by using the idea of open sets. We saw in Theorem 7.11, for example, how to characterize continuous functions $f : X \to Y$ between two metric spaces in terms of the open sets in X and Y. Also, Exercise 9 in Section 7.1 shows that different metrics on the same space may lead to the same open sets, and hence to the same continuous functions.

The definition of an abstract topological space was first given by Felix Hausdorff in his famous 1914 textbook *Grundzüge der Mengenlehre* (*Foundations of Set Theory*). Hausdorff's definition was different from, but only slightly less general than, the definition widely used today.

Definition 7.14 *A **topological space** is a set X together with a family $\mathcal{T}$ of subsets of X such that*

1. $\emptyset, X \in \mathcal{T}$;
2. *If A is any index set and $U_\alpha \in \mathcal{T}$ for all $\alpha \in A$, then $\bigcup_{\alpha \in A} U_\alpha \in \mathcal{T}$; and*
3. *If $U, V \in \mathcal{T}$, then $U \cap V \in \mathcal{T}$.*

*The subsets of X in $\mathcal{T}$ are called the **open** sets of X, and $\mathcal{T}$ is called a **topology** on X.*

In other words, we make a set into a topological space not by saying how to measure distances, but by saying which subsets are open. We can do this in any way we wish, subject to the three properties of Definition 7.14. You should recognize these as the properties listed in Theorem 7.8. They are properties of the open sets in a metric space as specified by Definition 7.7. Hence, with open sets defined in terms of the metric, any metric space is a topological space. Exercise 9 in Section 7.1 shows that different metrics on a set may lead to the same topology on the set. On the other hand, Exercises 10 and 11 show that different metrics on a set may lead to different topologies. As we shall soon see, there are also topologies that do not arise from any metric. The idea of a topological space is a true generalization of the idea of a metric space.

On any set, there are two topologies at the extreme ends of the spectrum.

Definition 7.15 *For any set X, the* **discrete** *topology on X is the collection of all subsets of X; that is, all subsets of X are open in the discrete topology. The* **indiscrete** *topology on X is the collection $\{\emptyset, X\}$; that is, only the empty set and X itself are open in the indiscrete topology.*

In terms of the number of open sets, the discrete topology is the largest possible topology on a set and the indiscrete topology is the smallest. The discrete topology comes from a metric, namely from the discrete metric, which is the generalization to X of the discrete metric defined on $\mathbb{R}^2$ in Example 7.2. On the other hand, part (a) of Exercise 3 shows that if X has more than one point, the indiscrete topology cannot come from a metric.

Example 7.16 Here are two other examples of topologies that cannot arise from a metric (see part (b) of Exercise 3). Such topologies are called **non-metrizable**.

(a) Let $X = \{a, b, c\}$, a set with three points. Let $\mathcal{T} = \{\emptyset, X, \{a\}, \{b\}, \{a, b\}, \{a, c\}\}$. The three conditions of Definition 7.14 are satisfied, so $(X, \mathcal{T})$ is a topological space.

(b) Let $\mathbb{N} = \{1, 2, 3, \ldots\}$ denote the natural numbers. Let $\mathcal{T}$ consist of $\emptyset$ and all subsets of $\mathbb{N}$ that have finite complements. Exercise 1 asks you to check that this is a topology. It is called the **finite-complement** topology on $\mathbb{N}$.

Any subset A of a topological space X inherits a topology from the parent space. As specified in the following definition, the open sets are defined to be the intersection of A with open sets of X. You can easily check that conditions of Definition 7.14 are met.

Definition 7.17 *Suppose $(X, \mathcal{T})$ is a topological space and A is a subset of X. The* ***subspace topology*** *on A is*

$$\mathcal{T}|_A = \{U \cap A \mid U \in \mathcal{T}\}.$$

Theorem 7.11 characterizes continuous functions between metric spaces in terms of open sets. This theorem suggests the proper definition of continuity for functions between topological spaces. It also ensures that for spaces whose topologies arise from metrics, the following definition is consistent with the definition of a continuous function between metric spaces.

Definition 7.18 *Suppose* $(X, \mathcal{T}_X)$ *and* $(Y, \mathcal{T}_Y)$ *are topological spaces. A function* $f : X \to Y$ *is* ***continuous*** *if and only if*

$$U \in \mathcal{T}_Y \Longrightarrow f^{-1}(U) \in \mathcal{T}_X.$$

Example 7.19 Consider the topological space $(X, \mathcal{T}_X)$ where $X = \{a, b, c\}$ and $\mathcal{T}_X = \{\emptyset, X, \{a\}, \{b\}, \{a, b\}, \{a, c\}\}$ as in Example 7.16(a). Also consider the topological space $(Y, \mathcal{T}_Y)$ where $Y = \{d, e\}$ and $\mathcal{T}_Y = \{\emptyset, Y, \{e\}\}$. Let $f : X \to Y$ be given by $f(a) = d$ and $f(b) = f(c) = e$. Determine whether f is continuous.

Solution. We simply compute the inverse images of the open sets in Y and see if they are open in X.

$$\begin{aligned} f^{-1}(\emptyset) &= \emptyset, \text{ which is in } \mathcal{T}_X; \\ f^{-1}(Y) &= X, \text{ which is in } \mathcal{T}_X; \\ f^{-1}(\{e\}) &= \{b, c\}, \text{ which is not in } \mathcal{T}_X. \end{aligned}$$

If you think about it, the inverse image of the empty set under any function will always be the empty set, and the inverse image of the entire range under any function will always be the entire domain. So it is not really necessary to check these two sets. However, the inverse images of all the other sets must be open for f to be continuous. Here the inverse image of the third set fails to be open, so f is not continuous. ❄

We can now give a very simple reformulation of the key concept of equivalence between topological spaces.

Definition 7.20 *If* $(X, \mathcal{T}_X)$ *and* $(Y, \mathcal{T}_Y)$ *are topological spaces, a function* $f : X \to Y$ *is a* ***homeomorphism*** *if and only if*

1. f *is a bijection* (f *is one-to-one and onto), and*
2. $U \in \mathcal{T}_X$ *if and only if* $f(U) \in \mathcal{T}_Y$ *(both* f *and* f^{-1} *are continuous).*

The complements of open sets in a topological space are also interesting.

Definition 7.21 *A subset* A *of a topological space* X *is* ***closed*** *if and only if its complement* $X - A$ *is open.*

This definition is consistent with terminology for intervals on the real line: an open interval is open, and a closed interval $[a, b]$ is closed because its complement $(-\infty, a) \cup$

(b, ∞) is open. Warning: "closed" does not mean "not open." On the real line for example, $[a, b)$ is not open, but it is also not closed. On the other hand, $\mathbb{R}$ itself is open and it is also closed, since its complement is the empty set, which is open. To decide if a subset A of a topological space X is closed, we really do have to check whether $X - A$ is open.

Example 7.22 What are the closed sets in the topological spaces of Example 7.16?

Solution.

(a) The closed sets are X, $\emptyset$, $\{b, c\}$, $\{a, c\}$, $\{c\}$, and $\{b\}$. In particular, $\{a\}$ is open but not closed, $\{c\}$ is closed but not open, and $\{b\}$ is both open and closed.

(b) The closed sets are $\mathbb{N}$, $\emptyset$, and all finite sets. Many sets, for instance the set of all even numbers, are neither open nor closed. ❄

The properties of open sets given in Definition 7.14 can be reformulated in terms of their complementary closed sets. Exercise 6 asks you to work through the details.

Theorem 7.23 *The collection of closed sets in a topological space satisfies*

1. *$\emptyset$ and X are closed,*
2. *any intersection of closed sets is closed, and*
3. *any finite union of closed sets is closed.*

With open and closed sets in hand, we can define for general topological spaces a number of other ideas that are familiar for subsets of Euclidean space.

Definition 7.24 *Suppose A is a subset of a topological space X.*

*The **closure** of A, denoted $\overline{A}$, is the intersection of all closed subsets of X that contain A.*

*The **interior** of A, denoted A°, is the union of all open subsets of X that are contained in A.*

*The **boundary** of A is* $\text{bd}(A) = \overline{A} - A^\circ$.

By part 2 of Theorem 7.23, we see that the closure of a set A is a closed set. Thus, $\overline{A}$ is the smallest closed set containing A in the sense that any closed set containing A also contains $\overline{A}$. Similarly, by part 2 of Definition 7.14, the interior of A is an open set. Thus, A° is the largest open set contained in A in the sense that any open subset of A is contained in A°.

We will see in the following sections that open sets also can be used to give elegant definitions of connectedness, compactness, and topological gluing. The abstraction of a topological space is exactly what we need to put these ideas on a firm logical foundation.

Exercises 7.2

1. Verify that the topologies given in Example 7.16 meet the three conditions of Definition 7.14 of a topology.

2. Suppose $X = \{a, b, c\}$. There are 29 possible topologies on X. How many can you describe? Rather than listing them, you might want to devise an easy and elegant way to describe these possible topologies.

3. **(a)** Show that the indiscrete topology on a space of more than one point cannot arise from a metric. You may want to try an indirect proof: suppose that a space X has more than one point and the indiscrete topology on X arises from a metric d. Let x and y be two distinct points in X and consider $B_\delta(x)$ where $\delta = d(x, y)$. Derive a contradiction using Theorem 7.9.

(b) Use similar ideas to verify that the topologies in Example 7.16 are non-metrizable.

4. In his definition of a topological space, Hausdorff included one additional condition, equivalent to the following:

> **Hausdorff Axiom 7.25** *Given any two distinct points x and y in X, there are open sets U and V with $x \in U$ and $y \in V$ and $U \cap V = \emptyset$. A topological space that satisfies this additional axiom is called a* ***Hausdorff space****.*

(a) Prove that any metric space is a Hausdorff space. Hence a topological space that does not satisfy the Hausdorff Axiom cannot be metrizable.

(b) Prove that if X has a finite number of points, the discrete topology is the only topology on X that satisfies the Hausdorff Axiom.

5. **(a)** Suppose that X has the discrete topology and Y is any topological space. Prove that any function $f : X \to Y$ is continuous.

(b) Suppose that Y has the indiscrete topology and X is any topological space. Prove that any function $f : X \to Y$ is continuous.

(c) Suppose X and Y are any topological spaces. Prove that any constant function $f : X \to Y$ that maps all points in X to a single point $y \in Y$ is continuous.

6. Prove Theorem 7.23.

7. Prove that a closed ball in a metric space is closed.

8. Consider the metric space $\mathbb{R}^2$ with the discrete metric.

(a) What are the open sets? What are the closed sets?

(b) What is $B_1((0, 0))$? What is $B_1[(0, 0)]$?

(c) What is the closure of $B_1((0,0))$?

(d) In a metric space, is the closure of an open ball always equal to the closed ball with the same center and radius?

9. Consider the real line $\mathbb{R}$ with the standard topology, that is, with the topology arising from the standard metric.

(a) Show that the set $A = \left\{\frac{1}{n} \middle| n \in \mathbb{N}\right\} = \left\{1, \frac{1}{2}, \frac{1}{3}, \frac{1}{4}, \ldots\right\}$ is not closed. What is $\overline{A}$?

(b) In a general topological space X, a point x is said to be a **limit point** of a set $A \subseteq X$ if and only if every open set containing x intersects A in a point other than x itself. Prove that a set A is closed if and only if it contains all of its limit points.

10. In $\mathbb{R}^2$ with the standard topology, let A be the half-open disk

$$\{(x, y) \in \mathbb{R}^2 \mid x^2 + y^2 \leq 1, x \geq 0\} \cup \{(x, y) \in \mathbb{R}^2 \mid x^2 + y^2 < 1, x < 0\}.$$

What are $\overline{A}$, A°, and $\mathrm{bd}(A)$?

11. Let $X = \{a, b, c, d, e\}$ with open sets $\emptyset$, X, $\{a\}$, $\{a, b\}$, $\{a, b, c\}$, $\{a, b, c, d\}$, $\{a, c, d\}$, $\{c, d\}$.

(a) As given, the open sets do not satisfy the conditions for a topology. What two sets must be added to the list of open sets to make it a topology?

(b) In the resulting topological space consider the set $A = \{a, d, e\}$. Find $\overline{A}$, A°, $\mathrm{bd}(A)$, $\mathrm{bd}(\overline{A})$, $\mathrm{bd}(A^\circ)$. It might help to start by listing all the closed sets of X.

12. In $\mathbb{R}$ with the standard topology, consider

$$A = (0, 1) \cup (1, 2) \cup \{x \in (2, 3) \mid x \text{ is rational}\} \cup \{4\}.$$

If we start with A and successively apply the closure and interior operators, how many different sets do we get? That is, how many different sets are there in the list

$$A, \quad A^-, \quad A^\circ, \quad A^{-\circ}, \quad A^{\circ-}, \quad A^{-\circ-}, \quad A^{\circ-\circ}, \quad \ldots \quad ?$$

Here we have written A^- instead of $\overline{A}$ to make clear the order in which the operators are applied.

7.3 Connectedness

The intuitive idea of a connected topological space is a set that hangs together as a single piece. It cannot be divided into two subsets A and B that are separated from each other. The sets are separated if somehow neither set contains points close to the other set. Now if points close to A are not in B, then they must be in A. In a metric space at least, this means that A is an open set. Similarly, B is an open set. This discussion leads to the following as a reasonable way to define connectedness for a topological space.

Definition 7.26 *A* ***disconnection*** *of a topological space X is a pair of subsets A and B such that*

1. $A \cup B = X$,
2. $A \cap B = \emptyset$,
3. $A \neq \emptyset$ *and* $B \neq \emptyset$, *and*
4. *A and B are open sets.*

A topological space X is ***connected*** *if and only if it does not have a disconnection.*

The first two conditions (that the space X is the union of two disjoint subsets A and B) say that A and B **partition** X. With this terminology, we can restate the definition: a connected space is one that cannot be partitioned into two nonempty open sets. Since $B = X - A$, the requirement that B is open means that A is closed. So any set other than $\emptyset$ and X that is both open and closed gives a disconnection. This gives us another formulation of connectedness: X is connected if and only if $\emptyset$ and X are the only subsets of X that are both open and closed. Thus, for example, any space with the indiscrete topology is connected, and any space of more than one point with the discrete topology is disconnected.

Example 7.27 Are the topological spaces in Example 7.16 connected?

Solution. The set X is not connected since $\{b\}$ and $\{a, c\}$ form a disconnection.

Because $\mathbb{N}$ is infinite, no subset of $\mathbb{N}$ can be both finite and have a finite complement. Hence $\emptyset$ and $\mathbb{N}$ are the only subsets of $\mathbb{N}$ that are both closed and open. It follows that $\mathbb{N}$ with the finite-complement topology is connected. ❄

Definition 7.26 applies to subsets of topological spaces. A subset S of a topological space X is connected if and only if it is connected in the subspace topology inherited from X. Referring to Definition 7.17, this means we cannot find two sets A and B that are open in X, both intersect S, and satisfy $S \subseteq A \cup B$ and $A \cap B \cap S = \emptyset$. In this situation, we will refer to the pair of sets A and B as a disconnection of the subspace S. For example, the subset $S = [1, 3] \cup [4, 7]$ of $\mathbb{R}$ (see Exercise 2 of Section 1.5) is disconnected: take $A = (0, 3.1)$ and $B = (3.9, 8)$ as a disconnection of S.

The definition of connectedness in terms of open sets gives an elegant connection between connectedness and continuity.

Theorem 7.28 *The continuous image of a connected space is connected. That is, if X is a connected topological space and $f : X \to Y$ is a continuous surjection onto a space Y, then Y is connected.*

Proof. Suppose Y has a disconnection formed by subsets A and B. We simply check the four properties of a disconnection to show that the subsets $f^{-1}(A)$ and $f^{-1}(B)$ form a disconnection of X. This contradicts the hypothesis that X is connected.

1. Since every point in X maps to $Y = A \cup B$, every point in X is either in $f^{-1}(A)$ or $f^{-1}(B)$. That is, $f^{-1}(A) \cup f^{-1}(B) = X$.
2. Since A and B are disjoint, no point in X maps to both A and B. Thus, $f^{-1}(A) \cap f^{-1}(B) = \emptyset$.
3. Let a denote a point in the nonempty set A. Since f is onto, there is a point in X that maps to a. Hence $f^{-1}(A)$ is nonempty. Similarly, $f^{-1}(B)$ is nonempty.
4. Because A and B are open and f is continuous, $f^{-1}(A)$ and $f^{-1}(B)$ are open. ❄

As an immediate consequence of this theorem, we have that connectedness is a topological property: if Y is homeomorphic to a connected space X, then Y is also connected. There is a general and useful principle lurking here. Any property that can be entirely defined in terms of points, sets, and open sets must be a topological invariant. Indeed, by Definition 7.20 homeomorphisms are exactly those bijections that induce an exact correspondence between the open sets in the two spaces.

The next theorem describes the connected subsets of the real line.

Theorem 7.29 *A nonempty subset of $\mathbb{R}$ with the standard topology is connected if and only if it is a single point or an interval.*

Proof. An interval can be open, closed, or half-open; it can be bounded or unbounded: (a,b), $[a,b]$, $[a,b)$, $(a,b]$, $(-\infty,b)$, $(-\infty,b]$, (a,∞), $[a,\infty)$, or $\mathbb{R}$ itself. A property that allows us to recognize a subset I of $\mathbb{R}$ as any of these types of intervals (and includes the empty set and singleton sets as degenerate cases) is

$$c, d \in I \quad \text{and} \quad c < x < d \quad \Longrightarrow \quad x \in I.$$

If a nonempty subset S of $\mathbb{R}$ is not an interval or a singleton set, then it does not satisfy the above property. Let $c < x < d$ be points such that $c, d \in S$ but $x \notin S$. Then $(-\infty, x)$ and (x, ∞) form a disconnection of S.

Conversely, we must show that every interval is connected. So suppose to the contrary that an interval I is disconnected by subsets A and B. Since both A and B intersect I, we can find $c \in A \cap I$ and $d \in B \cap I$. By interchanging the names of A and B if necessary, we can assume that $c < d$. The Completeness Property of the Real Numbers (Property 1.42) gives a least upper bound x for the bounded set $A \cap [c,d]$. Since $c \leq x \leq d$, we have that $x \in I$. We will derive a contradiction by showing that x is not contained in either A or B. Hence I must be connected.

If $x \in A$, then $x \neq d$. Also, since A is open, all numbers in some ε-ball around x are contained in A. In particular, some numbers larger than x are in $A \cap [c,d]$. So x is not an upper bound for $A \cap [c,d]$. Similarly, if $x \in B$, then $x \neq c$. And since B is open, all numbers in some ε-ball around x are contained in B. In particular, some numbers smaller than x are upper bounds for $A \cap [c,d]$. So x is not a least upper bound for $A \cap [c,d]$. ❄

Recall from Definition 1.40 that space X is **path-connected** if and only if for every $a, b \in X$, there is a path in X from a to b.

Theorem 7.30 *A path-connected topological space is connected.*

Proof. Suppose X is path-connected, but has a disconnection A, B. Choose $a \in A$, $b \in B$, and let $\alpha : [0, 1] \to X$ be a path from a to b. Exercise 4 at the end of this section asks you to verify that $\alpha^{-1}(A)$ and $\alpha^{-1}(B)$ form a disconnection of $[0, 1]$. This contradiction to Theorem 7.29 gives that X is connected. ❄

You may be surprised to find that a connected topological space need not be path-connected.

Example 7.31 In $\mathbb{R}^2$ with the standard topology, let $S = \{(x, y) \mid x > 0,\ y = \sin(\frac{\pi}{x})\}$. See Figure 7.32. Show that the space $X = S \cup \{(0, 0)\}$ is connected, but not path-connected.

Solution. Suppose A, B is a disconnection of X. Since S is path-connected and hence connected, it must be entirely contained in one of these sets, say A. Then since B intersects X, we must have $(0, 0) \in B$. Since B is open, it contains a ball $B_\varepsilon((0, 0))$. Now the sequence of points $(1, 0), (\frac{1}{2}, 0), (\frac{1}{3}, 0), (\frac{1}{4}, 0), \ldots$ is in $S \subseteq A$. But for $\frac{1}{n} < \varepsilon$ we also have $(\frac{1}{n}, 0) \in B_\varepsilon((0, 0)) \subseteq B$. This contradicts the condition that $A \cap B \cap X = \emptyset$. Hence X must be connected.

On the other hand, it is fairly clear that there cannot be a path in X from $(1, 0)$ to $(0, 0)$. As any such path travels toward $(0, 0)$, it would repeatedly have to move far away from $(0, 0)$. So it could not be continuous. A rigorous argument involves showing that if $\alpha : [0, 1] \to X$ were such a path, then $A = \alpha^{-1}(\{(0, 0)\})$ and $B = \alpha^{-1}(S)$ form a disconnection of $[0, 1]$. ❄

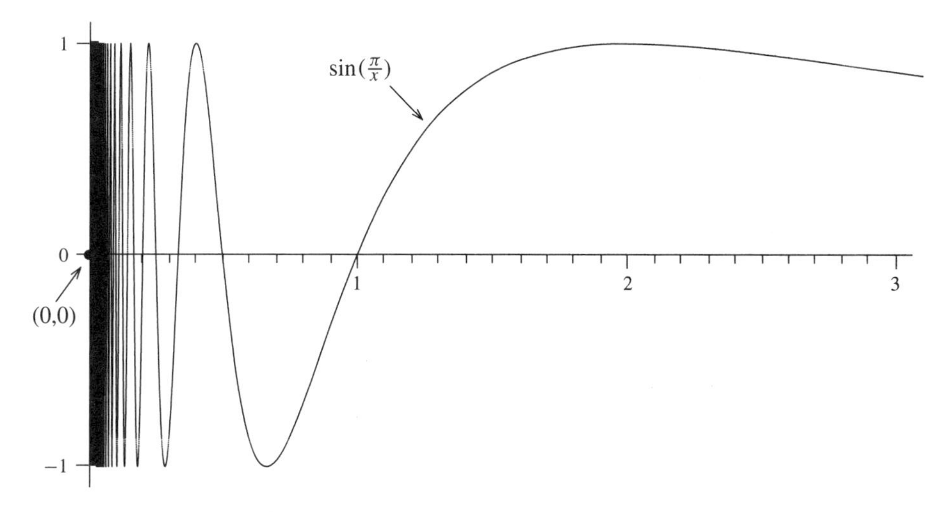

FIGURE 7.32
A space that is connected but not path-connected.

Our conclusion from Theorem 7.30 and Example 7.31 is that connectedness and path-connectedness are closely related, but not identical concepts: path-connectedness is slightly stronger. Both concepts are useful in topology.

Exercises 7.3

1. Let $X = \{a, b, c, d, e\}$ with open sets $\emptyset$, X, $\{a\}$, $\{c\}$, $\{a, b\}$, $\{a, c\}$, $\{c, d\}$, $\{a, b, c\}$, $\{a, c, d\}$, $\{a, b, c, d\}$. This should look familiar from Exercise 11 of Section 7.2. Is X connected?

2. Let X be an infinite set containing a point p. Let the open sets of X be X and any subset of X that does not contain p.

(a) Show that this is a topology on X.

(b) With this topology, is X connected? Why or why not?

3. In Exercise 13 of Section 1.5, you showed that the cone $C = \{(x, y, z) \in \mathbb{R}^3 \mid x^2+y^2 = z^2\}$ is not homeomorphic to $\mathbb{R}^2$ by showing that $C - \{(0, 0, 0)\}$ is not path-connected. Show that $C - \{(0, 0, 0)\}$ is not connected.

4. Verify in the proof of Theorem 7.30 that $\alpha^{-1}(A)$ and $\alpha^{-1}(B)$ form a disconnection of $[0, 1]$. Use the proof of Theorem 7.28 as a model for checking the four conditions of Definition 7.26.

5. In the discussion that the space in Example 7.31 is not path-connected, show that $B = \alpha^{-1}(S)$ is open in $[0, 1]$.

6. Prove that the continuous image of a path-connected topological space is path-connected.

7. Consider the space $\{0, 1\}$ with the discrete topology. Prove that a topological space X is disconnected if and only if there is a continuous surjection f from X onto $\{0, 1\}$.

7.4 Compactness

In Chapters 3 and 4, we defined compactness for surfaces and 3-manifolds in terms of the spaces having a finite triangulation. One of the nicest consequences of Hausdorff's definition of a topological space is that it allows us to define compactness in a more general setting. The general definition of compactness in terms of open sets is subtle—probably not among the first three or four generalizations you might think of.

Definition 7.33 *An **open cover** of a topological space X is a collection $\mathcal{F}$ of open sets such that*

$$\bigcup_{U \in \mathcal{F}} U = X.$$

*A topological space X is **compact** if and only if every open cover of X has a finite subcover.*

This means that whenever any collection $\mathcal{F}$ of open sets covers X, some finite number of them $U_1, U_2, \ldots, U_n$ already cover X. This is clearly a kind of finiteness condition, but a subtle one. It does not require that X have only a finite number of points, although if X does have only a finite number of points it will certainly be compact (see Exercise 1 at the end of this section). The fact that this definition of compactness is equivalent to being closed and bounded if X is a subset of $\mathbb{R}^n$ with the standard topology, is far from obvious. It is the content of a famous theorem in analysis, which was first noticed for subsets of $\mathbb{R}$ by Eduard Heine in the 1870s and was formulated precisely by Emile Borel in 1894. (Of course, Heine and Borel did not use the term compact.) The following Heine-Borel theorem, with its elegant bisection argument leading to a contradiction, appears on most mathematicians' lists of the ten most beautiful theorems in mathematics. It is worth careful study.

Heine-Borel Theorem 7.34 *A subset of $\mathbb{R}^n$ is compact if and only if it is closed and bounded.*

Proof. Assume first that X is a closed and bounded subset of $\mathbb{R}^n$. We want to show that X is compact. So let $\mathcal{F}$ denote an arbitrary open cover of X. Let us suppose that $\mathcal{F}$ has no finite subcover, and derive a contradiction.

Since X is bounded, it is contained in some large cube C. Let d be the diameter (maximum distance between any pair of points) of C. Since X is closed, $\mathbb{R}^n - X$ is open, and $\mathcal{F}$ together with $\mathbb{R}^n - X$ forms an open cover of C. According to our assumption that $\mathcal{F}$ does not have a finite subcover of X, we see that $\mathcal{F} \cup \{\mathbb{R}^n - X\}$ does not have a finite subcover of C. Using the cube C instead of X will make it easier for us to reach a contradiction.

Divide each side of C in half, cutting C into 2^n subcubes each of diameter $d/2$. If each of these subcubes were covered by a finite number of sets in our open cover of C, we could combine these sets to form a finite subcover of all of C. Hence at least one of these subcubes, call it C_1, is not covered by any finite number of sets in our cover of C.

Divide the sides of C_1 in half. Apply the argument of the previous paragraph to choose a subcube C_2 of diameter $d/2^2$ that cannot be covered by a finite number of sets in the open cover of C. Repeat this process to obtain a nested sequence of subcubes

$$C \supseteq C_1 \supseteq C_2 \supseteq \cdots \supseteq C_k \supseteq \cdots$$

such that C_k has diameter $d/2^k$ and no C_k is covered by a finite number of sets in our original open cover of C.

Since the diameters of the cubes C_k go to 0, the intersection of this sequence of cubes is a set consisting of a single point $x \in C$. The existence of the point x is essentially a multidimensional extension of the Completeness Property of the Real Numbers (Property 1.42). Now x is contained in some set U in our open cover of C. Since U is open, there is an $\varepsilon > 0$ such that $B_\varepsilon(x) \subseteq U$. Now choose k so large that $d/2^k < \varepsilon$. Then $C_k \subseteq B_\varepsilon(x) \subseteq U$. The cube C_k is contained in one of the sets in the cover. This contradicts the claim that C_k is not covered by any finite number of these sets.

The proof of the converse implication is a little more mundane. If X is either not bounded or not closed, we can use our bare hands to construct open covers of X that have no finite subcovers. Exercise 4 at the end of this section asks you to verify that these covers do the job to show that X is not compact. If X is not bounded, the open cover $\mathcal{F} = \{B_r(\mathbf{0}) \mid r = 1, 2, 3, \ldots\}$ will do. If X is not closed, its complement is not open. So there is some point $y \notin X$ such that for every $\varepsilon > 0$, the ball $B_\varepsilon(y)$ intersects X. Then

$$\mathcal{F} = \left\{\mathbb{R}^n - B_{1/r}[y] \mid r = 1, 2, 3, \ldots\right\}$$

is an open cover of X that has no finite subcover. ❄

Compactness has the same nice connection with continuity that connectedness has.

Theorem 7.35 *The continuous image of a compact space is compact. That is, if X is a compact topological space and $f : X \to Y$ is a continuous surjection onto a space Y, then Y is compact.*

Proof. Suppose $\mathcal{F}$ is an open cover of Y. Our mission is to find a finite subcover. To accomplish this, notice that $\mathcal{F}' = \{f^{-1}(U) \mid U \in \mathcal{F}\}$ is an open cover of X. Indeed, all of these sets are open because f is continuous; and they cover X because for every $x \in X$, $f(x)$ is in some U in $\mathcal{F}$, and so $x \in f^{-1}(U)$. Because X is compact, $\mathcal{F}'$ has a finite subcover, say $f^{-1}(U_1), \ldots, f^{-1}(U_n)$. Now we can easily check that the corresponding sets $U_1, \ldots, U_n$ are a finite subcover of Y. Indeed, since f is a surjection, every point in Y is $f(x)$ for some $x \in X$. Since x is contained in some $f^{-1}(U_i)$, we have $f(x) \in U_i$. ❄

It follows immediately that compactness, like connectedness, is a topological property. This is not surprising, since compactness is defined in terms of open sets. For surfaces and 3-manifolds, compactness is a crucial property that guarantees the existence of a finite triangulation. This in turn allows us to use combinatorial methods to investigate the structure of these spaces.

Exercises 7.4

1. Show that any finite set with any topology is compact.

2. Is the space in Exercise 2 of Section 7.3 compact? Why or why not?

3. **(a)** Show that Roman road space is not compact by finding an open cover of Roman road space that has no finite subcover. Notice that you cannot use the Heine-Borel Theorem here. Although the set is $\mathbb{R}^2$, it does not have the standard $\mathbb{R}^2$ topology.

(b) Show that the closed ball $B_1[(0, 0)]$ in Roman road space is not compact even though it is closed and bounded.

4. Consider a subset X of $\mathbb{R}^n$.

(a) If X is unbounded, show that $\mathcal{F} = \{B_r(\mathbf{0}) \mid r = 1, 2, 3, \ldots\}$ is a cover of X that does not have a finite subcover.

(b) If X is not closed, let $y \notin X$ be a point such that every ball $B_\varepsilon(y)$ with $\varepsilon > 0$ intersects X. Show that $\mathcal{F} = \{\mathbb{R}^n - B_{1/r}[y] \mid r = 1, 2, 3, \ldots\}$ is an open cover of X that does not have a finite subcover.

5. Prove that any closed subset of a compact space is compact.

6. **(a)** Prove that any compact subset of a metric space is closed.

(b) Prove that any compact subset of a Hausdorff space is closed.

7. Suppose $f : X \to Y$ is a continuous bijection from a compact space X to a Hausdorff space Y.

(a) For any open subset U of X, show that the image $f(U)$ is an open subset of Y.

(b) Show that f is a homeomorphism.

8. **(a)** Find an example of a function $f : \mathbb{R} \to \mathbb{R}$ to show that the continuous image of a closed subset need not be closed.

(b) Find an example of a function $f : \mathbb{R} \to \mathbb{R}$ to show that the continuous image of a bounded subset need not be bounded.

(c) What about the continuous image of a closed and bounded subset?

7.5 Quotient Spaces

In Chapters 3 and 4 we started with a topological space, for instance a polygonal disk in $\mathbb{R}^2$ or a polyhedral solid in $\mathbb{R}^3$, and glued parts of it together to get a new topological space, for instance T^2 or T^3. What exactly are we doing when we glue? It is clear how we get the points of the new space—they are equivalence classes of points in the old space under the relation that two points are equivalent if they are identical or if we wish to glue them together. However, a topological space is more than a set of points. To make the idea of gluing precise, we need to specify the open sets in the new space. Here is the natural way to do it.

Definition 7.36 *If $\sim$ is an equivalence relation on a topological space X, then the* ***quotient space*** *$X/\sim$ is the space whose points are the equivalence classes of $\sim$. The* ***projection function*** *$\pi : X \to X/\sim$ takes any $x \in X$ to the equivalence class containing x. The* ***quotient topology*** *on $X/\sim$ is defined by*

$$U \text{ is open in } X/\sim \quad \text{if and only if} \quad \pi^{-1}(U) \text{ is open in } X.$$

Example 7.37 Suppose we start with the square disk $\{(x, y) \mid |x| \leq 1, |y| \leq 1\}$ and glue the right edge to the left edge with a half-twist. What do the open sets in the resulting quotient space look like? What is this space?

Solution. The equivalence relation that accomplishes the desired gluing is

$$(x_1, y_1) \sim (x_2, y_2) \text{ if and only if}$$
$$(x_1, y_1) = (x_2, y_2), \text{ or } |x_1| = |x_2| = 1 \text{ and } (x_1, y_1) = (-x_2, -y_2).$$

A subset U of the quotient space is open if and only if its inverse image under the projection function is an open subset of the square. Let's look carefully at what this means. Suppose U contains the equivalence class $P = \{(1, y_0), (-1, -y_0)\}$. Then $\pi^{-1}(U)$ contains both $(1, y_0)$ and $(-1, -y_0)$. Hence for $\pi^{-1}(U)$ to be open, it must contain all points in the square that are within some positive distance ε_1 of $(1, y_0)$, and also all points within some distance ε_2 of $(-1, -y_0)$. See Figure 7.38. Letting ε be the smaller of ε_1 and ε_2, we see that $\pi^{-1}(U)$ must contain all points within ε of either $(1, y_0)$ or $(-1, -y_0)$. When we do the gluing, we can think of these two half-balls as being glued together into one ε-ball about the glued point P. The quotient space is, of course, the Möbius band M^2. ❄

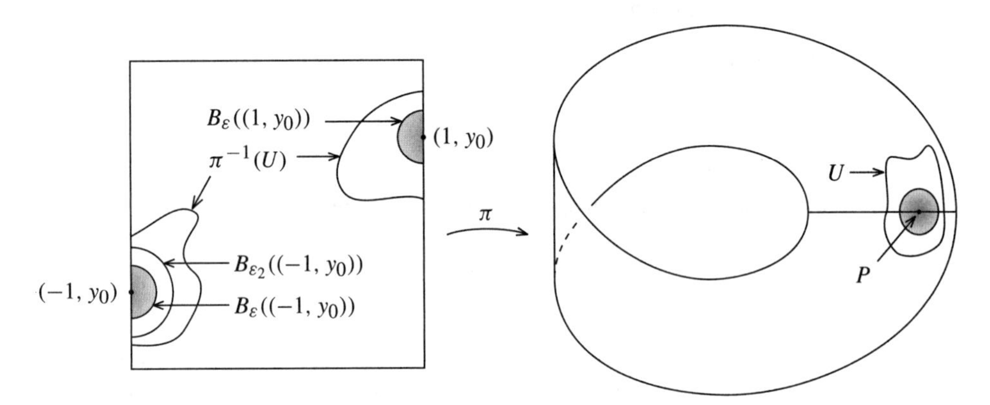

FIGURE 7.38
Gluing the square.

Notice that the definition of the quotient topology makes $\pi : X \to X/\sim$ continuous, and that π is always onto. Hence, by Theorem 7.35, if X is compact, $X/\sim$ will also be compact. Thus, when we take a polygonal region in $\mathbb{R}^2$, which is compact by the Heine-Borel Theorem (Theorem 7.34), and glue its edges together in pairs by identifying points on them, we are guaranteed that the resulting quotient space will be compact. Similarly, if we start with a polyhedral region in $\mathbb{R}^3$ and glue its faces together in pairs, the resulting three-dimensional pseudo-manifold is always compact.

Exercises 7.5

1. Let X be the space of Exercise 1 in Section 7.3. Form a quotient space from X by identifying a with d and identifying b with c.
   - **(a)** Set up some notation for the points of the quotient space. Explain your notation clearly.
   - **(b)** What is the quotient topology? In particular, list the open sets in the quotient space.

2. Let $X = \{(x, y) \mid x^2 + y^2 \leq 1\}$. Consider an equivalence relation on X defined by $(x_1, y_1) \sim (x_2, y_2)$ if and only if either $(x_1, y_1) = (x_2, y_2)$ or $x_1^2 + y_1^2 = x_2^2 + y_2^2 = 1$.

(a) Describe the equivalence classes of the quotient space $X/\sim$.

(b) Describe the open sets in the quotient topology on $X/\sim$.

(c) The quotient space $X/\sim$ is homeomorphic to what familiar space?

3. Consider the unit circle centered at the origin. For a chord across this circle, consider the line through the origin that is perpendicular to this chord. Let r denote the signed distance from the origin along this line to the chord (with positive values for points of intersection in the first or second quadrants). Let θ denote the angle from the positive x-axis to this line.

(a) Show that the pairs $(r, \theta) \in [-1, 1] \times [0, \pi]$ determine all possible chords (including single points on the boundary of the circle as degenerate cases).

(b) Which chords correspond to more than one pair $(r, \theta) \in [-1, 1] \times [0, \pi]$?

(c) What identification can be made to pairs of points in $[-1, 1] \times [0, \pi]$ so that there is a bijection from the quotient space to the set of all chords across the unit circle?

(d) What familiar object is formed by this quotient space?

4. Show that if $\sim$ is an equivalence relation on a connected topological space X, then $X/\sim$ is connected.

5. Suppose $\sim$ is an equivalence relation on a topological space X and $\pi : X \to X/\sim$ is the projection function. Suppose $f : X \to Y$ is a function to another topological space Y such that if $a \sim b$ in X, then $f(a) = f(b)$ in Y.

(a) Show that the function $\widetilde{f} : X/\sim \to Y$ defined by $\widetilde{f}([x]) = f(x)$ is well-defined.

(b) Show that f is continuous if and only if $\widetilde{f}$ is continuous.

(c) Exercise 16 of Section 1.2 established an equivalence relation on a square region and a bijection from the set of equivalence classes to a cylinder. Apply the result of part (b) to show that this bijection is continuous.

(d) Use Exercise 7 of Section 7.4 to show that this bijection is a homeomorphism.

(e) Formulate a general principle that will easily allow you to conclude that a pasting function defines a homeomorphism from the quotient space to the intended geometric object.

References and Suggested Readings for Chapter 7

53. M. A. Armstrong, *Basic Topology*, Springer-Verlag, New York, 1997.

An orderly development of continuity, compactness, connectedness, and other principles of topology leads to results about the fundamental group, surfaces, homology, and knots.

54. Bert Mendelson, *Introduction to Topology* (third edition), Dover, New York, 1990.

A presentation of the basic properties of metric spaces and topological spaces.

55. James Munkres, *Topology* (second edition), Prentice Hall, Upper Saddle River, NJ, 2000.

A classic introduction to general topology and algebraic topology.

Index

About the Authors

Robert Messer studied mathematics as an undergraduate at the University of Chicago. He wrote his thesis in geometric topology at the University of Wisconsin under D. Russell McMillan, receiving his PhD in 1975. He was a John Wesley Young Research Instructor at Dartmouth College and has taught at Western Michigan University and Vanderbilt University. He has been at Albion College since 1981 where he has served as chair of the Department of Mathematics and Computer Science from 1997 to 2002.

In addition to research in topology, he is the author of the textbook *Linear Algebra: Gateway to Mathematics* (1994) and one of the co-authors of *Learning by Discovery: A Lab Manual for Calculus* (MAA, 1993). He helped to organize and coach the Michigan All-Star Math Team for the American Regions Mathematics League Competitions and has served as director of the Michigan Mathematics Prize Competition for the Michigan Section of the Mathematical Association of America. He enjoys the combinatorics and symmetry of English change ringing as well as traditional American and English country dance.

Philip Straffin earned his undergraduate degree in mathematics from Harvard University. He learned knot theory from Ray Lickorish at Cambridge University on a Marshall Scholarship, and received his PhD from the University of California at Berkeley, with a thesis in algebraic topology under Emery Thomas. He has taught at Beloit College since 1970, and served as Chair of Mathematics and Computer Science from 1980 to 1990. He has twice been chosen as Beloit College's Teacher of the Year, and received the MAA's Haimo Award for Distinguished College Teaching of Mathematics in 1993.

Professor Straffin has published over 30 research and expository papers, and has won the Allendoerfer Award and the Trevor Evans Award for mathematical exposition from the MAA. His books include *Topics in the Theory of Voting* (1980) and *Game Theory and Strategy* (MAA, 1993), and edited collections *Political and Related Models* (with Steven

Brams and William Lucas, 1983) and *Applications of Calculus* (MAA, 1993). He is a member of the American Mathematical Society, the Mathematical Association of America and the Association for Women in Mathematics. For the MAA, he has been Chair of the Wisconsin Section, Editor of the Anneli Lax New Mathematical Library, and served on the MAA Notes Editorial Board, the Haimo Teaching Award Committee, the Beckenbach Book Prize Committee, the Council on Publications, and the Coordinating Council on Awards. He enjoys the challenge of scaling peaks in the mountains of Colorado.